Simulation-Driven Aerodynamic Design using Variable-Fidelity Models

Simulation-Driven Aerodynamic Design using Variable-Fidelity Models

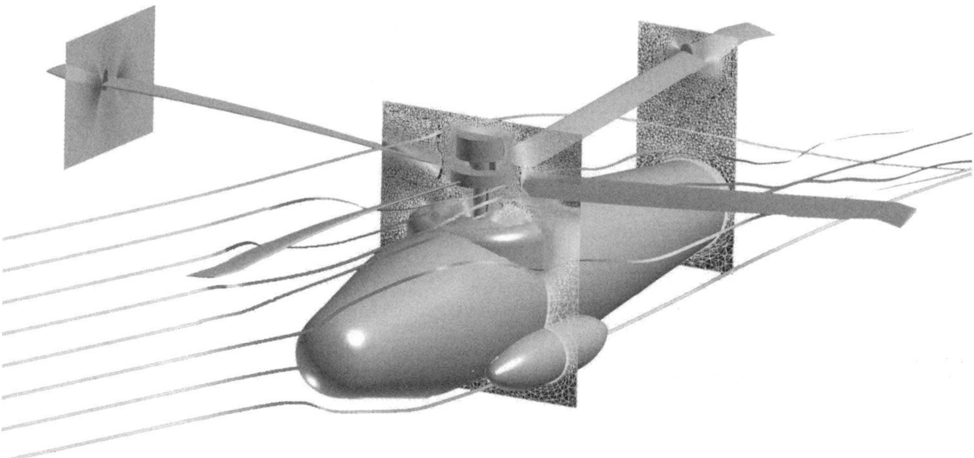

Leifur Leifsson
Slawomir Koziel
Reykjavik University, Iceland

ICP
Imperial College Press

Published by

Imperial College Press
57 Shelton Street
Covent Garden
London WC2H 9HE

Distributed by

World Scientific Publishing Co. Pte. Ltd.

5 Toh Tuck Link, Singapore 596224

USA office: 27 Warren Street, Suite 401-402, Hackensack, NJ 07601

UK office: 57 Shelton Street, Covent Garden, London WC2H 9HE

Library of Congress Cataloging-in-Publication Data
Leifsson, Leifur.
Simulation-driven aerodynamic design using variable-fidelity models / Leifur Leifsson, Reykjavik University, Iceland, Slawomir Koziel, Reykjavik University, Iceland.
pages cm
Includes bibliographical references and index.
ISBN 978-1-78326-628-9 (hardcover : alk. paper)
1. Airplanes--Design and construction--Data processing. 2. Airplanes--Computer simulation. 3. Aerodynamics--Data processing. 4. Tolerance (Engineering) I. Koziel, Slawomir. II. Title.
TL671.2.L375 2015
629.134'1--dc23

2014041980

British Library Cataloguing-in-Publication Data
A catalogue record for this book is available from the British Library.

Cover image: Simulation of the UH-60 rotor system in the National Full-Scale Aerodynamics Complex.
Credit: Elizabeth Lee-Rausch, Robert Biedron, NASA/Langley

Typeset by Stallion Press
Email: enquiries@stallionpress.com

Printed in Singapore by B & Jo Enterprise Pte Ltd

To our families: Sigrún, Eyrún Alda, Kolbrún Ósk, Bergrós María, Dorota, and Anna Halina.

Table of Contents

Preface

Design of contemporary engineering systems relies heavily on computer simulations. In particular, aircraft design requires the use of high-fidelity computational fluid dynamics (CFD) simulations to capture accurately nonlinear physics of the vehicle flow. On the other hand, direct use of high-fidelity CFD simulations in the design process, particularly for automated parameter optimization, is associated with high, often prohibitive, computational costs. Other issues, such as the presence of numerical noise, may results in a failure of optimization when using conventional methods. In this book, we demonstrate that numerically efficient design of aerodynamic surfaces, such as airfoils and wings, can be realized using surrogate-based optimization (SBO) methodologies. The essence of SBO techniques resides in shifting the optimization burden from an expensive high-fidelity CFD model to a fast replacement model: the surrogate. A properly created and handled surrogate serves as a reliable prediction tool so that satisfactory designs can be found at the cost of a limited number of simulations of the high-fidelity CFD model. The book is focused on so-called physics-based surrogates, constructed from underlying low-fidelity models, often involving relaxed-accuracy CFD simulations. The specific SBO techniques covered here include space mapping, shape-preserving response prediction, adaptive response correction, and multi-level optimization. We illustrate the approaches using numerous application examples, including transonic airfoil and wing shape design, and subsonic shape design. Recommendations concerning practical aspects of SBO in aerodynamic design are also presented.

Leifur Leifsson and Slawomir Koziel
2014, Reykjavík, Iceland

Acknowledgements

We would like to acknowledge the efforts of all those students, researchers, and colleagues who have helped us during the research work presented in this book. In particular, we would like to thank the following individuals for their efforts: Stanislav Ogurtsov, Yonatan Afework Tesfahunegn, and Eirikur Jonsson.

Part I

Motivation and Problem Formulation

Chapter 1

Introduction

1.1 Simulation-Driven Design

Designer experience, analytical models, and physical experiments were the fundamental components of engineering design for many decades. Nowadays, the majority of engineering design is carried out using computational analysis, especially in the early phases. In fact, due to the complexity of contemporary engineering systems, computer simulations are necessary to provide reliable evaluations of system performance and are thus indispensable in the design process. This is commonly referred to as *simulation-driven* (or *simulation-based*) design. Physical experiments and testing are of course still important and necessary, however — because of cost and time constraints — they are mostly used for validation purposes.

Modern aircraft design is a good example of an area in which computer simulations play a key role in achieving high efficiency and robustness. Figure 1.1 shows a computer rendering of a typical transport aircraft. In an aircraft, there is a strong coupling between the major disciplines governing the dynamics and function (see Fig. 1.2). The coupling of aerodynamics with structures and weights can be illustrated by the following example: a 1% improvement in performance during takeoff of a generic large twin-engine transport aircraft translates, for a given range, into an increase in payload of 14 passengers or 2,800 lbs (1,270 kg) (van Dam, 2002).

Traditionally, the design of high-lift systems has been carried out using detailed wind-tunnel testing, which is both challenging and expensive due to complicated flow features. With improvements in computer technology, computational fluid dynamics (CFD)

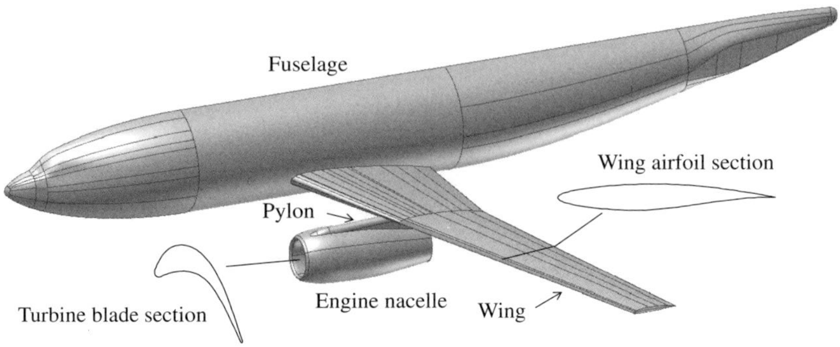

Fig. 1.1. A computer-aided design (CAD) drawing of a typical transport aircraft with a turbofan jet engine. The aircraft wings and the turbine blades of the turbofan engines are streamlined aerodynamic surfaces defined by airfoil sections.

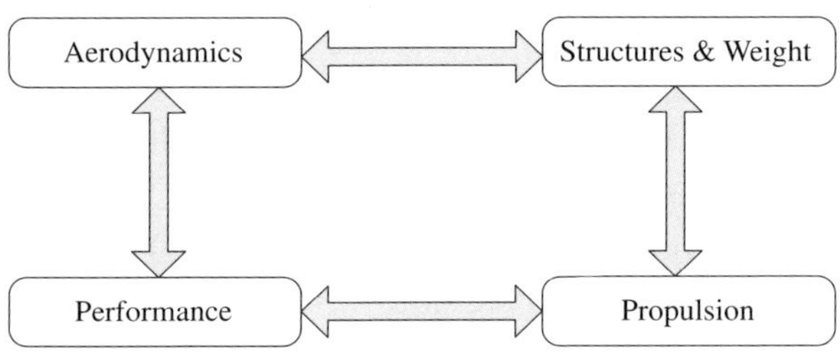

Fig. 1.2. The major disciplines governing the aircraft function and performance are highly coupled.

simulations have replaced wind-tunnel tests during the early stages of the design cycle, allowing designers to solve the flow accurately around complex shapes.

As a consequence, the design cycle time can be reduced significantly, which leads to a reduction of the development costs. In this respect, i.e. the replacement of expensive and time-consuming experimental techniques, computer simulations can be seen as a major revolution in the design process.

An additional benefit of exploiting computer simulations in the process creates the possibility of developing the best design for a given situation. Typically, a given engineering system, such as an aircraft, has a large number of designable parameters. It can be an impossible task to make even slight improvements to a design by manually manipulating the parameters. Such tasks can, however, be tackled by numerical procedures. In fact, optimization algorithms are often the only alternative when dealing with a highly nonlinear system and a large design space. In that sense, the coupling of computer simulations with optimization algorithms can be another revolution in the design process.

There are numerous examples in which computer simulations and optimization algorithms play key roles in design, such as vehicles (e.g. cars, ships, aircraft) or electronic devices (e.g. antennas, microwave filters).

In this book, we will focus on *aerodynamic shape optimization*, which involves the design of shapes in fluid flow. A special emphasis is given to the optimization of airfoil and wing shapes (see Fig. 1.1).

1.2 Aerodynamic Shape Optimization

This section provides a very brief description of aerodynamic analysis by CFD models. We discuss the design optimization process. Some of the key challenges related to coupling CFD models with numerical optimization algorithms are highlighted. Following that, an overview of the state-of-the-art in aerodynamic shape optimization is presented.

1.2.1 *Aerodynamic analysis*

Aerodynamic analysis of streamlined shapes through CFD models contains, in general, the following elements: geometry modeling and parameterization, computational grid generation, flow solution, and a calculation of the measures of merit. The entire process is automated and integrated within an optimization framework.

A wing geometry is typically parameterized using planform variables (such as the span and sweep) and section airfoil shapes (see Fig. 1.3).

A computational grid is needed for a solution to the governing flow equations. The grid is generated by partitioning the solution domain into small elements (or cells), such as the one shown in Fig. 1.4. The grid needs to be sufficiently dense to capture the physics of the flow

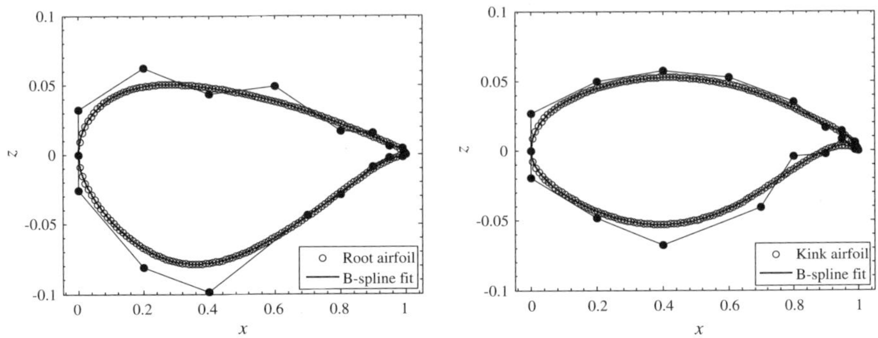

Fig. 1.3. Approximations of the NASA Common Research Model airfoil profiles at the wing root and kink with a B-spline curve fit. The filled dots are control points that determine the profile shape.

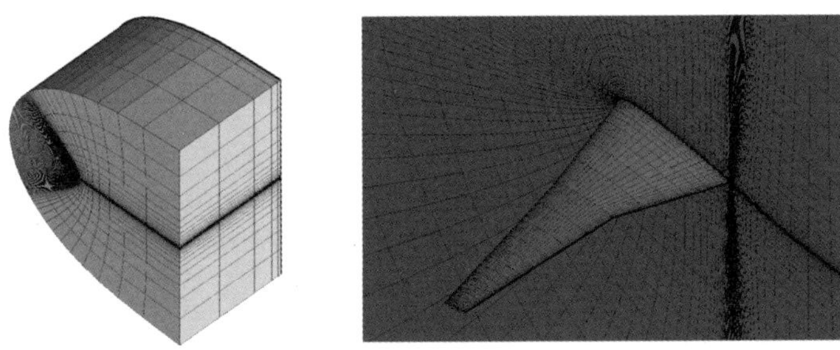

Fig. 1.4. A structured computational grid for the flow past the NASA Common Research Model wing.

around the wing. A typical transport wing needs in the order of 10 million grid cells to represent the flow accurately.

The simulation of the flow about such a wing could take around 24 hours (assuming a parallel computation on eight processors). Obviously, a more powerful hardware will reduce the computational time. A computation of the two-dimensional flow past an airfoil needs around 0.5 million grid cells and takes around half an hour (on a similar machine).

From the output of the CFD model the measures of merit are calculated. Examples of measures include the lift, drag, and moment coefficients. Figure 1.5 shows an example output from a CFD analysis of a typical transport wing.

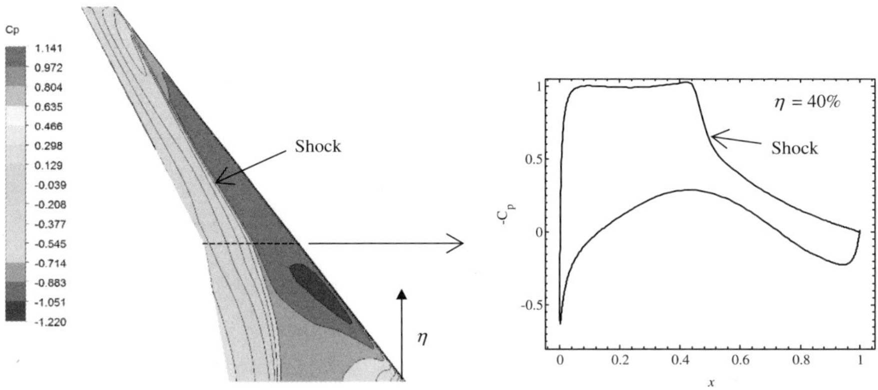

Fig. 1.5. CFD results of the flow past the NASA Common Research Model wing at a Mach number of 0.85, a lift coefficient $C_L = 0.50$, and a drag coefficient $C_D = 0.0364$: contours of pressure coefficient (C_p) on the upper surface (left), and a pressure distribution at the 40% spanwise location (right).

The fidelity of the computational methods used in design has been steadily increasing. Over 40 years ago, the CFD tools were only capable of simulating potential flow past simplified wing configurations. Today's commercial CFD tools are capable of simulating three-dimensional viscous flows past full aircraft configurations using the Reynolds-averaged Navier–Stokes (RANS) equations with the appropriate turbulence models.

1.2.2 *Design optimization*

The objective of optimization is to obtain the best possible design for a given operating condition(s) and constraints. The best design maximizes/minimizes the measure of merit. In practice, the optimization is realized in the following way. Consider Fig. 1.6, which shows a simplified optimization loop concept. The optimizer (optimization algorithm) suggests the candidate design, represented by a variable vector $\mathbf{x}$. It is subsequently verified by evaluation of the aerodynamic (simulation) model that returns the values of the objective function ($f(\mathbf{x})$) and the constraints ($g(\mathbf{x})$). Based on those values, the optimizer decides on the acceptance/rejection of $\mathbf{x}$ and suggests another

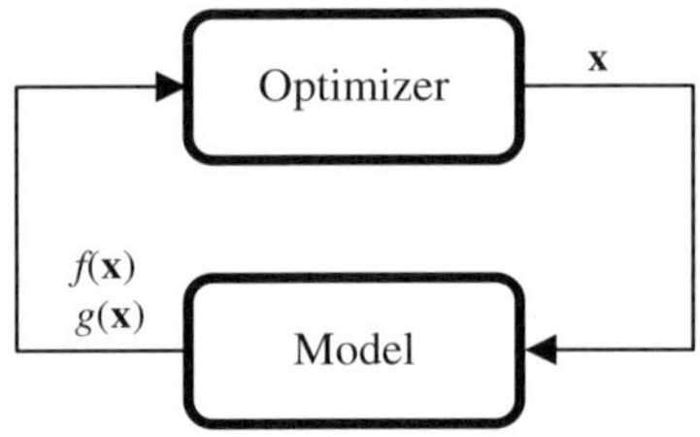

Fig. 1.6. A flowchart of an optimization loop.

candidate design. This iterative process is continued until satisfying, given termination criteria (e.g. convergence).

1.2.3 *Key challenges*

Simulation-driven aerodynamic design optimization involves numerous challenges, including the following:

- High-fidelity CFD simulations are computationally expensive. A three-dimensional CFD simulation can take on the order of one day on a parallel computing cluster.
- A large number of design variables. Typically, an airfoil shape can be described accurately with 10–15 design variables. The entire transport wing shape might require a few airfoils at various spanwise locations, leading to at least 30 design variables, aside from the planform variables (e.g. span, sweep, twist).
- Design optimization normally requires a large number of simulations. For example, even in the case of a two-dimensional airfoil shape optimization with a few design variables, a gradient-based optimization method can require over 100 function evaluations, and the optimization process could take as long as one week. For higher-dimensional problems, the required number of function evaluations may be substantially larger.
- Multiple operating conditions (e.g. a range of Mach numbers and lift coefficients) and multiple objectives (e.g. minimum takeoff gross weight, minimum drag, minimum noise) may need to be considered in the design process simultaneously.

- Uncertainty in the operating conditions and in the airfoil shape may need to be taken into account, leading to the need to carry out stochastic analysis, which is always more time-consuming than a deterministic one.
- The simulation results normally include numerical noise. This can be due to partially converged solutions or badly generated computational meshes.
- The objectives can be numerically sensitive to the grid resolution.
- Coupling of the aerodynamics with other disciplines needs to be considered as well, for example the coupling of the aerodynamic load with the wing structure by means of structural analysis. This is referred to as multidisciplinary design and optimization.

Considering only the first three of the aforementioned points, it is clear that high-fidelity CFD simulations are computationally far too expensive to be used in a direct, simulation-driven design optimization. The overall cost is simply too high for the process to be practical.

One way of tackling this challenge is to speed up the CFD simulations as much as possible. This can be realized using, for example, more powerful computers, parallel or distributed computations. Another way is to reduce the number of simulations required by the optimization algorithm. This can be achieved by reducing the dimensionality of the design space and by using efficient optimization algorithms. Addressing the latter issue is the topic of this book.

1.2.4 *State of the art*

Aerodynamic shape optimization dates back to 1687, when Newton claimed to have solved the problem of minimizing hydrodynamic resistance on a revolving surface (Gunzburger, 1997). Hicks *et al.* (1974) are generally credited for the first practical application of aerodynamic shape optimization. They used a conjugate-gradient method to design transonic airfoils with a potential flow solver. Later, Hicks and Henne (1978) extended the work to three-dimensional wing design with a steepest-descent gradient method.

Nowadays, gradient-based methods are the most widely used approaches for aerodynamic shape optimization. Depending on how the constraints (such as the geometry or the flow) are handled, the application of gradient-based methods can be categorized as either being constrained or unconstrained.

The steepest-descent method is most popular for unconstrained problems (Hicks and Henne, 1978; Nadarajah and Jameson, 2000; Jameson *et al.*, 2004; Mavriplis, 2007). An alternative are quasi-Newton methods, particularly those with BFGS (Broyden–Fletcher–Goldfarb–Shanno) Hessian update (Elliott and Peraire, 1997; Nemec and Zingg, 2002; Rumpfkeil and Zingg, 2008).

Sequential quadratic programming belongs to the most popular methods for constrained aerodynamic shape optimization problems. In particular, the SNOPT (Sparse Nonlinear OPTimizer) (Gill *et al.*, 2005) algorithm is widely used (Kim *et al.*, 2005; Hicken and Zingg, 2007). Other algorithms include the KSOPT (Kreisselmeier–Steinhauser Optimizer) (Wrenn, 1989), which is used, for example, by Anderson and Venkatakrishnan (1997), and the NPSOL (Nonlinear Programming Systems Optimization Laboratory) (Gill *et al.*, 1986), used by Cliff *et al.* (2001).

All of these methods require the availability of the gradients of both the objective function and constraints. Early work used finite differentiation. Its implementation is straightforward; however, the cost of a gradient evaluation scales linearly with the number of design variables. This restricts the use of finite differencing to problems with few variables since the cost of each CFD model evaluation is high.

The development of the adjoint method represents a breakthrough in this area. The main advantage of the adjoint approach is that the cost of a gradient calculation can be made nearly independent of the number of design variables. This opens the gateway for applying aerodynamic shape optimization to problems with a large design space. It was Pironneau (1983) who initially developed the method for optimization of fluid flows, whereas it was Jameson (1988) who developed the method for aerodynamic shape optimization. Other methods of calculating the gradient include the complex-step method

(Martins *et al.*, 2003) and automatic differentiation (see, for example, Bischof *et al.* (1998)).

The adjoint methods can be categorized into continuous and discrete approaches. In the continuous approach, the governing equations are first differentiated to formulate the adjoint equation, and then discretized. In the discrete approach, the governing equations are first discretized and then differentiated to formulate the adjoint equation.

Both approaches have been successfully implemented in aerodynamic shape optimization. Applications of the continuous adjoint approach include Pironneau (1983), Anderson and Venkatakrishnan (1997), Jameson *et al.* (1998), and Reuther *et al.* (1999a, 1999b). Applications of the discrete adjoint approach include Jameson *et al.* (2004) and Leoviriakit and Jameson (2005).

Various other types of optimization algorithms are used for aerodynamic shape optimization, such as derivative-free methods, one-shot methods, and surrogate-based methods. Performance of these methods varies in terms of computational effort and speed.

Evolutionary algorithms, such as genetic algorithms, are the most popular derivative-free methods for aerodynamic shape optimization. These methods are based on mimicking natural or social phenomena, such as the process of natural selection, recombination (data exchange), mutation, and survival of the fittest. Some of the recent application examples can be found in Obayashi (1997), Holst and Pulliam (2003), and Epstein and Peigin (2006).

The fundamental advantage of evolutionary algorithms (or, more broadly, population-based metaheuristics) over gradient-based ones is their ability to perform global searches. However, this comes at a price as a large number of model evaluations are needed, especially for a large design space. Zingg *et al.* (2008) provide a comparison between genetic algorithms and gradient-based methods. Other derivative-free methods include, for example, the Nelder–Mead simplex method (Duvigneau and Visonneau, 2001).

One-shot methods are based on the same Lagrangian formulation as the gradient-based methods, but the flow equations and the first-order optimality conditions are solved simultaneously, thus avoiding repeated flow and gradient evaluations. The linear system from the coupled system is twice the size of the flow equations plus the number of design variables. This system scales linearly with the number of objective functions considered. An overview of the approach can be found in Gunzburger (1997). Applications can be found in Gatsis and Zingg (2003), Iollo *et al.* (2005), and Hazra and Jameson (2007). According to Frank and Shubin (1992), the one-shot method is considerably more efficient than gradient-based methods, but less robust.

The overall objective of surrogate-based optimization (SBO) is to reduce the number of evaluations of the high-fidelity models, thereby making the optimization process more efficient (Queipo *et al.*, 2005; Forrester and Keane, 2009). This is achieved by using computationally cheap surrogate functions in lieu of the computationally more expensive high-fidelity models. The surrogate models can be created by: approximating the sampled high-fidelity model data using, e.g., regression (so-called function-approximation surrogates); or by correcting physics-based low-fidelity models that are less accurate but computationally cheap representations of the high-fidelity models.

A variety of techniques are available to create function-approximation surrogate models. These include, among others, polynomial regression (Queipo *et al.*, 2005), radial basis function interpolation (Forrester and Keane, 2009), kriging (Simpson *et al.*, 2001), and support vector regression (Gunn, 1998). Function-approximation models are versatile; however, they normally require substantial amounts of data samples to ensure good accuracy. In the context of optimization, function approximation techniques (particularly kriging) with adaptive sampling are often used (Forrester and Keane, 2009), where the initial surrogate is updated by adding new training samples that can be allocated either to improve the global modeling accuracy (design space exploration) or to improve the current design (design space exploitation). The allocation strategies (so-called infill criteria) are often based on a combination of the aforementioned goals. Examples of SBO with various function-approximation models related to aerodynamic design can be found in Forrester *et al.* (2007), Jouhoud *et al.* (2007), Brooks *et al.* (2011), March *et al.* (2011), March and Willcox (2012), and Allaire and Willcox (2014).

Physics-based surrogates are typically constructed by correcting low-fidelity models. Popular techniques include response correction (Alexandrov and Lewis, 2001) and space mapping (SM) (Bandler *et al.*, 2004a). The low-fidelity models can be obtained using one of, or a combination of, the following: simplified physics models (also called variable-fidelity physics models (Alexandrov *et al.*, 2000), the high-fidelity model with a coarser computational mesh discretization (called variable-resolution models (Alexandrov *et al.*, 2000)), relaxed convergence criteria (called variable-accuracy models (Forrester *et al.*, 2007)). These models are typically more expensive to evaluate than the function-approximation surrogates. On the other hand, less high-fidelity model data are needed to obtain a given accuracy level. In SBO with physics-based low-fidelity models, called multi-fidelity SBO, only a single high-fidelity model evaluation is typically required per algorithm iteration. Consequently, multi-fidelity SBO methods are scalable to larger numbers of design variables (assuming that no derivative information is utilized).

One of the most popular SBO methods exploiting physics-based surrogates is SM (Bandler *et al.*, 2004a). SM surrogate is a composition of the low-fidelity model and simple, usually linear, transformations that reshape the model domain (input-like SM (Bandler *et al.*, 2004a)) to correct the model response (output-like SM (Koziel *et al.*, 2008b)) or change the overall model properties (implicit-like SM (Koziel *et al.*, 2008b)). SM is able to yield a satisfactory design after a few evaluations of the high-fidelity model, but its performance depends on the proper selection of the SM transformations and their parameters. The surrogate management framework (SMF) algorithm (Booker *et al.*, 1999) is a pattern-based technique that uses the surrogate model as a predictive tool, while retaining the robust convergence properties of pattern search methods for local searches. Typically the surrogate model is constructed using kriging (Simpson *et al.*, 2001) and it is updated in each iteration using accumulated high-fidelity data. Another generic SBO approach is the approximation and model management optimization (AMMO)

(Alexandrov *et al.*, 2000; Alexandrov *et al.*, 2001; Forrester *et al.*, 2007). AMMO is based on ensuring zero- and first-order consistency conditions between the high-fidelity model and the surrogate (i.e. agreement between the function values and first-order derivatives at the current iteration point) by using a suitable correction term. AMMO requires high-fidelity model sensitivity; therefore it is efficient if cheap (e.g. adjoint) derivatives are available. Examples of SBO with physics-based models related to aerodynamic design can be found in Barrett *et al.* (2006) and Robinson *et al.* (2008).

Several SBO techniques exploiting variable-fidelity models have recently been developed for aerodynamic shape optimization. These techniques have been shown to yield significantly better computational efficiency compared to SBO with standard functional surrogate models. The variable-fidelity techniques share the commonality of constructing the surrogate using an underlying low-fidelity model or models. They differ in the correction process and/or how the low-fidelity models are utilized during the optimization run.

The shape-preserving response prediction (SPRP) technique (Koziel and Leifsson, 2013a) predicts the high-fidelity model pressure distribution using low-fidelity model data. The adaptive response correction (ARC) technique (Koziel and Leifsson, 2012b) also predicts the high-fidelity pressure distribution, but unlike SPRP it does not rely on a one-to-one correspondence of the high- and low-fidelity characteristics. The SM technique (Koziel and Leifsson, 2012a) corrects directly the low-fidelity model measures of merit, such as the lift and drag coefficients, to align them with the high-fidelity model. The multi-level optimization (MLO) technique (Koziel and Leifsson, 2013b) constructs a family of low-fidelity models of increasing fidelity that are sequentially optimized, with the optimal design of the lower-fidelity model being the initial design of the higher-fidelity model.

1.3 Book Outline

The main topic of this book is surrogate-based optimization methods for simulation-driven aerodynamic shape design with the focus on techniques exploiting variable-fidelity models and physics-based surrogates. The book is organized in five parts.

The first part includes the motivation (this chapter) and the problem formulation, Chapter 2, in which we formulate aerodynamic shape design as an optimization task. We also briefly discuss the designable parameters of airfoils and wings.

The second part is an exposition of optimization techniques. In Chapter 3, we give an overview of conventional optimization techniques, including gradient-based and derivative-free methods, such as metaheuristics. Adjoint sensitivity techniques are also briefly described. The basic concepts of SBO are introduced in Chapter 4. SBO with approximation-based surrogates is described in Chapter 5, and SBO with physics-based surrogates is described in Chapter 6. Several techniques are presented, including AMMO, SM, SPRP, ARC, and MLO.

In the third part, an overview of aerodynamic modeling is provided. Chapter 7 describes several airfoil geometry parameterization methods. High-fidelity aerodynamics modeling using CFD is described in Chapter 8. We give examples of two-dimensional airfoil CFD models and three-dimensional wing CFD models. In Chapter 9, we describe ways of formulating low-fidelity models appropriate for constructing physics-based surrogates.

In the fourth part, we provide applications of the presented optimization techniques to transonic airfoil shape design (Chapter 10), transonic wing shape design (Chapter 11), subsonic shape design (Chapter 12), as well as design problems in other areas (Chapter 13), such as microwave filter design and antenna design. Chapter 14 gives examples of basic SBO implementations with MATLAB.

The fifth and final part concludes the book. Chapter 15 discusses practical aspects of variable-fidelity design and optimization. Here we formulate recommendations for the readers interested in applying the presented algorithms and techniques in their aerodynamic shape design and discuss possible future developments concerning mostly automation of simulation-driven design. A summary of the book is given in Chapter 16.

Chapter 2

Aerodynamic Shape Optimization

This chapter describes the aerodynamic shape optimization (ASO) problem. In particular, objectives and constraints for typical deterministic design problems are formulated, including one-objective and multi-objective, inverse design, as well as one-point and multi-point approaches.

2.1 Introduction

The overall goal of ASO is to find a shape (or adjust the existing one) that improves a given aerodynamic measure of merit while adhering to appropriate constraints. The first step of the ASO process is to appropriately define the shape, the measure of merit, and the design constraints. With those definitions at hand, the optimization problem is formulated and suitable numerical methods can be applied to yield the solution.

Aerodynamic shapes are typically described by a set of parameters and shape functions. For example, the aircraft wing shown in Fig. 2.1 can be described on one hand by planform variables, such as the span (b), sweep (Λ), and chord lengths (c) shown in Fig. 2.1(a) and, on the other hand, by the airfoil shapes at each span station, such as the root, kink, and tip. Each airfoil section, such as the one in Fig. 2.1(b), can be described by a set of shape functions and parameters. An overview of several airfoil parameterization methods is provided in Chapter 7.

The measure of merit is a quantity that characterizes the performance of the aerodynamic surface. The characteristics of an aerodynamic surface are typically represented with the non-dimensional

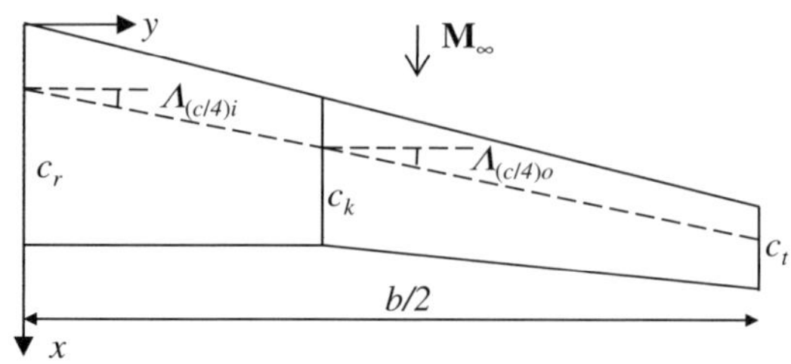

(a) Planform view

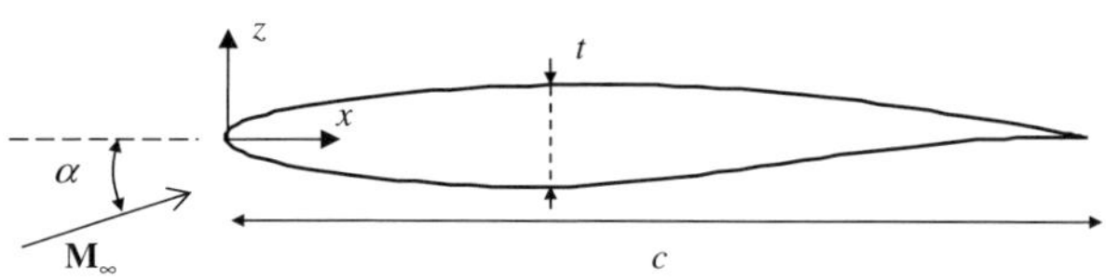

(b) A typical transonic airfoil shape

Fig. 2.1. Nomenclature for typical transonic aircraft wing geometry.

coefficients of forces, pressures, and moments acting on it, such as the coefficients of lift (C_L), drag (C_D), pressure (C_p), and pitching moment (C_M). Typically, the objective is to minimize the drag. For example, in the design of transonic wings, one of the main objectives is to minimize the drag induced by a pressure shock, such as the one shown in Fig. 2.2. Often, in low-speed design the objective is to maximize the lift. Maximization of the lift-to-drag ratio can also be considered. Anderson (2010) provides an excellent discussion of aerodynamic design of transonic airfoils and wings.

There can be constraints pertaining to various aspects of the engineering system under consideration. Aircraft design is highly multidisciplinary and the constraints can be related to the aerodynamics, structures, propulsion system, and control systems. As all the disciplines are highly coupled, ASO needs to account for that in some way. For example, the wings of an aircraft have many structural, mechanical, and electrical components. There needs to be space for these components. This is often accounted for in ASO by thickness constraints.

Section 2.2 gives an overview of typical ASO problem formulations.

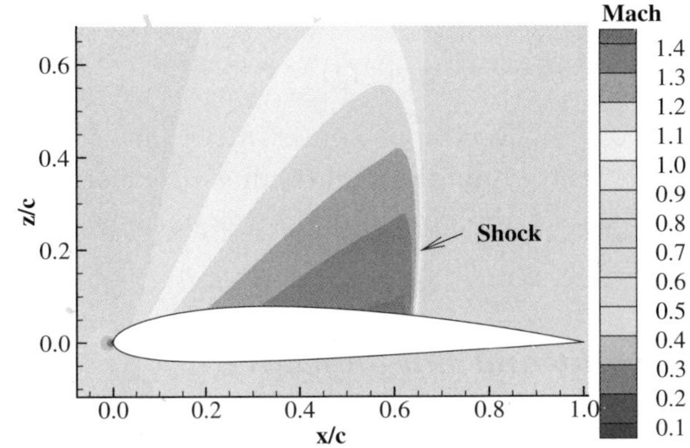

(a)

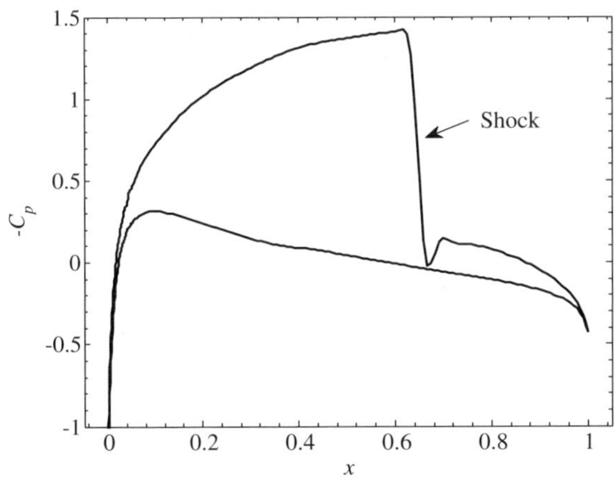

(b)

Fig. 2.2. Inviscid transonic flow analysis of NACA 2412 at Mach 0.75 and an angle of attack of 1 degree with the compressible Euler equations: (a) contours of Mach number and (b) surface pressure distribution. The lift coefficient is $C_l = 0.67$ and drag coefficient $C_d = 0.0261$.

2.2 Problem Formulation

In more rigorous terms, the ASO problem consists of determining the values of design variables $\boldsymbol{x} \in R^n$, such that the objective function J: $R^n \rightarrow R$ is minimized:

$$\min_{\boldsymbol{x}} J(\boldsymbol{x}, \boldsymbol{Q}), \tag{2.1}$$

subject to constraint equations:

$$\boldsymbol{g}(\boldsymbol{x}, \boldsymbol{Q}) \leq 0, \tag{2.2}$$

where $\mathbf{Q}$ denotes the vector of conservative flow variables, and $\mathbf{g}$: $R^n \rightarrow R^m$ is a vector function containing m constraints. The flow variables must satisfy the governing flow equations, $\mathbf{R}$:

$$\boldsymbol{R}(\boldsymbol{x}, \boldsymbol{Q}) = 0. \tag{2.3}$$

2.2.1 *One-point and one-objective*

The problem formulation Equations (2.1)–(2.3) are general and can be applied to different design scenarios. The *one-point* and *one-objective* approach is widely adopted, in which the aerodynamic surface is optimized for one operating condition with a single merit function. The most common of this type is the *lift-constrained drag minimization problem.* Here, the goal is to improve the aerodynamic efficiency while maintaining a required lift. The objective function is set as:

$$J = C_D, \tag{2.4}$$

where C_D is the drag coefficient, and the lift constraint is:

$$g_1 = C_L^* - C_L \leq 0, \tag{2.5}$$

where C_L is the lift coefficient obtained for design $\mathbf{x}$, and C_L^* is the required lift coefficient. Parameters of the operating condition include the Mach number, M_∞, and the Reynolds number, Re. The angle-of-attack, α, can be set as a design variable, or it can be considered a state variable that is adjusted during the flow solve to satisfy Equation (2.5).

Another approach is to maximize the lift for a given drag value. The objective function is then set as:

$$J = -C_L, \tag{2.6}$$

and the drag constraint is:

$$g_1 = C_D - C_D^* \leq 0, \tag{2.7}$$

where C_D^* is the target drag coefficient.

The above objectives can be combined into one by using a weighted sum approach. For example, Nemec *et al.* (2004) defines:

$$J = \omega_L \left(1 - \frac{C_L}{C_L^*}\right)^2 + \omega_D \left(1 - \frac{C_D}{C_D^*}\right)^2, \tag{2.8}$$

if $C_D > C_D^*$, otherwise:

$$J = \omega_L \left(1 - \frac{C_L}{C_L^*}\right)^2. \tag{2.9}$$

The weights ω_L and ω_D are user-assigned constants and specify the relative importance of maintaining lift versus drag reduction. In a similar way, Leung (2010) writes:

$$J = \omega_L \left(1 - \frac{L}{L^*}\right)^2 + \omega_D \left(1 - \frac{D}{D^*}\right)^2, \tag{2.10}$$

where L is the lift (which is equal to the aircraft weight for level flight), $L = C_L \cdot S$, D is the drag, $D = C_D \cdot S$, and S is the wing's reference surface area. The designer specifies the targets in lift and drag, L^* and D^*, as well as the weights, ω_L and ω_D. If the target lift is attainable and target drag is not, this objective function represents lift-constrained drag minimization with the lift constraint appearing as a penalty term. The advantage of having the objective function in the form of Equations (2.8)–(2.10) is that additional terms (which may be considered as constraints) can be added easily (Leung, 2010). We will describe this approach below.

2.2.2 *Constraints*

Additional constraints can be added depending on the design goals. For example, the pitching moment, C_M, can be constrained to be below a certain value, C_M^*, i.e.:

$$g_2 = C_M - C_M^* \leq 0. \tag{2.11}$$

Geometric constraints can be added to enforce limitations related to volume (e.g. due to fuel) or structural (e.g. wing box) requirements. A volume constraint can be expressed as:

$$g_3 = |v - v^*| - v_f v^* \leq 0, \tag{2.12}$$

where v is the volume enclosed by the geometry, v^* is a reference volume, and v_f is a user-specified factor that the volume v is allowed to deviate from v^*. The thickness of an airfoil section at a given span station can be constrained as:

$$g_4 = t^* - t \leq 0, \tag{2.13}$$

where t is the thickness of the airfoil and t^* is the target minimum thickness.

Often, constraints are implemented through penalty terms in the objective function. Then the objective function is written as:

$$J = J_0 + \sum_j (J_p)_j, \tag{2.14}$$

where J_0 is the design objective described above, and $(J_p)_\mathrm{j}$ is the penalty term for each constraint violation. Quadratic penalty terms are common. For example, if the lift constraint Equation (2.5) is violated, the penalty term added to J_0 would be:

$$J_{p,L} = \omega_L \left(1 - \frac{C_L}{C_L^*}\right)^2, \tag{2.15}$$

where ω_L is the user-supplied penalty weight. In a similar fashion, the volume constraint penalty factor is:

$$J_{p,v} = \omega_v \frac{(|v - v^*| - v_f v^*)^2}{(v^*)^2}, \tag{2.16}$$

where ω_v is the penalty weight, and the thickness constraint penalty factor is:

$$J_{p,t} = \omega_t \left(1 - \frac{t}{t^*}\right)^2. \tag{2.17}$$

2.2.3 *Multi-point and one-objective*

The above formulations can be applied to the simultaneous optimization at multiple operating conditions by using the weighted-sum approach by writing the objective as:

$$J_m = \sum_{i=1}^{N_m} \omega_i J_i, \tag{2.18}$$

where N_m is the number of design points, ω_i a user-assigned weight for each design point, and J_i is any of the above-mentioned objective functions. This is termed as *multi-point* optimization. Typically, the different design points are represented by different Mach numbers, which are usually close to the drag rise at transonic speeds.

2.2.4 *Inverse design*

The inverse design problem involves finding a shape that attains a predefined flow behavior. Typically, a target pressure distribution, C_p^*, is specified at each span station. The airfoil shape is recovered by minimizing, in a least-squares sense:

$$J = \frac{1}{2} \sum_{i=1}^{N_p} [(C_p)_i - (C_p^*)_i]^2, \tag{2.19}$$

where C_p is the local pressure coefficient, and N_p is the number of points on the distribution.

It should be noted that the specified target pressure distribution may not be attainable. Minimization of Equation (2.19) will, however, yield a shape that recovers the specified distribution as closely as possible in a least square sense.

Moreover, inverse design only yields the airfoil shape pertaining to the specified pressure distribution, which may or may not be optimal in terms of other measures of merit, such as the drag and lift. Direct design, in contrast, may yield a truly optimized design in terms of those measures.

2.2.5 *Multi-objective*

By nature, aerodynamic shape optimization is a multi-objective problem. In many cases, *a priori* preference articulation is possible, so that only one specific objective (e.g. drag coefficient) is to be explicitly optimized, whereas the others (e.g. lift coefficient) can be handled through design constraints. This allows for solving the problem with the one-objective formulation. Sometimes, however, this is neither possible nor convenient, e.g. when gaining knowledge about possible trade-offs between competing objectives is important. In those situations genuine *multi-objective* optimization is necessary. Typically, a solution to multi-objective design is represented as a so-called Pareto front, which represents the set of the best possible designs that are non-commensurable in the conventional (one-objective) sense.

An example multi-objective formulation can involve minimizing the drag coefficient, as well as minimizing the negative lift coefficient (which is equivalent to maximizing the lift coefficient). These objectives are conflicting because a reduction in drag will, typically, result in a reduction in lift. Consequently, this problem does not have a unique solution. A possible way of addressing this issue is objective function aggregation using, e.g., a weighted sum method, where:

$$J = \omega_D J_D + (1 - \omega_D) J_L. \tag{2.20}$$

Here, ω_D is a weighing constant, $J_D = C_D$, and $J_L = -C_L$. Constraints, such as a thickness constraint, can be added to this formulation as discussed above.

It should be emphasized that combining objectives into a single scalar function only allows for finding a single solution, i.e. one point on the Pareto front. Other solutions can be obtained (with some restrictions) by optimizing an aggregated objective function such as Equation (2.20) with different set ups of the weighting factors.

A discussion of genuine multi-objective optimization that aims at identifying the entire set of solutions representing the Pareto front exceeds the scope of this book. Some information, in the context of two-dimensional airfoil design, can be found in Section 10.6.

Part II

Optimization Techniques

Chapter 3

Simulation-Driven Design: Direct Methods

Computational fluid dynamics (CFD) simulation plays an important and still increasing role in contemporary aerodynamic design. Perhaps the most straightforward method of using CFD analysis in the design process is to handle the design task as an optimization problem. This is normally done by formulating design goals (usually as an explicit objective function) and identifying a set of parameters to be adjusted, as well as defining the design space, usually by assigning lower and upper bounds for the parameters and/or additional linear or nonlinear constraints. The problem is subsequently solved using a suitable numerical optimization routine. In the optimization process, CFD analysis is used to evaluate the candidate designs suggested by the algorithm. Figure 3.1 shows the flowchart of a simulation-driven design process exploiting direct optimization of a CFD-evaluated objective function.

In this chapter, we give a brief overview of conventional optimization techniques, including gradient-based and derivative-free methods, as well as metaheuristics. We also discuss adjoint sensitivity techniques that allow us to obtain derivatives of the objective function with little or no extra computational effort, which is of fundamental importance when using gradient-based methods for solving complex aerodynamic shape optimization problems, particularly if large numbers of design variables are involved.

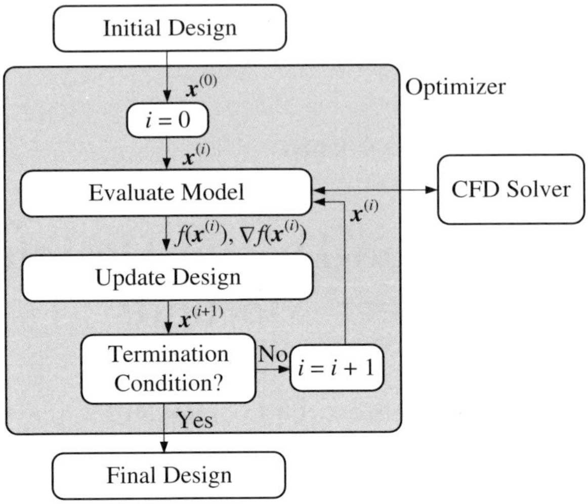

Fig. 3.1. Simulation-driven design through optimization. The generic optimization scheme is an iterative process in which the candidate designs generated by the optimization algorithm are evaluated by a CFD solver for verification purposes and to provide the optimizer with information that allows for the searching of better designs. Depending on the type of the algorithm, the search process may be guided by the model response or (if available) by the response and its derivatives (gradient).

3.1 Design Optimization Problem

The design optimization problem is formulated as:

$$\boldsymbol{x}^* = \arg\min_{\boldsymbol{x} \in X} f(\boldsymbol{x}), \tag{3.1}$$

where $f(\boldsymbol{x})$ is a scalar objective function, whereas $\boldsymbol{x} \in X \subseteq R^n$ is the objective function domain. For the sake of simplicity, in this chapter we use the symbol f for what is otherwise in the book considered to be a composition $H(f(\cdot))$ of the high-fidelity model f (usually a vector-valued function) and a scalar merit function H. The objective function represents the performance of the system under design. The problem can also be formulated using explicit constraints, such as lower/upper bounds for design variables, as well as other (both linear and nonlinear) inequality or equality constraints (cf. Chapter 2).

In Sections 3.2 and 3.3, we present the outline of the conventional optimization methods, both gradient-based and derivative-free, focusing on unconstrained optimization. The reader interested in in-depth description of the methods is referred to the literature (e.g. Michalewicz (1996) for evolutionary methods; Nocedal and Wright (2006) for gradient-based optimization; and Conn *et al.* (2009) for derivative-free methods).

3.2 Gradient-Based Optimization

Gradient-based methods are probably the most widely used optimization techniques in engineering and science (Nocedal and Wrigth, 2006). The main driving force when searching for a lower value of the objective function $f(\boldsymbol{x})$ (i.e. a better design) is the gradient:

$$\nabla f(\boldsymbol{x}) = \left[\frac{\partial f}{\partial x_1}(\boldsymbol{x})\frac{\partial f}{\partial x_2}(\boldsymbol{x})\ldots\frac{\partial f}{\partial x_n}(\boldsymbol{x})\right]^T, \tag{3.2}$$

which gives information about descent of f in the vicinity of $\boldsymbol{x}$, assuming that f is sufficiently smooth (i.e. at least continuously differentiable). More specifically, we have:

$$f(\boldsymbol{x}+\boldsymbol{h}) \cong f(\boldsymbol{x}) + \nabla f(\boldsymbol{x})^T \cdot \boldsymbol{h} < f(\boldsymbol{x}), \tag{3.3}$$

for sufficiently small $\boldsymbol{h}$ as long as $\nabla f(\boldsymbol{x})^T \cdot \boldsymbol{h} < 0$. In particular, $\boldsymbol{h} = -\nabla f(\boldsymbol{x})$ determines the direction of the (locally) steepest descent. Gradient-based methods find a new approximation of the optimum design by either moving along a descent direction or utilizing the local model of f constructed using its derivatives. Some of the methods also exploit exact or approximate second-order derivatives; in particular, the Hessian of f, defined as:

$$H(\boldsymbol{x}) = \left[\frac{\partial^2 f}{\partial x_j \partial x_k}(\boldsymbol{x})\right]_{j,k=1,\ldots,n}. \tag{3.4}$$

In Sections 3.2.1–3.2.3, we outline selected gradient-based algorithms for unconstrained optimization, and also make a few comments about constrained optimization.

3.2.1 *Descent methods*

Here we consider iterative algorithms, i.e. the methods that generate a sequence of vectors $\boldsymbol{x}^{(i)}$, $i = 0, 1, \ldots,$ being approximate solutions to Equation (3.1). For all methods, we require that $f(\boldsymbol{x}^{(i+1)}) < f(\boldsymbol{x}^{(i)})$. The generic structure of a descent method is the following ($\boldsymbol{x}^{(0)}$ is the initial solution; Nocedal and Wright (2006)):

1. set $i = 0$;
2. find a search direction, $\boldsymbol{h}_d$;
3. find a step length, α;
4. set $\boldsymbol{x}^{(i)} = \boldsymbol{x}^{(i)} + \alpha \boldsymbol{h}_d$;
5. set $i = i + 1$;
6. if the termination condition is not satisfied, go to 2; and
7. END.

In the algorithm above, the search is performed along a descent direction $\boldsymbol{h}_d$ found in Step 2. A descent direction is a vector $\boldsymbol{h}_d$ for which $\nabla f(\boldsymbol{x})^T \cdot \boldsymbol{h}_d < 0$ (cf. Equation (3.3)). The length of the step, i.e. $\alpha \boldsymbol{h}_d$, is determined by a separate algorithm to ensure that $f(\boldsymbol{x}^{(i+1)}) < f(\boldsymbol{x}^{(i)})$. The termination condition may be, among others, of the form $\|\boldsymbol{x}^{(i+1)} - \boldsymbol{x}^{(i)}\| \leq \varepsilon_1$ (convergence in argument), $\|\nabla f(\boldsymbol{x}^{(i)})\| \leq \varepsilon_2$ (vanishing of gradient), $f(\boldsymbol{x}^{(i)}) - f(\boldsymbol{x}^{(i+1)}) \leq \varepsilon_3$ (convergence in function value), or a combination of the above.

Line search (Nocedal and Wright, 2006) is a procedure of finding the length of the step taken along the descent direction. We consider a function:

$$\varphi(\alpha) = f(\boldsymbol{x} + \alpha \boldsymbol{h}_d). \tag{3.5}$$

In *exact line search*, the step length α_e is found as $\alpha_e = \arg\min\{\alpha > 0\colon f(\boldsymbol{x} + \alpha \boldsymbol{h}_d)\}$. *Soft line search* only requires the satisfaction of certain (usually quite loose) conditions, such as the Wolfe conditions (Nocedal and Wright, 2006):

$$\varphi(\alpha_s) \leq \lambda(\alpha_s) \quad \text{where } \lambda(\alpha) = \varphi(0) + \rho \varphi'(0) \cdot \alpha \quad \text{with } 0 < \rho < 0.5 \tag{3.6}$$

and

$$\varphi'(\alpha_s) \geq \beta \cdot \varphi'(0) \quad \text{with } \rho < \beta < 1, \tag{3.7}$$

which ensure a sufficient decrease of f while not being too short. Descent methods with line search governed by Equations (3.6) and (3.7) are normally convergent to at least a first-order stationary point of f. It should be mentioned that a soft line search is generally more advantageous than an exact line search in terms of the overall computational cost of the optimization process (the use of exact line search results in a smaller number but much more expensive iterations).

It should be mentioned that selecting the search direction to be the steepest descent direction, i.e. $\boldsymbol{h} = -\nabla f(\boldsymbol{x})$, in each iteration (the so-called steepest descent method), leads to a poorly performing algorithm when close to the optimum. On the other hand, it is useful in the initial stages of the optimization process, which leads to various hybrid methods.

In general, a better choice of the search direction is a combination of the previous direction, $\boldsymbol{h}_{prev}$, and the current gradient, i.e.

$$\boldsymbol{h} = -\nabla f(\boldsymbol{x}^{(i)}) + \gamma \boldsymbol{h}_{prev}, \tag{3.8}$$

which is characteristic to the so-called conjugate gradient methods (Yang, 2010). An example method of selecting the coefficient γ is a Fleecher–Reeves method, with:

$$\gamma = \frac{\nabla f(\boldsymbol{x})^T \nabla f(\boldsymbol{x})}{\nabla f(\boldsymbol{x}_{prev})^T \nabla f(\boldsymbol{x}_{prev})}, \tag{3.9}$$

or a Polak–Ribiére method, with:

$$\gamma = \frac{(\nabla f(\boldsymbol{x}) - \nabla f(\boldsymbol{x}_{prev}))^T \nabla f(\boldsymbol{x})}{\nabla f(\boldsymbol{x}_{prev})^T \nabla f(\boldsymbol{x}_{prev})}. \tag{3.10}$$

In both cases, the new search direction $\boldsymbol{h}$ obtained from Equation (3.8) is not guaranteed to be downhill (e.g. when the soft line search is used to find the iteration step), so safeguards such as defaulting to $\boldsymbol{h} = -\nabla f(\boldsymbol{x})$ are necessary in case $\nabla f(\boldsymbol{x})\boldsymbol{h}^T \geq 0$.

Instead of using a line search, descent methods can also be realized with a trust region framework (Conn *et al.*, 2000). Here, the iteration step is determined by the properties of the (local) model of the objective function f created using first-order (and, optionally, higher-order) derivatives. The first-order Taylor expansion of f is defined as:

$$f(\boldsymbol{x} + \boldsymbol{h}) \simeq q(\boldsymbol{h}) \quad \text{with } q(\boldsymbol{h}) = f(\boldsymbol{x}) + \nabla f(\boldsymbol{x})\boldsymbol{h}^T. \tag{3.11}$$

Similarly, the second-order Taylor expansion is given by:

$$f(\boldsymbol{x}+\boldsymbol{h}) \simeq q(\boldsymbol{h}) \quad \text{with } q(\boldsymbol{h}) = f(\boldsymbol{x}) + \nabla f(\boldsymbol{x}) \cdot \boldsymbol{h}^T + \frac{1}{2}\boldsymbol{h}^T \boldsymbol{H}(\boldsymbol{x})\boldsymbol{h}. \tag{3.12}$$

In both cases, the model $q(\boldsymbol{h})$ is a good approximation of $f(\boldsymbol{x}+\boldsymbol{h})$ for sufficiently small vectors $\boldsymbol{h}$. In a trust region framework, the iteration step size $\boldsymbol{h}_{tr}$ is determined as follows:

$$\boldsymbol{h}_{tr} = \arg \min_{\boldsymbol{h},\, \|\boldsymbol{h}\| \le \delta} q(\boldsymbol{h}), \tag{3.13}$$

where $\delta > 0$ is the trust region radius. If $f(\boldsymbol{x}+\boldsymbol{h}_{tr}) < f(\boldsymbol{x})$, $\boldsymbol{h}_{tr}$ becomes the next optimization step, otherwise it is rejected. The size of the trust region radius is adjusted based on the *gain ratio*:

$$r = \frac{f(\boldsymbol{x}) - f(\boldsymbol{x}+\boldsymbol{h})}{q(0) - q(\boldsymbol{h})}. \tag{3.14}$$

A typical adjustment scheme would be $\delta \leftarrow 2\delta$ if $r > 0.75$ and $\delta \leftarrow \delta/3$ if $r < 0.25$. It should be noted that for both Equations (3.11) and (3.12) the condition $f(\boldsymbol{x}+\boldsymbol{h}_{tr}) < f(\boldsymbol{x})$ would be satisfied if δ were sufficiently small.

3.2.2 *Newton and quasi-Newton methods*

Assuming the f is at least twice continuously differentiable, the function f can be represented locally using its second-order Taylor Equation (3.12). If the Hessian $\boldsymbol{H}(\boldsymbol{x})$ is positive definite, then the model $q(\boldsymbol{h})$ has a unique minimizer for $\boldsymbol{h}$, such that $\nabla q(\boldsymbol{h}) = 0$, i.e. where

$$\nabla f(\boldsymbol{x}) + \boldsymbol{H}(\boldsymbol{x})\boldsymbol{h} = 0. \tag{3.15}$$

This is the origin of the Newton method, in which the next iteration point is obtained by solving Equation (3.15) as follows:

$$\boldsymbol{x}^{(i+1)} = \boldsymbol{x}^{(i)} - [\boldsymbol{H}(\boldsymbol{x}^{(i)})]^{-1}\nabla f(\boldsymbol{x}^{(i)}). \tag{3.16}$$

Equation (3.16) is well defined, assuming that $\boldsymbol{H}(\boldsymbol{x})$ is non-singular. Moreover, it is positive definite for all iterations, the starting point is sufficiently close to the optimum, and the methods usually converge very quickly (quadratically) to a minimum.

Unfortunately, the basic Newton algorithm has serious disadvantages, making it not very practical. These include the lack of global convergence for many problems, the possibility of converging to a maximum or a saddle point, and the possibility of Equation (3.16) being ill-conditioned. Moreover, the Newton method requires analytical second-order derivatives.

The main difficulty, i.e. the Hessian not being positive definite, can be alleviated by using the so-called damped Newton method, where Equation (3.16) is replaced with:

$$\boldsymbol{x}^{(i+1)} = \boldsymbol{x}^{(i)} - [\boldsymbol{H}(\boldsymbol{x}^{(i)}) + \mu\boldsymbol{I}]^{-1}\nabla f(\boldsymbol{x}^{(i)}), \quad (3.17)$$

where $\boldsymbol{I}$ is the identity matrix. If $\mu > 0$ is sufficiently large, the matrix $\boldsymbol{H}(\boldsymbol{x}) + \mu\boldsymbol{I}$ is positive definite. Then the step $\boldsymbol{h}_\mu$, found as a solution to the problem $[\boldsymbol{H}(\boldsymbol{x}) + \mu\boldsymbol{I}]\boldsymbol{h}_\mu = -\nabla f(\boldsymbol{x})$, is a minimizer of the following model:

$$\begin{aligned} q_\mu(\boldsymbol{h}) &= q(\boldsymbol{h}) + \frac{1}{2}\mu\boldsymbol{h}^T\boldsymbol{h} \\ &= f(\boldsymbol{x}) + \nabla f(\boldsymbol{x}) \cdot \boldsymbol{h}^T + \frac{1}{2}\boldsymbol{h}^T[\boldsymbol{H}(\boldsymbol{x}) + \mu\boldsymbol{I}]\boldsymbol{h}. \end{aligned} \quad (3.18)$$

Moreover, $\boldsymbol{h}_\mu$ is a descent direction for f at $\boldsymbol{x}$. It can be observed that the term $\mu\boldsymbol{h}^T\boldsymbol{h}/2$ penalizes large steps. If μ is very large then the algorithm defaults to the steepest descent method, as $\boldsymbol{h}_\mu \approx -\nabla f(\boldsymbol{x})/\mu$.

In practice, large values of μ are used at the early stages of the iteration process (Equation (3.17)), and μ is reduced as the algorithm converges, which allows for finding the minimizer of f. A popular method exploiting this approach is the Levenberg–Marquardt (LM) algorithm, in which the value of μ is increased at the beginning of each iteration until $\boldsymbol{H}(\boldsymbol{x}) + \mu\boldsymbol{I}$ becomes positive definite, and then updated based on the value of the gain ratio r (Equation (3.14)). The typical updating formula is: $\mu \leftarrow \mu \cdot \max\{1/3, 1 - (2r - 1)^3\}$ if $r > 0$, and $\mu = 2\mu$ otherwise (Nielsen, 1999).

Damped Newton methods allow for overcoming all of the aforementioned disadvantages of the Newton algorithm but one: the potentially high cost of obtaining second-order derivatives of f. In quasi-Newton methods, exact Hessian (or, even better, its inverse) is replaced by approximations obtained using suitable updating formulas. One of the best and most popular updating formulas is the Broyden–Fletcher–Goldfarb–Shanno (BFGS) one (Nocedal and Wright, 2006). BFGS preserves the positive symmetry and definiteness of the approximation. It is defined as follows for Hessian approximation $\boldsymbol{B}$:

$$\boldsymbol{B}_{new} = \boldsymbol{B} + \frac{1}{\boldsymbol{h}^T\boldsymbol{y}}\boldsymbol{y}\boldsymbol{y}^T - \frac{1}{\boldsymbol{h}^T\boldsymbol{u}}\boldsymbol{u}\boldsymbol{u}^T,$$

$$\text{where } \boldsymbol{h} = \boldsymbol{x}_{new} - \boldsymbol{x}, \quad \boldsymbol{y} = \nabla f(\boldsymbol{x}_{new}) - \nabla f(\boldsymbol{x}), \quad \boldsymbol{u} = \boldsymbol{B}\boldsymbol{h}. \tag{3.19}$$

Approximation of the inverse of Hessian $\boldsymbol{D}$ is defined as:

$$\boldsymbol{D}_{new} = \boldsymbol{D} + \kappa_1\boldsymbol{h}\boldsymbol{h}^T - \kappa_2(\boldsymbol{h}\boldsymbol{v}^T + \boldsymbol{v}\boldsymbol{h}^T),$$

$$\text{where } \boldsymbol{h} = \boldsymbol{x}_{new} - \boldsymbol{x}, \quad \boldsymbol{y} = \nabla f(\boldsymbol{x}_{new}) - \nabla f(\boldsymbol{x}), \quad \boldsymbol{v} = \boldsymbol{B}\boldsymbol{y}$$

$$\kappa_2 = \frac{1}{\boldsymbol{h}^T\boldsymbol{y}}, \quad \kappa_1 = \kappa_2(1 + \kappa_2(\boldsymbol{y}^T\boldsymbol{v})). \tag{3.20}$$

The quasi-Newton algorithm with BFGS formula always uses a soft-line search with very loose conditions (e.g. $\rho = 0.0001$, and $\beta = 0.9$, cf. Equations (3.6) and (3.7)).

Quasi-Newton algorithms belong to the most efficient methods for unconstrained optimization. On the other hand, conjugate gradient methods may be better when the number of design variables is large: Newton-like methods rely on matrix operations while conjugate gradient methods only use vector operations, which are cheaper; also, Newton-type methods require more storage ($n \times n$ matrix versus a few vectors for conjugate gradient methods).

3.2.3 *Constrained optimization*

The methods discussed in Sections 3.2.1 and 3.2.2 are for unconstrained optimization. Most practical problems are constrained, with typical constraints being lower/upper bounds for design variables, as well as linear/nonlinear equality and inequality constraints. In this section, we make a few comments on constrained optimization methods. The reader interested in more detail is referred to one of numerical optimization textbooks (e.g. Nocedal and Wright, 2006).

Formulation of a constrained aerodynamic shape optimization problem has been presented in Section 2.2. Here, we consider Equation (3.1) assuming N inequality and M equality constraints:

$$g_k(\boldsymbol{x}) \leq 0, \quad k = 1, \ldots, N; \tag{3.21}$$

$$h_k(\boldsymbol{x}) = 0, \quad k = 1, \ldots, M. \tag{3.22}$$

The set of points satisfying the above constraints is called a feasible region.

The first-order necessary conditions (so-called Karush–Kuhn–Tucker, or KKT, conditions; Kuhn and Tucker, 1951) for an optimum of a constrained problem (Equations (3.1), (3.21), and (3.22)) state that — at the local optimum $\boldsymbol{x}^*$ of f — there exist constants $\mu_1, \ldots, \mu_N$, and $\lambda_1, \lambda_2, \ldots, \lambda_M$, such that:

$$\nabla f(\boldsymbol{x}^*) + \sum_{k=1}^{N} \mu_k \nabla g_k(\boldsymbol{x}^*) + \sum_{k=1}^{M} \lambda_k \nabla h_k(\boldsymbol{x}^*) = 0, \tag{3.23}$$

and

$$g_k(\boldsymbol{x}^*) \leq 0, \quad \mu_k g_k(\boldsymbol{x}^*) = 0, \quad k = 1, \ldots, N, \tag{3.24}$$

where

$$\mu_k \geq 0, \quad k = 1, \ldots, N, \tag{3.25}$$

and provided that all the functions (both objective and constraints) are continuously differentiable. The function at the left-hand side of Equation (3.25) is called the Lagrangian function, whereas coefficients μ and λ are called Lagrange multipliers. It should be noted that these conditions are more complex than for unconstrained optimization, in which the first-order necessary simply states that $\nabla f(\boldsymbol{x}^*) = 0$.

There are two general approaches to handling constrained optimization. Assuming that the unconstrained minimum of f is not in the interior of the feasible region, the constrained minimum must be on the boundary of the feasible region. So-called active set methods search for the optimum along the boundary. Inequality constraints are handled by keeping track of the set of active constraints while moving downhill along the boundary. Another group of methods is based on approaching the optimal solution in an iterative manner, either from within the feasible region (so-called interior-point methods) or by the possible use of infeasible points (however, not by moving along the feasible region boundary). In the latter approaches, the objective function is modified and corresponding unconstrained optimization problems are solved in each iteration.

Perhaps the simplest method of handling constraints is through penalty functions. In this approach, the original problem (Equation (3.1)) with constraints (Equations (3.21) and (3.22)) is replaced with:

$$\begin{aligned}&\arg\min_{\boldsymbol{x}} \phi(\boldsymbol{x}, \boldsymbol{\alpha}, \boldsymbol{\beta}) \\ &\quad = \arg\min_{\boldsymbol{x}} \left\{ f(\boldsymbol{x}) + \sum_{k=1}^{N} \alpha_k \bar{g}_k^2(\boldsymbol{x}) + \sum_{k=1}^{M} \beta_k h_k^2(\boldsymbol{x}) \right\}, \qquad (3.26)\end{aligned}$$

where $\boldsymbol{\alpha} = [\alpha_1 \ldots \alpha_N]$, $\boldsymbol{\beta} = [\beta_1 \ldots \beta_M]$, α_k, $\beta_k \gg 1$, are penalty factors, and $\bar{g}_k(\boldsymbol{x}) = \max\{0, g_k(\boldsymbol{x})\}$ (inequalities only contribute to ϕ if they are active). In practice, one solves a sequence of problems (Equation (3.26)) for increasing values of penalty factors using the solution of the previous iteration as the starting points for the next one.

The barrier method replaces the original constrained problem (Equation (3.1)) by the following formulation (for simplicity, only inequality-type constraints are assumed here):

$$\arg\min_{\boldsymbol{x}} \phi(\boldsymbol{x}, \mu) = \arg\min_{\boldsymbol{x}} \left\{ f(\boldsymbol{x}) - \mu \sum_{k=1}^{N} \ln(-g_k(\boldsymbol{x})) \right\}, \qquad (3.27)$$

where $\mu > 0$ is a barrier parameter. The search starts from a feasible point. A sequence of unconstrained problems is solved, with μ decreasing to zero so that the minimum of $\phi(\boldsymbol{x}, \mu)$ converges to a solution of f.

Augmented Lagrangian methods are a class of algorithms that are similar to penalty methods, in the sense that the original constrained problem is solved as a sequence of suitably formulated unconstrained ones. The unconstrained objective is the Lagrangian of the constrained problem with an additional penalty term (so-called augmentation). Assuming for simplicity equality-only constraints, $h_k(\boldsymbol{x})$, $k = 1, \ldots, M$, the augmented Lagrangian algorithm generates a new approximation of the constrained solution to Equation (3.1) as:

$$\begin{aligned}\boldsymbol{x}^{(i+1)} &= \arg\min_{\boldsymbol{x}} \phi^{(i)}(\boldsymbol{x}) \\ &= \arg\min_{\boldsymbol{x}} \left\{ f(\boldsymbol{x}) + \frac{\mu^{(i)}}{2} \sum_{k=1}^{M} h_k^2(\boldsymbol{x}) - \sum_{k=1}^{M} \lambda_k h_k(\boldsymbol{x}) \right\}. \end{aligned} \tag{3.28}$$

The starting point is the previous approximation, $\boldsymbol{x}^{(i)}$. In each iteration, the coefficients μ_k are updated as follows: $\lambda_k \leftarrow \lambda_k - \mu^{(i)} h_k (\boldsymbol{x}^{(i+1)})$. The coefficients λ_k are estimates of the Lagrange multipliers (cf. Equation (3.23)), and their accuracy increases as the optimization process progresses. The value of μ is increased in each iteration; however, unlike in the penalty method, it is not necessary to take $\mu \rightarrow \infty$.

One of the most popular methods for solving constrained optimization problems is sequential quadratic programming (Han, 1977; Nocedal and Wright, 2006). At each iteration, the following quadratic programming subproblem is used to compute a search direction $\boldsymbol{h}^{(i)}$:

$$\boldsymbol{h}^{(i)} = \arg\min_{\boldsymbol{h}} \left\{ f(\boldsymbol{x}^{(i)}) - \nabla f(\boldsymbol{x}^{(i)})^T \boldsymbol{h} + \frac{1}{2} \boldsymbol{h}^T \boldsymbol{H}(\boldsymbol{x}^{(i)}) \boldsymbol{h} \right\}, \tag{3.29}$$

so that:

$$g_k(\boldsymbol{x}^{(i)}) + \nabla g_k(\boldsymbol{x}^{(i)})^T \boldsymbol{h} \leq 0, \quad k = 1, \ldots, N, \tag{3.30}$$

and:

$$h_k(\boldsymbol{x}^{(i)}) + \nabla h_k(\boldsymbol{x}^{(i)})^T \boldsymbol{h} = 0, \quad k = 1, \ldots, M. \tag{3.31}$$

Here, $\boldsymbol{x}^{(i)}$ is a current iteration point, whereas $\boldsymbol{H}$ is a symmetric, positive definite matrix (preferably an approximation of the Hessian of the Lagrangian of f, cf. Equation (3.23)). The new approximation $\boldsymbol{x}^{(i+1)}$ is subsequently obtained using line search.

3.3 Derivative-Free Optimization

In many situations, a gradient-based search may not be a good option. This is particularly the case when derivative information is not available or too expensive to compute (e.g. through finite differentiation of an expensive objective function). On the other hand, a gradient-based search does not perform well then the objective function is noisy, which is often the case for objective functions evaluated using computer simulation, including CFD analysis.

Optimization methods that do not rely on derivative data are referred to as derivative-free methods (Conn *et al.*, 2009). These techniques include various types of pattern search algorithms (Kolda *et al.*, 2003) and global optimization methods such as metaheuristics (Yang, 2010), as well as surrogate-based optimization (SBO) algorithms (Queipo *et al.*, 2005; Koziel *et al.*, 2011b). SBO is the main focus of this book and, therefore, its more detailed exposition will be presented in Chapters 3–6. Here, we briefly present the concept of the pattern search (Section 3.3.1), Nelder–Mead algorithm (Section 3.3.2), population-based search methods for global optimization, including genetic/evolutionary algorithms (Section 3.3.3), and a few popular metaheuristics, including particle swarm optimizers and differential evolution (Sections 3.3.4 and 3.3.5).

3.3.1 *Pattern search*

Pattern search algorithms are optimization methods based on evaluating the objective function on a stencil determined by a set of directions that are suitable from geometric or algebraic points of view. The initial grid is subsequently modified during the optimization run.

Figure 3.2 illustrates the concept of the pattern search methodology (Kolda *et al.*, 2003), in which the search for the objective function minimum is restricted to the rectangular grid and explores a grid-restricted vicinity of the current design. Failure to make the step improving the current design leads to refining the grid size and allowing smaller steps. The termination criterion is typically a user-defined minimum stencil size.

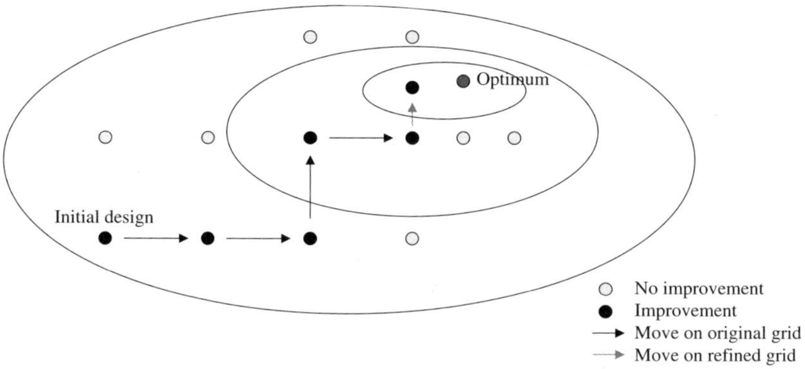

Fig. 3.2. The concept of pattern search using a rectangular grid. The search is based on exploratory movements restricted to the rectangular grid around the initial design. Upon the failure to make the successful move, the grid is refined to allow smaller steps. The actual implementations of pattern search routines also use more sophisticated strategies (e.g. grid-restricted line search).

Many variants of the pattern search methods are available (see, for example, Kolda *et al.* (2003)). Assuming a sufficiently large size of the initial grid, these techniques can be used to perform a quasi-global search.

Pattern search and similar methods are usually robust, although their convergence is relatively slow compared to gradient-based routines. Their fundamental advantage is in the fact that they do not use derivative information and, even more importantly, they are relatively immune to the numerical noise. Their recent popularity is not only due to the development of rigorous convergence theory (Kolda *et al.*, 2003; Conn *et al.*, 2009) but also because of the fact that pattern searches may easily benefit from the availability of parallel computing resources.

3.3.2 *Nelder–Mead algorithm*

The Nelder–Mead algorithm is a derivative-free downhill simplex procedure for unconstrained optimization (Nelder and Mead, 1965). The search process is based on manipulating a set of candidate solutions, forming a simplex. A simplex or n-simplex is a convex hull of a set of $n + 1$ affinely independent points, $\boldsymbol{x}^{(i)}$, in a vector space

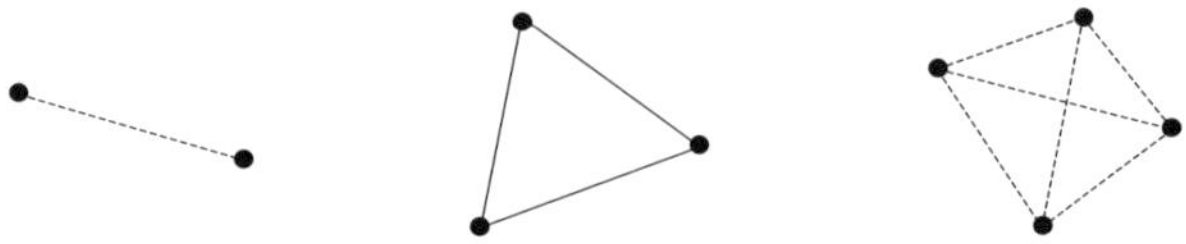

Fig. 3.3. The concept of simplex. (a) 1-simplex; (b) 2-simplex; (c) 3-simplex.

(normally, we consider n-simplex in an n-dimensional space). The set $X = \{\boldsymbol{x}^{(i)}\}_{i=1,\ldots,k}$ is affinely independent if the vectors $\boldsymbol{v}^{(i)} = \boldsymbol{x}^{(i+1)} - \boldsymbol{x}^{(1)}$, $i = 1, \ldots, k-1$, are linearly independent. A convex hull $H(X)$ of X is defined as a set of all convex combinations of the points from X, i.e. $H(X) = \{\sum_{i=1,\ldots,k} a_i \boldsymbol{x}^{(i)} \colon \boldsymbol{x}^{(i)} \in X, a_i \geq 0, \sum_{i=1,\ldots,k} a_i = 1\}$. 1-simplex, 2-simplex, and 3-simplex are, respectively, a line segment, a triangle, and a tetrahedron (see Fig. 3.3).

At any given iteration, the Nelder–Mead algorithm processes a set of simplex vertices and updates it based on the corresponding objective function values. The algorithm flow is the following (see also Fig. 3.4).

- Order the vertices so that $f(\boldsymbol{x}^{(1)}) \leq f(\boldsymbol{x}^{(2)}) \leq \ldots f(\boldsymbol{x}^{(n+1)})$. An auxiliary point, $\boldsymbol{x}^{(0)}$, is then defined to be the center of gravity of all vertices but $\boldsymbol{x}^{(n+1)}$.
- Compute a *reflection* $\boldsymbol{x}^r = (1+\alpha)\boldsymbol{x}^{(0)} - \alpha\boldsymbol{x}^{(n+1)}$ (a typical value of α is 1); if $f(\boldsymbol{x}^{(1)}) \leq f(\boldsymbol{x}^r) < f(\boldsymbol{x}^{(n)})$, reject $\boldsymbol{x}^{(n+1)}$ and update the simplex using $\boldsymbol{x}^r$.
- If $f(\boldsymbol{x}^r) < f(\boldsymbol{x}^{(1)})$, compute an *expansion* $\boldsymbol{x}^e = \rho\boldsymbol{x}^r + (1-\rho)\boldsymbol{x}^{(0)}$ (a typical value of ρ is 2); if $f(\boldsymbol{x}^e) < f(\boldsymbol{x}^r)$, reject $\boldsymbol{x}^{(n+1)}$ and update the simplex using $\boldsymbol{x}^e$.
- If $f(\boldsymbol{x}^{(n)}) \leq f(\boldsymbol{x}^r) < f(\boldsymbol{x}^{(n+1)})$, compute a *contraction* $\boldsymbol{x}^c = (1+\gamma)\boldsymbol{x}^{(0)} - \gamma\boldsymbol{x}^{(n+1)}$ (a typical value of γ is 0.5); if $f(\boldsymbol{x}^c) \leq f(\boldsymbol{x}^r)$, reject $\boldsymbol{x}^{(n+1)}$ and update the simplex using $\boldsymbol{x}^c$.
- If $f(\boldsymbol{x}^c) > f(\boldsymbol{x}^r)$ or $f(\boldsymbol{x}^r) \geq f(\boldsymbol{x}^{(n+1)})$, shrink the simplex: $\boldsymbol{x}^{(i)} = \boldsymbol{x}^{(1)} + \sigma(\boldsymbol{x}^{(i)} - \boldsymbol{x}^{(1)})$, $i = 1, \ldots, n+1$ (a typical value of σ is 0.5).

The above procedure constitutes one iteration of the algorithm, which is continued until the termination condition is met (typically, the reduction of the simplex size below a user-defined threshold).

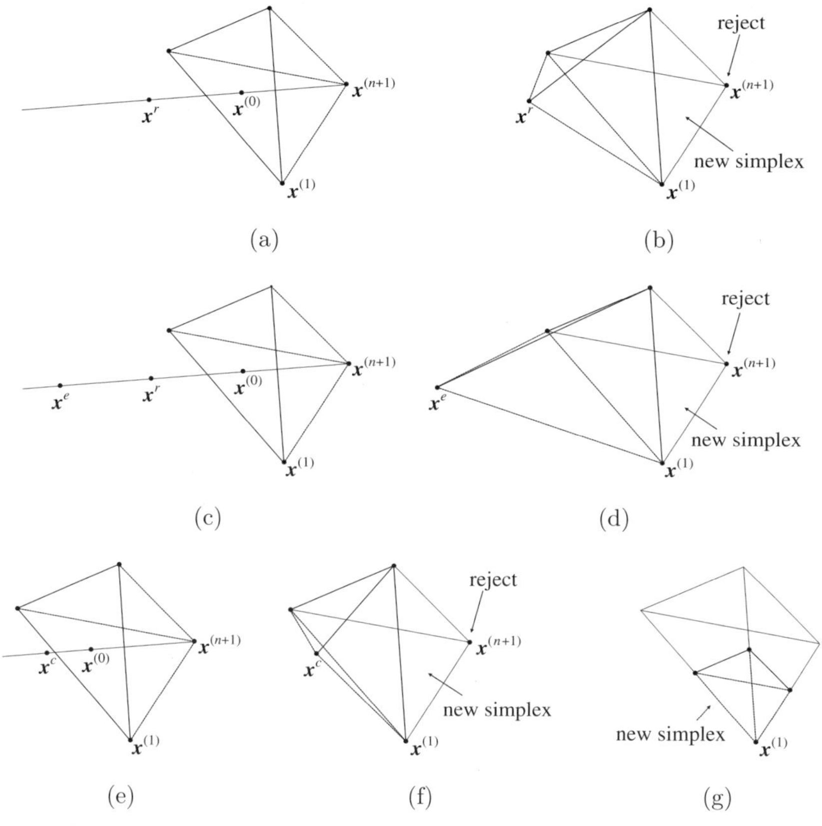

Fig. 3.4. Simplex operations performed by the Nelder–Mead algorithm. (a) reflection: $\boldsymbol{x}^r = (1+\alpha)\boldsymbol{x}^{(0)} - \alpha\boldsymbol{x}^{(n+1)}$; (b) acceptance of the reflection point $\boldsymbol{x}^r$ if $f(\boldsymbol{x}^{(1)}) \leq f(\boldsymbol{x}^r) < f(\boldsymbol{x}^{(n)})$; (c) expansion: $\boldsymbol{x}^e = \rho\boldsymbol{x}^r + (1-\rho)\boldsymbol{x}^{(0)}$; (d) acceptance of the expansion point $\boldsymbol{x}^e$ if $f(\boldsymbol{x}^e) < f(\boldsymbol{x}^r)$; (e) contraction: $\boldsymbol{x}^c = (1+\gamma)\boldsymbol{x}^{(0)} - \gamma\boldsymbol{x}^{(n+1)}$; (f) acceptance of the contraction point $\boldsymbol{x}^c$ if $f(\boldsymbol{x}^c) \leq f(\boldsymbol{x}^r)$; (g) shrinking of the simplex if neither of the above operations led to the objective function improvement.

The Nelder–Mead algorithm is very popular. It is characterized by a relatively low computational cost (only a few objective function evaluations per iteration); however, its convergence is rather slow.

3.3.3 *Metaheuristics and global optimization: genetic and evolutionary algorithms*

Metaheuristics are derivative-free global optimization methods capable of handling issues that are challenging for many conventional algorithms (including gradient-based ones), such as non-differentiable, discontinuous, or noisy objective functions, as well as the presence of multiple local optima. Metaheuristics are based on the observation or mimicking of natural processes (e.g. biological or social systems). The majority of these methods process sets (or populations) of potential solutions to the optimization problem at hand (often called individuals or agents). The optimization process progresses using method-specific interactions between individuals as well as selection methods, designed to both explore and exploit promising regions of the design space. Some older metaheuristic algorithms (e.g. simulated annealing; Kirkpatrick *et al.* (1983)) process single solutions rather than a set of individuals.

Probably the most popular varieties of metaheuristics are genetic algorithms (GA) (Goldberg, 1989) and evolutionary algorithms (EA) (Back *et al.*, 2000). Other methods include evolution strategies (ES) (Back *et al.*, 2000), particle swarm optimizers (PSO) (Kennedy *et al.*, 2001), differential evolution (DE) (Storn and Price, 1997), and, more recently, firefly algorithm (Yang, 2010).

Let P denote the set of individuals (population) processed by the algorithm. The typical flow of the metaheuristic algorithm is the following:

1. initialize population P;
2. evaluate population P;
3. select parent individuals S from P;
4. apply recombination operators to create a new population P from parent individuals S;
5. apply mutation operators to introduce local perturbations in individuals of P;
6. if termination condition is not satisfied go to 2; and
7. END.

Initialization of the population is usually random. In Step 2, the members of the population are evaluated, and the corresponding values of the objective function determine their "fitness". For many algorithms (GA, EA), an important step is the selection of the subset of individuals to form a new population. Depending on the algorithm, the selection can be deterministic (pick up the best ones only, ES) or partially random (probability of being selected depends on the fitness value but there is a chance to survive even for poor individuals, EA). A typical example of a selection procedure is a tournament selection, in which out of the set of candidates randomly picked from the current population, the one with the smallest value of the objective function (or the highest "fitness") will be picked up for the new population. This kind of procedure favors better individuals, but also gives a chance for worse ones to be selected. The selection pressure can be adjusted by the candidate set size.

In some algorithms, such as PSO or DE, no selection is used (i.e. individuals are modified from iteration to iteration but never die).

There are two major types of operations that are used to modify individuals: exploratory (e.g. crossover in EA or ES) and exploitative (e.g. mutation in GA). Exploratory operators combine information contained in the parent individuals to create an offspring. An example of such an operator used by evolutionary algorithms with natural (floating point) representation is an arithmetic crossover, in which a new individual $\boldsymbol{c}$ can be created as a convex combination of the parents $\boldsymbol{p}_1$ and $\boldsymbol{p}_2$, i.e. $\boldsymbol{c} = \alpha\boldsymbol{p}_1 + (1 - \alpha)\boldsymbol{p}_2$, where $0 < \alpha < 1$ is a random number. These types of operators allow the making of large "steps" in the design space and, therefore, explore new and promising regions.

Exploitative operators introduce small perturbations (e.g. $\boldsymbol{p} \leftarrow \boldsymbol{p} + \boldsymbol{r}$, where $\boldsymbol{r}$ is a random vector selected according to a normal probability distribution with zero mean and certain, problem-dependent variance). These operators allow the exploitation of a given region of the design space, improving local search properties of the algorithm. In some of the more recent algorithms, e.g. PSO, the difference between both types of operators is not as clear (modification of the individual may be based on the best solution found so far by that given individual as well as the best solution found by the entire population). On the other hand, the operators used by algorithms such as PSO should be interpreted as an exchange between individuals of information that affect their behavior, rather than the individuals themselves.

3.3.4 *Particle swarm optimization*

PSO is nowadays one of the most popular metaheuristic algorithms. PSO is based on mimicking swarm behavior, such as fish or bird schooling, in nature (Kennedy *et al.*, 2001). Since its first development in 1995, many of its variations have been applied to virtually every area of optimization, computational intelligence, and design applications.

The PSO algorithm simulates a set of individuals that interact with each other using social influence and social learning. The swarm is typically modeled by particles in a multidimensional space that are represented by their positions and velocities. Individual particles communicate good positions to each other and adjust their own velocity and position based on this data. The movement of a swarming particle consists of two major components: a stochastic component and a deterministic component. Each particle is attracted toward the position of the current global best, $\boldsymbol{g}$, and its own best location, $\boldsymbol{x}_i^*$, found during the optimization run. At the same time, it has a tendency to move randomly.

Let $\boldsymbol{x}_i$ and $\boldsymbol{v}_i$ denote the position of the velocity vectors of the ith particle. Both vectors are updated according to the following rules:

$$\boldsymbol{v}_i \leftarrow \chi[\boldsymbol{v}_i + c_1\boldsymbol{r}_1 \bullet (\boldsymbol{x}_i^* - \boldsymbol{x}_i) + c_2\boldsymbol{r}_2 \bullet (\boldsymbol{g} - \boldsymbol{x}_i)], \tag{3.32}$$

$$\boldsymbol{x}_i \leftarrow \boldsymbol{x}_i + \boldsymbol{v}_i, \tag{3.33}$$

where $\boldsymbol{r}_1$ and $\boldsymbol{r}_2$ are vectors with components being uniformly distributed random numbers between 0 and 1 and $\bullet$ denotes componentwise multiplication. The control parameter χ is typically equal to 0.7298 (Kennedy *et al.*, 2001); c_1 and c_2 are acceleration constants determining how much the particle is directed toward good positions (they represent a "cognitive" and a "social" component, respectively); usually, $c_1, c_2 \approx 2$. The sum of c_1 and c_2 should not be smaller than 4.0; as the values of these parameters increase, χ gets smaller, as does the damping effect.

The initial locations of all particles should distribute relatively uniformly so that they can sample over most regions, which is especially important for multimodal problems. The initial velocity of a particle can be taken as zero.

There are many different variants and extensions of the standard PSO algorithm (Kennedy *et al.*, 2001; Clerc and Kennedy, 2002).

3.3.5 *Differential evolution*

DE was developed in the late 1990s by Storn and Price (Storn and Price, 1997). It is a vector-based evolutionary algorithm, and can be considered a further development to genetic algorithms. It is a derivative-free stochastic search algorithm with self-organizing tendency. The design parameters are represented as n-dimensional vectors (also called agents). New agents are allocated in the search space by combining the positions of other existing agents. More specifically, an intermediate agent is generated from two agents randomly chosen from the current population. This temporary agent is then mixed with a predetermined target agent. The new agent is accepted for the next generation if and only if it yields reduction in an objective function.

The basic DE algorithm uses a random initialization. A new agent, $\boldsymbol{y} = [y_1 \dots y_n]$, is created from the exisiting one, $\boldsymbol{x} = [x_1 \dots x_n]$, as follows.

1. Three agents, $\boldsymbol{a} = [a_1 \dots a_n]$, $\boldsymbol{b} = [b_1 \dots b_n]$, and $\boldsymbol{c} = [c_1 \dots c_n]$, are randomly extracted from the population (all distinct from each other and from $\boldsymbol{x}$).
2. A position index, $p \in \{1, \dots, N\}$, is determined randomly (N is the population size).
3. The new agent, $\boldsymbol{y}$, is determined as follows (separately for each of its components $i = 1, \dots, n$):
   a. select a random number, $r_i \in (0, 1)$, with uniform probability distribution;
   b. if $i = p$ or $r_i < CR$, let $y_i = a_i + F(b_i - c_i)$, otherwise let $y_i = x_i$; here, $0 \leq F \leq 2$ is the differential weight and $0 \leq CR \leq 1$ is the crossover probability, both defined by the user; and
   c. if $f(\boldsymbol{y}) < f(\boldsymbol{x})$ then replace $\boldsymbol{x}$ with $\boldsymbol{y}$; otherwise reject $\boldsymbol{y}$ and keep $\boldsymbol{x}$.

It should be emphasized that although DE resembles some other stochastic optimization techniques, unlike traditional EA, DE perturbs the solutions in the current generation vectors with scaled differences of two randomly selected agents. As a consequence, no separate probability distribution has to be used, and thus the scheme presents some degree of self-organization. Additionally, DE is simple to implement, uses very few control parameters, and has been observed to perform satisfactorily in a number of multimodal optimization problems (Chakraborty, 2008).

Many research studies have been focused on the choice of the control parameters F and CR as well as the modification of the updating scheme of Step 3b above. The interested reader is referred to Price *et al.* (2005) for details.

3.3.6 *Other methods*

Metaheuristic optimization is a fast-growing research area and a number of new techniques, as well as improvements of the existing ones, are being developed almost every year. Examples include ant colony optimization (Dorigo and Stutzle, 2004), bee algorithm (Yang, 2005), harmony search (Geem *et al.*, 2001), or firefly algorithm (Yang, 2008). Most of these techniques are based on observations of specific natural or social phenomena, however the general concept (as described in Section 4.3.3) is usually similar. A good exposition of the recent methods can be found in Yang (2010).

3.4 Sensitivity Analysis: The Adjoint Method

Gradient-based optimization routines require the knowledge of the objective function derivatives with respect to the designable parameters, $\boldsymbol{x}$, i.e. $\partial f/\partial x_k$, $k = 1, \ldots, n$. The most straightforward method of estimating the derivatives is finite differentiation. For example, in a central finite difference scheme we have (for small positive ε):

$$\frac{df(\boldsymbol{x})}{dx_k} \cong \frac{f([x_1 x_2 \ldots x_k + \varepsilon \ldots x_n]) - f([x_1 x_2 \ldots x_k - \varepsilon \ldots x_n])}{2\varepsilon}, \tag{3.34}$$

which requires $2n$ objective function evaluations. A less accurate but cheaper forward scheme,

$$\frac{df(\boldsymbol{x})}{dx_k} \cong \frac{f([x_1x_2\ldots x_k+\varepsilon\ldots x_n]) - f([x_1x_2\ldots x_k\ldots x_n])}{\varepsilon}, \quad (3.35)$$

still requires n additional evaluations of f.

An alternative method of realizing finite differentiation is a complex variable method:

$$\frac{df(\boldsymbol{x})}{dx_k} \cong \frac{\mathrm{Im}(f([x_1x_2\ldots x_k+i\varepsilon\ldots x_n]))}{\varepsilon}, \quad (3.36)$$

where i is an imaginary factor ($i = (-1)^{1/2}$) and $\mathrm{Im}(z)$ is an imaginary part of the complex number z. It has been shown that Equation (3.36) outperforms Equation (3.34). In particular, the estimated derivatives are less sensitive to the value of ε (Papadimitriou and Giannakoglou, 2008).

An interesting alternative strategy in obtaining sensitivity information is automatic differentiation (AD) (Griewank, 2000; Bischof *et al.*, 2008). AD is applied directly to the code of the computational model, in which, in order to generate the derivative code (tangent linear or adjoint model), the model code is decomposed into elementary functions corresponding (more or less) to the individual statements in the code. These elementary functions are differentiated (the results are referred to as local Jacobians). The derivative code multiplies these local Jacobians, which, according to the chain rule, yields the derivative of the composite function. As opposed to finite differentiation, AD provides sensitivity information that is accurate within round-off error. Automatic differentiation has been applied to aerodynamic problems (see, for example, Sherman *et al.* (1996); Takemiya (2008); Takemiya and Mavris (2009)).

Let us assume that the objective functions (some of them may actually represent design constraints) are functions of the design variables, $\boldsymbol{x} = [x_1x_2\ldots x_n]^T$, and a set of state variables, $\boldsymbol{u} = [u_1u_2\ldots u_M]^T$, i.e. we have $f_l = f_l(\boldsymbol{x}, \boldsymbol{u}(\boldsymbol{x}))$, $l = 1,\ldots,L$. Direct differentiation is an approach in which the gradient of the objective function is obtained by calculating the gradients of the state variables, u_m, with respect to the design variables, and applying a chain rule:

$$\frac{df_l(\boldsymbol{x})}{dx_k} = \frac{\partial f_l(\boldsymbol{x}, \boldsymbol{u}(\boldsymbol{x}))}{\partial x_k} + \frac{\partial f_l(\boldsymbol{x}, \boldsymbol{u}(\boldsymbol{x}))}{\partial \boldsymbol{u}} \frac{\partial \boldsymbol{u}(\boldsymbol{x})}{\partial x_k}. \tag{3.37}$$

Here, $\partial f_l / \partial \boldsymbol{u}$ is a row matrix, and $\partial \boldsymbol{u} / \partial x_k$ is a column matrix.

In the discrete direct differentiation approach, the state equations are cast in discrete form, $R_m = 0$, $m = 1, \ldots, M$, and differentiated to yield:

$$\frac{dR_m(\boldsymbol{x})}{dx_k} = \frac{\partial R_m(\boldsymbol{x}, \boldsymbol{u}(\boldsymbol{x}))}{\partial x_k} + \frac{\partial R_m(\boldsymbol{x}, \boldsymbol{u}(\boldsymbol{x}))}{\partial \boldsymbol{u}} \frac{\partial \boldsymbol{u}(\boldsymbol{x})}{\partial x_k} = 0. \tag{3.38}$$

The system of $M \times n$ equations is then solved for $\partial \boldsymbol{u} / \partial x_k$, and the results are substituted into Equation (3.37), allowing the computation of df_l / dx_k. The details of the continuous direct differentiation method can be found in Papadimitriou and Giannakoglou (2008). The direct differentiation approach is an efficient way of calculating gradients of many (objective/constraint) functions of a few design variables.

Yet another method of calculating gradients is the adjoint approach, which, in many practical cases, may lead to substantial computational savings when compared to the direct approach. In the discrete adjoint method, the so-called augmented functionals, $f_{aug.l}$, that are defined by introducing the adjoint variables, ψ_{ml}, are used, as follows (Papadimitriou and Giannakoglou, 2008):

$$\frac{df_{aug.l}(\boldsymbol{x})}{dx_k} = \frac{df_l(\boldsymbol{x})}{dx_k} + \psi_{ml} \frac{dR_m(\boldsymbol{x})}{dx_k}. \tag{3.39}$$

As $dR_m / dx_k = 0$, the derivatives of $f_{aug.l}$ and f_l are identical. From Equations (3.37) and (3.38):

$$\frac{df_{aug.l}(\boldsymbol{x})}{dx_k} = \frac{\partial f_l(\boldsymbol{x}, \boldsymbol{u}(\boldsymbol{x}))}{\partial x_k} + \frac{\partial f_l(\boldsymbol{x}, \boldsymbol{u}(\boldsymbol{x}))}{\partial \boldsymbol{u}} \frac{\partial \boldsymbol{u}(\boldsymbol{x})}{\partial x_k} + \psi_{ml} \left(\frac{\partial R_m(\boldsymbol{x}, \boldsymbol{u}(\boldsymbol{x}))}{\partial x_k} + \frac{\partial R_m(\boldsymbol{x}, \boldsymbol{u}(\boldsymbol{x}))}{\partial \boldsymbol{u}} \frac{\partial \boldsymbol{u}(\boldsymbol{x})}{\partial x_k} \right), \tag{3.40}$$

which is:

$$\frac{df_{aug.l}(\boldsymbol{x})}{dx_k} = \frac{\partial f_l(\boldsymbol{x},\boldsymbol{u}(\boldsymbol{x}))}{\partial x_k} + \psi_{ml}\frac{\partial R_m(\boldsymbol{x},\boldsymbol{u}(\boldsymbol{x}))}{\partial x_k} + \left(\frac{\partial f_l(\boldsymbol{x},\boldsymbol{u}(\boldsymbol{x}))}{\partial \boldsymbol{u}} + \psi_{ml}\frac{\partial R_m(\boldsymbol{x},\boldsymbol{u}(\boldsymbol{x}))}{\partial \boldsymbol{u}}\right)\frac{\partial \boldsymbol{u}(\boldsymbol{x})}{\partial x_k}. \tag{3.41}$$

By satisfying the adjoint equations,

$$\frac{\partial f_l(\boldsymbol{x},\boldsymbol{u}(\boldsymbol{x}))}{\partial \boldsymbol{u}} + \psi_{ml}\frac{\partial R_m(\boldsymbol{x},\boldsymbol{u}(\boldsymbol{x}))}{\partial \boldsymbol{u}} = 0, \tag{3.42}$$

the derivatives of the augmented functionals can be obtained as:

$$\frac{df_{aug.l}(\boldsymbol{x})}{dx_k} = \frac{\partial f_l(\boldsymbol{x},\boldsymbol{u}(\boldsymbol{x}))}{\partial x_k} + \psi_{ml}\frac{\partial R_m(\boldsymbol{x},\boldsymbol{u}(\boldsymbol{x}))}{\partial x_k}. \tag{3.43}$$

As opposed to the discrete direct approach, the discrete adjoint method requires the solution of $M \times L$ equations, and, having the adjoint variables, the required derivatives are obtained from Equation (3.43). Consequently, the adjoint method is a very efficient method of calculating gradients if $n \gg L$, i.e. the number of designable variables is much larger than the number of functions, which is typically the case for aerodynamic problems. A compact description of the continuous adjoint approach can be found in Papadimitriou and Giannakoglou (2008).

3.5 Conclusion

The direct optimization methods described in this chapter are attractive solutions for solving various engineering design problems. Depending on the situation, some of those methods are better than others. For example, gradient-based optimizers are more suitable for optimizing smooth objective functions, particularly if it is possible to obtain cheap sensitivity information (e.g. through adjoints). Derivative-free methods are suitable for handling analytically intractable (noisy, discontinuous) functions. Among them, metaheuristics allow for global optimization as well as handling problems with multiple objectives.

Despite their usefulness, direct methods have an important disadvantage related to the computational cost of the optimization process. Conventional optimization algorithms (e.g. gradient-based schemes with numerical derivatives) require tens, hundreds, or even thousands of objective function calls per run, depending on the number of design variables. Numerical simulations can be very time-consuming; in some instances (e.g. high-fidelity three-dimensional wing or full aircraft simulations) the evaluation time can be as long as several hours, days, or even weeks per single design. Therefore, attempts to solve Equation (3.1) by embedding the simulator directly in the optimization loop often may be impractical, even when adjoint sensitivity is employed. On the other hand, the availability of massive computing resources is not always translated into computational speed-up due to a growing demand for simulation accuracy, both by including more accurate models (governing equations), multi-physics, and second-order effects, and by using finer discretization of the structure under consideration. An additional problem is numerical noise, which is always present in CFD-based objective functions.

The situation is even more challenging in the case of metaheuristics. Population-based algorithms normally require a large number of objective function evaluations to converge. Typical population size is anything between 10 and 100, whereas the number of iterations may be a few dozen to a few hundred. Also, their performance may be heavily dependent on the values of control parameters, which may not be easy to determine beforehand. Finally, because they are stochastic methods, a solution obtained in each run of the algorithm will be generally different from the previous one. Therefore, from the point of view of engineering design, metaheuristics is an attractive approach for problems in which the evaluation time of the objective function is not of primary concern and when multiple local optima are expected. A typical application would be finding infill samples for kriging-based surrogate models, which requires global optimization but with a computationally cheap objective function (Couckuyt, 2013).

One of the possible methods of handling computationally expensive design problems is SBO, in which direct optimization of the high-fidelity model is replaced by optimization of its cheap and analytically tractable representation, referred to as a surrogate model. As indicated in the following chapters, it is possible to construct and exploit such representations in such a way that — with occasional reference to the high-fidelity model — a satisfactory design can be found at a fraction of the computational effort required by direct optimization.

Chapter 4

Surrogate-Based Optimization

In this chapter, the basic concepts of surrogate-based optimization (SBO) are presented. In particular, we describe the SBO optimization flow and discuss important properties of the SBO process. As the fundamental component of SBO is the surrogate model itself, part of the chapter is focused on various surrogate modeling techniques. We also emphasize a distinction between function approximation and physics-based surrogates, as well as discussing an important issue of exploration and exploitation of the design space in the context of SBO. Formulation and discussion of various SBO techniques, exploiting both approximation- and physics-based surrogates, will be presented in Chapters 5 and 6, respectively.

4.1 Design Problem Formulation: Surrogate-Based Optimization Concept

The high computational costs of accurate, high-fidelity computational fluid dynamics (CFD) simulation, and the resulting difficulties of automated, simulation-driven aerodynamic design, exploiting numerical optimization techniques, are the main driving forces behind the development of more efficient design methods. One of the options is the use of cheap adjoint sensitivities (Pironneau, 1974; Jameson, 1988; Baysal and Eleshaky, 1991; Desai and Ito, 1994; Huan and Modi, 1994; Burgreen and Baysal, 1996; Anderson and Venkatakrishnan, 1999; Maute *et al.*, 2003; Martins *et al.*, 2005), making gradient-based optimization feasible even for reasonably expensive problems with relatively large numbers of designable parameters (Baysal and Eleshaky, 1992; Korivi *et al.*, 1998;

Kim *et al.*, 2001; Le Moigne and Qin, 2004). Arguably, gradient-based optimization with adjoints is currently the dominant and (theoretically) one of the best-founded approaches in CFD-based aerodynamic design.

SBO (Bandler *et al.*, 2004a; Queipo *et al.*, 2005; Forrester and Keane, 2009; Koziel *et al.*, 2011b) — the main topic of this book — is relatively recent but nevertheless one the most promising approaches to alleviating the problem of high computational costs of CFD-based design. Despite addressing the computational aspect of the design process, SBO may be helpful in handling other problems, such as the numerical noise that is always present in CFD-based objective functions and searching for globally optimum designs (Jones *et al.*, 1998; Jones, 2001; Couckuyt, 2013). While many SBO techniques do not rely on derivative information, it is possible — as indicated later in this book — to incorporate sensitivity data (if available) into the SBO process, with the benefit of improved performance and further reductions of the design cost (see also Alexandrov *et al.* (1998); Koziel *et al.* (2012a)).

Let us recall the formulation of the aerodynamic design problem as a nonlinear minimization task:

$$\boldsymbol{x}^* = \arg\min_{\boldsymbol{x}} H(f(\boldsymbol{x})), \tag{4.1}$$

where $f(\boldsymbol{x})$ is the high-fidelity model of system of interest (e.g. an airfoil) and $\boldsymbol{x} \in R^n$ is a vector of designable variables, e.g. those parameterizing the airfoil geometry. $f(\boldsymbol{x})$ may represent aerodynamic forces (wave drag, lift coefficient, etc.) or certain intermediate data such as pressure distribution. H is the objective function that is defined, so that a better design corresponds to a lower value of $H(f(\boldsymbol{x}))$. An optimal design vector is denoted by $\boldsymbol{x}^*$. Here, it is assumed that f is obtained through computationally expensive CFD simulations.

Conventional optimization algorithms (e.g. gradient-based schemes with numerical derivatives) require tens, hundreds, or even thousands of objective function calls per run, depending on the number of design variables. Numerical simulations can be very time-consuming; in some instances (e.g. high-fidelity three-dimensional wing or full aircraft simulations) the evaluation time can be as long as several hours, days, or even weeks per single design. Therefore, attempts to solve Equation (4.1) by embedding the simulator directly into the optimization loop may often be impractical, even when adjoint sensitivity is employed. On the other hand, the availability of massive computing resources is not always translated into computational speed-up, due to a growing demand for simulation accuracy, both by including more accurate models (governing equations), multi-physics, and second-order effects and by using finer discretization of the structure under consideration. An additional problem is numerical noise, which is always present in CFD-based objective functions.

SBO (Bandler *et al.*, 2004a; Queipo *et al.*, 2005; Forrester and Keane, 2009) has been developed from the practical necessity of handling computationally expensive or otherwise difficult engineering design problems. SBO is potentially one of the most effective approaches when time-consuming computer simulations are involved. The fundamental idea behind SBO is to replace the direct optimization of the computationally expensive model with an iterative process that involves the creation, optimization, and updating of a fast and analytically tractable surrogate model. Surrogate modeling is a fast-evolving research area in itself (Gunn (1998); Simpson *et al.* (2001), and references therein; and Schölkopf (2004); Rasmussen and Williams (2006); Couckuyt (2013), and references therein). In the context of SBO, it is normally sufficient that the surrogate is a locally accurate representation of the high-fidelity model, whereas global accuracy (i.e. across the entire design space) is not essential (Bandler *et al.*, 2004a). The surrogate model serves as a prediction tool, supposedly guiding the SBO process toward a better design. The design obtained through optimizing the surrogate model is verified by evaluating the high-fidelity model. These data are subsequently used to update the surrogate. SBO proceeds in this predictor-corrector fashion iteratively until some termination criterion is met. The practical advantage of SBO stems from the fact that most of the operations are performed on the surrogate model with the high-fidelity model referred to only occasionally. As a result, SBO can often significantly reduce the computational cost of the optimization process when compared to optimizing the high-fidelity model directly, without resorting to any surrogate.

In this chapter, we review the basics of SBO. We briefly describe the SBO flow and main properties and discuss the surrogate modeling process, as well as various ways of constructing the surrogate models. In particular, we emphasize the distinction between models based on function approximations of sampled high-fidelity model data and models constructed from physically based low-fidelity models, the latter being the main focus of this book. A discussion of various SBO methods will be presented in Chapters 5 and 6.

4.2 Surrogate-Based Optimization Basics and Formulation

As mentioned in Section 4.1, the fundamental idea of SBO is that direct optimization of an expensive model — here, accurate high-fidelity CFD simulation of the aerodynamic component/structure of interest — is replaced by an iterative process, in which the estimated location of the optimum design is obtained by optimizing a fast representation of the high-fidelity model, referred to as a surrogate. Normally, the high-fidelity model is evaluated at such a candidate design to verify it. Depending on the particular algorithm implementation (Koziel *et al.*, 2011b), the new design may or may not be accepted. The surrogate model is subsequently enhanced using the accumulated high-fidelity model data. Formally, the SBO design process can be represented as (Koziel *et al.*, 2011b):

$$\boldsymbol{x}^{(i+1)} = \arg\min_{\boldsymbol{x}} s^{(i)}(\boldsymbol{x}), \tag{4.2}$$

where $\boldsymbol{x}^{(i)}$, $i = 0, 1, \ldots,$ is a sequence of approximated solutions to the original Equation (4.1), whereas $s^{(i)}$ is the surrogate model at iteration i. Here, $\boldsymbol{x}^{(0)}$ is the initial design. The surrogate model is supposedly much cheaper in computational terms than the high-fidelity model f. On the other hand, it is assumed to be sufficiently accurate, particularly in the vicinity of the current design $\boldsymbol{x}^{(i)}$. If these conditions are satisfied, Equation (4.2) is likely to quickly approach the high-fidelity optimum $\boldsymbol{x}^*$. Typically, the high-fidelity model is

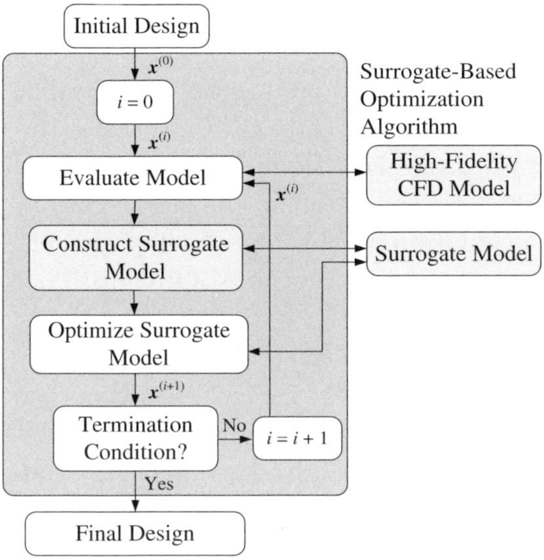

Fig. 4.1. SBO concept and flow diagram. An approximate high-fidelity model optimum is found in an iterative manner by optimizing the surrogate model. The high-fidelity model is evaluated at each new design for verification purposes. If the termination condition is not satisfied, the surrogate model is updated and the search continues. In many cases the high-fidelity model is evaluated only once per iteration. The number of iterations needed in SBO is often substantially fewer than those in conventional (direct) optimization techniques.

evaluated only once per iteration (at every new design $\boldsymbol{x}^{(i+1)}$). As mentioned before, these data are used for design verification and also to update the surrogate model. Figure 4.1 shows the flow diagram of the generic SBO process.

It should be emphasized that — in the optimization context — global accuracy of the surrogate (i.e. across the entire design space) is not of primary concern, unless globally optimum design is sought (Koziel *et al.*, 2011b; Couckuyt, 2013). For the sake of computational savings, it is normally more advantageous to take care of the surrogate model accuracy only in the most promising parts of the design space, in particular along the optimization path. Many SBO algorithms are implemented so that — upon the algorithm convergence — accumulated high-fidelity data allow very good agreement between the surrogate and the high-fidelity model in the optimum design vicinity, whereas the surrogate quality in other areas is usually not important.

One of the aforementioned assumptions concerning the surrogate model is that it is computationally cheap. If this is the case, the optimization cost associated with Equation (4.2) can be viewed as small or even negligible, so the total optimization cost is mostly determined by the evaluation of the high-fidelity model. Normally, the number of iterations needed by the SBO algorithm is substantially fewer than that needed by the majority of methods that optimize the high-fidelity model directly (e.g. gradient-based schemes with numerical derivatives) (Koziel *et al.*, 2006). Unfortunately, the cost of evaluating the surrogate model cannot be neglected in some cases. A typical example would be physics-based surrogates constructed from underlying low-fidelity CFD simulations, in which the time evaluation ratio between the high- and low-fidelity models may be as low as 20–50 (Koziel and Leifsson, 2013a). This poses an additional challenge for designing SBO algorithms: ensuring that not only the number of high-fidelity but also surrogate-model evaluations is as small as possible.

From the algorithm convergence standpoint, the following consistency conditions between the surrogate and the high-fidelity models are important: zero-order consistency, $s^{(i)}(\boldsymbol{x}^{(i)}) = f(\boldsymbol{x}^{(i)})$, and first-order consistency, $\nabla s^{(i)}(\boldsymbol{x}^{(i)}) = \nabla f(\boldsymbol{x}^{(i)})$ (Alexandrov and Lewis, 2001), where ∇ stands for the model gradient (or Jacobian in the case of vector-valued models). If both conditions are satisfied, the SBO Equation (4.2) is provably convergent to a local optimum of f, under some extra assumptions regarding model smoothness that are also necessary (Alexandrov *et al.*, 1998; Echeverría and Hemker, 2008) and upon enhancing the algorithm by a trust region mechanism (Conn *et al.*, 2000):

$$\boldsymbol{x}^{(i+1)} = \arg\min_{\boldsymbol{x}} U(\boldsymbol{R}_s^{(i)}(\boldsymbol{x})) \quad \text{subject to } \|\boldsymbol{x} - \boldsymbol{x}^{(i)}\| \leq \delta^{(i)}, \tag{4.3}$$

where $\delta^{(i)}$ denotes the trust region radius at iteration i. The trust region is updated at every iteration based on the quality of prediction produced by the surrogate model (Conn *et al.*, 2000). The idea of the trust region approach is that if the vicinity of $\boldsymbol{x}^{(i)}$ is sufficiently small, the first-order consistent surrogate becomes, in this vicinity, a sufficiently suitable representation of f to generate — through Equation (4.3) — a design that reduces the high-fidelity objective function value.

In order to satisfy the first-order consistency condition, both high-fidelity and surrogate-model sensitivity data are required. Convergence of the SBO algorithm can also be guaranteed under various other settings, see, for example, Booker *et al.* (1999) (surrogate management framework), Koziel *et al.* (2006) and Koziel *et al.* (2008b) (space mapping (SM)), or Echeverría and Hemker (2008) (manifold mapping). For methods exploiting physics-based surrogates, the quality of the underlying low-fidelity model may be critical (e.g. Koziel *et al.* (2008b)).

Another important aspect of SBO is the exploration and exploitation of the design space. In general, the basic SBO scheme (Equation (4.2)) can be implemented as a local or global optimization. In many situations, particularly if the trust-region-like convergence safeguards (Equation (4.3)) are used and the surrogate model is first-order consistent, the SBO scheme is typically executed as a local search. On the other hand, if the surrogate model is constructed to represent the high-fidelity model over the entire design space, it is possible to turn the process (Equation (4.2)) into a global optimization algorithm. In this case, promising designs are obtained by optimizing the surrogate using global methods, e.g. evolutionary algorithms (Jones *et al.*, 1998; Couckuyt, 2013). The surrogate model itself is updated using certain statistical infill criteria based on the expected improvement of the objective function or minimization of the overall modeling error.

4.3 Surrogate Model Construction: Approximation- and Physics-Based Surrogates

SBO is a relatively young area. Still, there have been a number of SBO methods developed over the last two decades or so (Alexandrov and Lewis, 2001; Bandler *at al.*, 2004a; Queipo *et al.*, 2005; Koziel, 2010a). Many of these techniques are quite recent and have been developed out of practical necessity, specifically to solve problems otherwise unmanageable at the time. A good example is SM (Bandler *et al.*, 2004a), developed to handle optimization problems in microwave engineering. Another, relatively fresh, example is shape-preserving response prediction (Koziel, 2010a), applied both to electrical engineering and aerodynamic shape optimization problems (Koziel, 2010a; Koziel and Leifsson, 2013a).

The majority of SBO methods rely on the same principles as discussed in Section 4.1. The differences between particular techniques are not only in the way they operate but also, more importantly, in the method of constructing the surrogate model. As explained earlier, the surrogate is a fundamental component of all SBO algorithms. There are several characteristics of the surrogate that are important from the point of view of the SBO performance: it should be computationally cheap, analytically tractable (smooth), and reasonably accurate for reliable prediction of the high-fidelity model minimizer. The last feature is often understood in the local sense (Bandler *et al.*, 2004a).

Surrogate models can be roughly divided into two categories: approximation (or functional); and physics-based. Functional surrogate models are constructed through approximations of the high-fidelity model data obtained in the process of sampling the design space using appropriate design of experiments (DOE) methodologies (Queipo *et al.*, 2005). In Section 5.4 we briefly discuss the modeling flow for this type of surrogate model, various strategies for allocating samples (Simpson *et al.*, 2001), generating approximations (Simpson *et al.*, 2001; Queipo *et al.*, 2005; Forrester and Keane, 2009), and validating the models. Approximation-based surrogates can be considered generic models that are independent of the physics or any other knowledge of the problem at hand. The other class of models is based on exploitation of some knowledge about the system under consideration, usually embedded in the physics-based low-fidelity model. In the case of aerodynamic problems, the low-fidelity model can be obtained using a simplified analytical description of the structure under design or from coarse-discretization CFD simulations. The surrogate itself is constructed by the appropriate correction of the low-fidelity model, usually based on a limited amount of high-fidelity model data. The concept and discussion of physics-based surrogates is presented in Section 4.5.

4.4 Approximation-Based Surrogates

Approximation-based models are probably the most popular class of surrogates due to a large variety of techniques that are available in the literature and implemented as ready-to-use toolboxes, particularly in MATLAB (2012). The models from this group have the following characteristics (Søndergaard, 2003; Forrester and Keane, 2009).

- They can be constructed without previous knowledge of the physical system of interest.
- They are generic, and therefore applicable to a wide class of problems.
- They are usually based on algebraic (often explicitly formulated) models.
- They are often very cheap to evaluate.
- They usually require a considerable amount of data to ensure reasonable accuracy; the number of training samples quickly grows with the dimensionality of the design space.

All but the last item in the above list are advantages of approximation models. Large numbers of training samples may be a problem, particularly in the optimization context and particularly for one-time *ad hoc* optimization of a given structure: the computational effort necessary to set up a decent surrogate model may be higher than the cost of even direct optimization of the high-fidelity model.

In this section, we discuss the surrogate modeling flow for approximation-based surrogate construction. We briefly describe various designs of experiment methods and approximation techniques, as well as model validation.

4.4.1 *Surrogate modeling flow*

Approximation-based surrogates are constructed from the sampled high-fidelity model data. Figure 4.2 shows the modeling flowchart for this type of model. The first stage is DOE, which is essentially

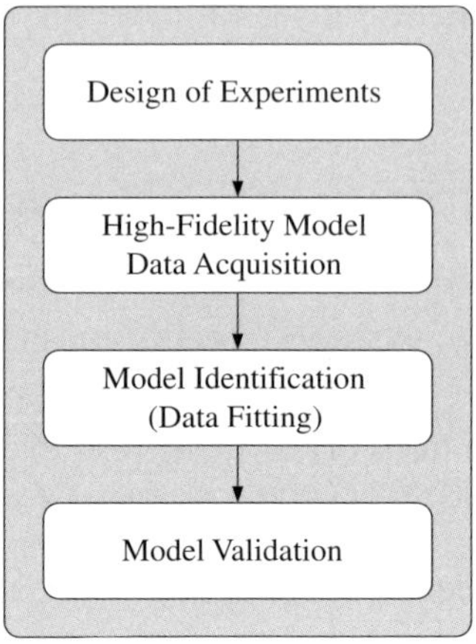

Fig. 4.2. Approximation-based surrogate model construction flowchart. In case the accuracy of the model is not sufficient, the procedure can be iterated by adding more training data points and updating the surrogate.

a strategy allocating the required number of samples in the design space. In many cases, the number of training samples is restricted by the available computational budget (in certain situations, only a few points may be allowed). Nowadays, uniform sampling is usually preferred (Simpson *et al.*, 2001). A brief review of the most popular DOE techniques is provided in Section 4.4.2.

Having acquired high-fidelity model data, the model identification is performed using a selected approximation method (see Section 4.4.3 for an overview of available methods). Although in some cases the surrogate can be found using explicit formulas by solving an appropriate regression problem (e.g. polynomial approximation (Queipo *et al.*, 2005)), in most situations it is computed by means of a separate minimization problem (e.g. when using kriging (Kleijnen, 2009) or neural networks (Rayas-Sanchez, 2004)).

The surrogate has to be verified to ensure sufficient accuracy, particularly its generalization capability, i.e. the quality of model predictions at the designs not seen during the identification stage. For that reason, model testing is normally performed using a separate set of testing samples. One of the important issues is to keep a balance between accuracy in the known and unknown data (training and testing set, respectively).

In practice, the modeling process may be iterative, with the flow depicted in Fig. 4.2 constituting a single iteration. Upon validation of the surrogate, new set-up samples, together with the corresponding high-fidelity model data, are then used to re-identify the model, and such an adaptive sampling scheme is continued until the accuracy goals are met. In the optimization context, the surrogate model update may also be oriented toward finding better designs rather than toward ensuring global accuracy (Forrester and Keane, 2009).

4.4.2 *Design of experiments*

DOE (Koehler and Owen, 1996; Giunta *et al.*, 2003; Santner *et al.*, 2003) is a strategy for allocating samples (points) in the design space. The high-fidelity model is evaluated at the selected locations to create the training data set that subsequently is used to construct the functional surrogate model. The purpose is to maximize — in a given sense — the amount of information acquired from the high-fidelity model data. It is clear that there is a trade-off between the number of training samples used and the amount of information that can be extracted from these points. If no prior knowledge about the problem is available, the samples are typically spread apart as much as possible in order to capture global trends in the design space. In some cases (adaptive sampling; Devebhaktuni *et al.* (2001); Couckuyt (2013)), information about the model performance may be fed back to the sampling algorithm so that more emphasis is put (and, consequently, more samples are allocated) in the design space regions in which the behavior of the system is more nonlinear.

Chronologically, the oldest DOE techniques are factorial designs (Giunta *et al.*, 2003), in which the samples are allocated in the extreme points of the design space (e.g. center, corners, etc.). Spreading them apart had the effect of minimizing possible errors of estimating the main trends as well as the interactions between design variables, if the data about the system were coming from physical measurements. On the other hand, some of the factorial designs are rather "economical" in terms of the number of samples, which is important if the computational budget for data acquiring is very

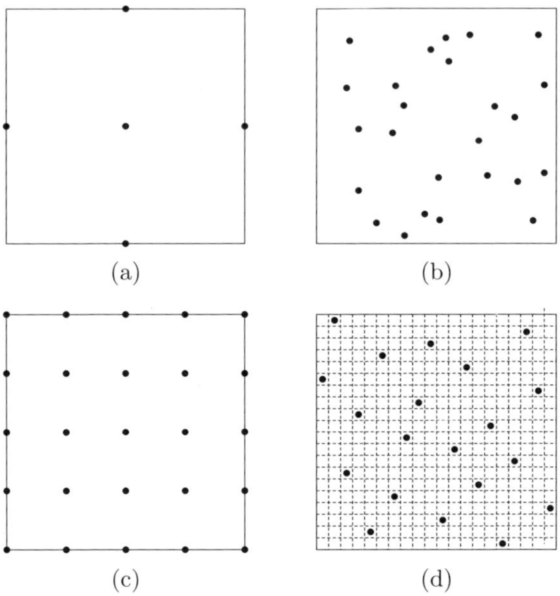

Fig. 4.3. Popular design of experiments techniques. (a) One of the possible factorial designs (star distribution); (b) random sampling; (c) uniform grid sampling; (d) LHS.

limited. One of such designs, so-called star distribution (McKay *et al.* (1979); see also Fig. 4.3(a)), is frequency used in combination with SM (Cheng *et al.*, 2006).

In cases in which no prior knowledge about the system at hand is available (typical while constructing the initial surrogate), most recent DOE approaches attempt to allocate the samples uniformly within the design space (Queipo *et al.*, 2005). A number of such space-filling DOE are available. The simplest ones either do not ensure sufficient uniformity (e.g. pseudo-random sampling (Guinta *et al.*, 2003), see also Fig. 4.3(b)) or are not practical (e.g. stratified random sampling, in which the number of samples needed is on the order of 2^n). Uniform grid sampling may be a good solution for low-dimensional spaces (Fig. 4.3(c)): it provides the best possible uniformity, however the number of samples is restricted to $N_1 \cdot N_2 \cdot \cdots \cdot N_n$, where N_j is the number of samples along jth axis of the design space.

Probably the most popular DOE for (relatively) uniform sample distributions is Latin hypercube sampling (LHS) (McKay *et al.*, 1979). In order to allocate p samples with LHS, the range for each parameter is divided into p bins, which, for n design variables, yields a total number of p^n bins in the design space. The samples are randomly selected in the design space so that: each sample is randomly placed inside a bin, and for all one-dimensional projections of the p samples and bins, there is exactly one sample in each bin. Figure 4.3(d) shows an example LHS realization of 20 samples for two design variables ($n = 2$). It is worth noting that the standard LHS may lead to non-uniform distributions (for example, samples allocated along the design space diagonal satisfy both of the conditions above). Various improvements of standard LHS have been developed which provide more uniform sampling distributions (Ye, 1998; Palmer and Tsui, 2001; Beachkofski and Grandhi, 2002; Leary *et al.*, 2003).

Other uniform sampling techniques that are commonly used include orthogonal array sampling (Queipo *et al.*, 2005), quasi-Monte Carlo sampling (Guinta *et al.*, 2003), or Hammersley sampling (Guinta *et al.*, 2003). It should also be mentioned that uniform sampling can be approached as an optimization problem. In this case, starting from an initial set of samples, their uniformity can be improved by minimizing a specific non-uniformity measure, e.g. $\sum_{i=1,\ldots,p} \sum_{j=i+1,\ldots,p} d_{ij}^{-2}$ (Leary *et al.*, 2003), where d_{ij} is the Euclidean distance between samples i and j.

4.4.3 *Approximation techniques: polynomial regression*

One of the simplest methods of approximating sampled high-fidelity model data is polynomial regression (Queipo *et al.*, 2005). The surrogate model assumes the following relation between the function of interest f and K polynomial basis functions v_j using p samples $f(\boldsymbol{x}^{(i)})$, $i = 1, \ldots, p$:

$$f(\boldsymbol{x}^{(i)}) = \sum_{j=1}^{K} \beta_j v_j(\boldsymbol{x}^{(i)}). \tag{4.4}$$

These equations can be represented in a matrix form:

$$\boldsymbol{f} = \boldsymbol{X}\boldsymbol{\beta}, \tag{4.5}$$

where $\boldsymbol{f} = [f(\boldsymbol{x}^{(1)})\ f(\boldsymbol{x}^{(2)})\ \ldots\ f(\boldsymbol{x}^{(p)})]^T$, $\boldsymbol{X}$ is a $p \times K$ matrix containing the basis functions evaluated at the sample points, and $\beta = [\beta_1 \beta_2 \ldots \beta_K]^T$. The number of sample points, p, should be consistent with the number of basis functions considered, K (typically $p \geq K$). If the sample points and basis functions are taken arbitrarily, some columns of $\boldsymbol{X}$ can be linearly dependent. If $p \geq K$ and $\text{rank}(\boldsymbol{X}) = K$, a solution to Equation (4.5) in the least-squares sense can be computed through $\boldsymbol{X}^+$, the pseudoinverse (or generalized inverse) of $\boldsymbol{X}$ (Golub and Van Loan, 1996):

$$\boldsymbol{\beta} = \boldsymbol{X}^+ = (\boldsymbol{X}^T\boldsymbol{X})^{-1}\boldsymbol{X}^T\boldsymbol{f}. \tag{4.6}$$

The simplest examples of regression models are the first- and second-order polynomial models:

$$s(\boldsymbol{x}) = s([x_1\ x_2\ \ldots\ x_n]^T) = \beta_0 + \sum_{j=1}^{n} \beta_j x_j, \tag{4.7}$$

$$s(\boldsymbol{x}) = s([x_1\ x_2\ \ldots\ x_n]^T) = \beta_0 + \sum_{j=1}^{n} \beta_j x_j + \sum_{i=1}^{n} \sum_{j \leq i}^{n} \beta_{ij} x_i x_j. \tag{4.8}$$

It should be mentioned that the modeling scheme (Equations (4.4)–(4.6)) can be used for other types of basis functions, not necessarily monomials, as long as the surrogate model is linear with respect to its parameter set, $\boldsymbol{\beta}$. Polynomial interpolation/regression appears naturally and is crucial in developing robust and efficient derivative-free optimization algorithms. Interested readers are referred to Conn *et al.* (2009) for more details.

4.4.4 *Approximation techniques: radial basis functions*

Radial basis function interpolation/approximation (Wild *et al.*, 2008; Forrester and Keane, 2009) exploits linear combinations of K radially symmetric functions, ϕ:

$$s(\boldsymbol{x}) = \sum_{j=1}^{K} \lambda_j \phi(\|\boldsymbol{x} - \boldsymbol{c}^{(j)}\|), \tag{4.9}$$

where $\lambda = [\lambda_1 \lambda_2 \ldots \lambda_K]^T$ is the vector of model parameters and $\boldsymbol{c}^{(j)}$, $j = 1, \ldots, K$ are the (known) basis function centers.

The model parameters, λ, can be calculated similarly to polynomial regression:

$$\boldsymbol{\lambda} = \boldsymbol{\Phi}^{+} = (\boldsymbol{\Phi}^T \boldsymbol{\Phi})^{-1} \boldsymbol{\Phi}^T \boldsymbol{f}, \tag{4.10}$$

where again $\boldsymbol{f} = [f(\boldsymbol{x}^{(1)}) f(\boldsymbol{x}^{(2)}) \ldots f(\boldsymbol{x}^{(p)})]^T$ and the $p \times K$ matrix Φ is defined as:

$$\boldsymbol{\Phi} = [\Phi_{kl}]_{k=1,\ldots,p;\ l=1,\ldots,K}, \tag{4.11}$$

where

$$\Phi_{kl} = \phi(\|\boldsymbol{x}^{(k)} - \boldsymbol{c}^{(l)}\|). \tag{4.12}$$

If the number of basis functions is equal to the number of samples, i.e. $p = K$, the centers of the basis functions coincide with the data points and are all different, and Φ is a regular square matrix. In such a case, we have $\lambda = \Phi^{-1} \boldsymbol{f}$.

Typical choices for the basis functions are $\phi(r) = r$, $\phi(r) = r^3$, or $\phi(r) = r^2 \ln r$ (thin plate spline). More flexibility can be obtained by using parametric basis functions such as $\phi(r) = \exp(-r^2/2\sigma^2)$ (Gaussian), $\phi(r) = (r^2 + \sigma^2)^{1/2}$ (multi-quadric), or $\phi(r) = r^2 + \sigma^2)^{-1/2}$ (inverse multi-quadric).

4.4.5 *Approximation techniques: kriging*

Kriging is one of the most popular techniques for interpolating deterministic noise-free data (O'Hagan, 1978; Journel and Huijbregts, 1981; Simpson *et al.*, 2001; Kleijnen, 2009). It is a Gaussian process-based modeling method, which is compact and cheap to evaluate (Rasmussen and Williams, 2006). Kriging has applications in numerous areas (e.g. Jones *et al.* (1998) and Forrester and Keane (2009) for applications in optimization).

In its basic formulation, kriging (Journel and Huijbregts, 1981; Simpson *et al.*, 2001) assumes that the function of interest f is of the form:

$$f(\boldsymbol{x}) = \boldsymbol{g}(\boldsymbol{x})^T \boldsymbol{\beta} + Z(\boldsymbol{x}), \tag{4.13}$$

where $\boldsymbol{g}(\boldsymbol{x}) = [g_1(\boldsymbol{x}) g_2(\boldsymbol{x}) \ldots g_K(\boldsymbol{x})]^T$ are known (e.g. constant) functions, usually low-order polynomials and $\boldsymbol{\beta} = [\beta_1 \beta_2 \ldots \beta_K]^T$ are the unknown model parameters, whereas $Z(\boldsymbol{x})$ is a realization of a normally distributed Gaussian random process with zero mean and variance σ^2.

The regression part, $\boldsymbol{g}(\boldsymbol{x})^T\ \beta$, is a trend function that globally approximates f. $Z(\boldsymbol{x})$ takes into account localized variations. The covariance matrix of $Z(\boldsymbol{x})$ is given as:

$$Cov[Z(\boldsymbol{x}^{(i)})Z(\boldsymbol{x}^{(j)})] = \sigma^2 \boldsymbol{R}([R(\boldsymbol{x}^{(i)}, \boldsymbol{x}^{(j)})]), \tag{4.14}$$

where $\boldsymbol{R}$ is a $p \times p$ correlation matrix with $R_{ij} = R(\boldsymbol{x}^{(i)}, \boldsymbol{x}^{(j)})$. Here, $R(\boldsymbol{x}^{(i)}, \boldsymbol{x}^{(j)})$ is the correlation function between sampled data points $\boldsymbol{x}^{(i)}$ and $\boldsymbol{x}^{(j)}$. One of the most popular choices is the Gaussian correlation function:

$$R(\boldsymbol{x}, \boldsymbol{y}) = \exp\left[-\sum_{k=1}^{n} \theta_k |x_k - y_k|^2\right], \tag{4.15}$$

where θ_k are unknown correlation parameters (or hyperparameters) and x_k and y_k are the kth component of the vectors $\boldsymbol{x}$ and $\boldsymbol{y}$, respectively.

The kriging predictor (Journel and Huijbregts, 1981; Simpson *et al.*, 2001) is defined as:

$$s(\boldsymbol{x}) = \boldsymbol{g}(\boldsymbol{x})^T \boldsymbol{\beta} + \boldsymbol{r}^T(\boldsymbol{x})\boldsymbol{R}^{-1}(\boldsymbol{f} - \boldsymbol{G}\boldsymbol{\beta}), \tag{4.16}$$

where $\boldsymbol{r}(\boldsymbol{x}) = [R(\boldsymbol{x}, \boldsymbol{x}^{(1)}) \ldots R(\boldsymbol{x}, \boldsymbol{x}^{(p)})]^T$, $\boldsymbol{f} = [f(\boldsymbol{x}^{(1)}) f(\boldsymbol{x}^{(2)}) \ldots f(\boldsymbol{x}^{(p)})]^T$, and $\boldsymbol{G}$ is a $p \times K$ matrix with $G_{ij} = g_j(\boldsymbol{x}^{(i)})$. The vector of model parameters β can be computed as $\beta = (\boldsymbol{G}^T\boldsymbol{R}^{-1}\boldsymbol{G})^{-1} \boldsymbol{G}^T\boldsymbol{R}^{-1}\boldsymbol{f}$. An estimate of the variance σ^2 is given by $\hat{\sigma}^2 = p^{-1} (\boldsymbol{f} - \boldsymbol{G}\boldsymbol{\beta})^T\boldsymbol{R}^{-1}(\boldsymbol{f} - \boldsymbol{G}\boldsymbol{\beta})$. Model fitting is accomplished by maximum likelihood for θ_k (Journel and Huijbregts, 1981). In particular, the n-dimensional unconstrained nonlinear maximization problem with cost function $-[p \ln(\hat{\sigma}^2) + \ln |\boldsymbol{R}|]/2$, where the variance σ^2 and $|\boldsymbol{R}|$ are both functions of θ_k, is solved for positive values of θ_k as optimization variables.

One of the important advantages of kriging is that once the surrogate has been obtained, the random process, $Z(\boldsymbol{x})$, gives valuable information regarding the approximation error that can be used for improving the surrogate by allocating extra training samples at the locations where the estimated model error is the highest (Journel and Huijbregts, 1981; Forrester and Keane, 2009). This feature is also utilized in various global optimization methods (see Couckuyt (2013) and references therein).

4.4.6 *Approximation techniques: neural networks*

Artificial neural networks (ANN) is a large area of research in itself (Haykin, 2008). In the context of this book, ANN can be viewed as yet another method of approximating sampled high-fidelity model data to create a computationally cheap surrogate model.

The fundamental component of a neural network (Minsky and Papert, 1969; Haykin, 1998) is the neuron (or single-unit perceptron). A neuron performs an affine transformation followed by a nonlinear operation, as shown in Fig. 4.4. Denoting the inputs to a neuron as $x_1, \ldots, x_n$, the neuron output y can be computed as:

$$y = \frac{1}{1 + \exp(-\eta/T)}, \tag{4.17}$$

where $\eta = w_1x_1 + \ldots + w_nx_n + \gamma$, with $w_1, \ldots, w_n$ being regression coefficients. Here, γ is the bias value of a neuron, and T is a user-defined (slope) parameter. Neurons can be combined in multiple ways (Haykin, 1998). The most common neural network architecture is the multi-layer feed-forward network (see Fig. 4.4(b)).

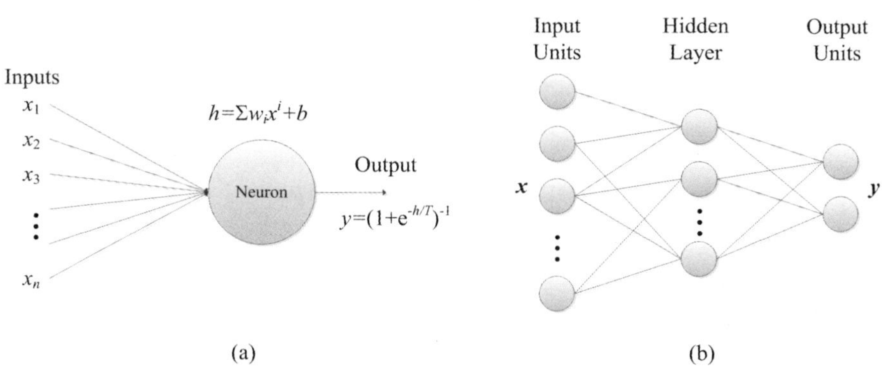

Fig. 4.4. Neural networks. (a) Neuron basic structure; (b) two-layer feed-forward neural network architecture.

The following two steps are involved in the construction of a neural network model: selection of the network architecture and network training, i.e. assigning values to the weight coefficient of the neurons. The network training is normally formulated as a nonlinear least-squares regression problem. The objective is to obtain as good a match as possible between the network outputs and given training data. As the optimization cost function is nonlinear in all the optimization variables (neurons coefficients), it is not possible to obtain the solution in a closed-form expression, as was the case of linear regression Equations (4.6) and (4.10). A popular technique for solving this regression problem is the error back-propagation algorithm (Haykin, 1998; Simpson *et al.*, 2001). For large networks with complex architectures, the use of global optimization methods might be necessary (Alba and Marti, 2006).

ANN have universal approximation capability, i.e. they can approximate any set of training data with any required accuracy, provided the network is sufficiently complex (Simpson *et al.*, 2001). However, in complicated cases (e.g. nonsmooth functions with a large number of variables) the underlying regression problem may be significantly involved. Another issue is possible network overtraining: while trained, an approximation error of a sufficiently complex network can be very small, whereas its generalization capability can be poor. Obtaining the balance between approximation and generalization is mostly controlled by the network architecture. In some areas, quite sophisticated software tools have been developed, allowing simultaneous determination of the optimum network architecture and weight coefficient values (Zhang and Gupta, 2000).

4.4.7 *Approximation techniques: support vector regression*

Support vector regression (SVR) (Gunn, 1998) is one of the relatively recent techniques that is characterized by good generalization capability (Angiulli *et al.*, 2007) and easy training through quadratic programming, resulting in a global optimum for the model parameters (Smola and Schölkopf, 2004). SVR is a variant of the support vector machines (SVM) methodology developed by Vapnik (1995), which was originally applied to solve classification problems. SVM exploits the structural risk minimization (SRM) principle, which has been shown to be superior (Gunn, 1998) to traditional empirical risk minimization (ERM) principles employed by conventional methods used in empirical data modeling, e.g. neural networks. ERM is based on minimizing an error function for the set of training points. When the model structure is complex (e.g. higher order polynomials), ERM-based surrogates often result in overfitting. SRM incorporates the model complexity in the regression, and therefore yields surrogates that may be more accurate outside the training set. SVR is gaining popularity in various areas, including aerodynamic design (Ceperic and Baric, 2004; Rojo-Alvarez *et al.*, 2005; Yang *et al.*, 2005; Xia *et al.*, 2006; Meng and Xia, 2007; Andrés *et al.*, 2012; Zhang and Han, 2013).

Here, we formulate the SVR surrogate model for vector-valued functions. Let $r^k = f(\boldsymbol{x}^k)$, $k = 1, 2, \ldots, N$ denote the sampled high-fidelity model responses. We want to use support vector regression to approximate r^k at all base points, $\boldsymbol{x}^k$, $k = 1, 2, \ldots, N$. We shall also use the notation $r^k = [r_1^k r_2^k \ldots r_m^k]^T$ to denote components of the vector r^k. In the case of linear regression, we want to approximate a given set of data, in our case the data pairs $D_j = \{(\boldsymbol{x}^1, R_j^1), \ldots, (\boldsymbol{x}^N, R_j^N)\}$, $j = 1, 2, \ldots, m$, by a linear function, $f_j(\boldsymbol{x}) = \boldsymbol{w}_j^T \boldsymbol{x} + b_j$. The optimal regression function is given by the minimum of the functional (Smola and Schölkopf, 2004):

$$\Phi_j(\boldsymbol{w}, \xi) = \frac{1}{2}\|\boldsymbol{w}_j\|^2 + C_j \sum_{i=1}^{N} (\xi_{j.i}^+ + \xi_{j.i}^-), \tag{4.18}$$

where C_j is a user-defined value, and $\xi_{j.i}^+$ and $\xi_{j.i}^-$ are slack variables representing upper and lower constraints on the output of the system. The typical cost function used in support vector regression is the so-called ε-insensitive loss function:

$$L_\varepsilon(y) = \begin{cases} 0 & \text{for } |f_j(\boldsymbol{x}) - y| < \varepsilon \\ |f_j(\boldsymbol{x}) - y| & \text{otherwise} \end{cases}. \tag{4.19}$$

The value of C_j determines the trade-off between the flatness of f_j and the amount up to which deviations larger than ε are tolerated (Gunn, 1998).

Here we describe nonlinear regression employing the kernel approach, in which the linear function $\boldsymbol{w}_j^T\boldsymbol{x} + b_j$ is replaced by the nonlinear function $\Sigma_i \gamma_{j.i} K(\boldsymbol{x}^k, \boldsymbol{x}) + b_j$, where K is a kernel function. Thus, the SVR model is defined as:

$$s(\boldsymbol{x}) = \begin{bmatrix} \sum_{i=1}^{N} \gamma_{1.i} K(\boldsymbol{x}^i, \boldsymbol{x}) + b_1 \\ \vdots \\ \sum_{i=1}^{N} \gamma_{m.i} K(\boldsymbol{x}^i, \boldsymbol{x}) + b_m \end{bmatrix}, \tag{4.20}$$

with parameters $\gamma_{j.i}$ and b_j, $j = 1, \ldots, m$, $i = 1, \ldots, N$ obtained according to a general support vector regression methodology. In particular, Gaussian kernels of the form

$$K(\boldsymbol{x}, \boldsymbol{y}) = \exp\left(-\frac{\|\boldsymbol{x} - \boldsymbol{y}\|^2}{2c^2}\right), \quad c > 0 \tag{4.21}$$

can be used, where c is the scaling parameter. Both c and parameters C_j and ε can be adjusted to minimize the generalization error calculated using a cross-validation method (Queipo *et al.*, 2005) and exponential grid search (Ceperic and Baric, 2004).

4.4.8 *Other approximation techniques*

One of other approximation methods worth mentioning here is moving least squares (MLS) (Levin, 1998), which is a technique particularly popular in aerospace engineering. MLS is formulated as weighted least squares (WLS) (Aitken, 1935). In MLS, the error contribution from each training point, $\boldsymbol{x}^{(i)}$, is multiplied by a weight, ω_i, that depends on the distance between $\boldsymbol{x}$ and $\boldsymbol{x}^{(i)}$. A common choice for the weights is:

$$\omega_i(\|\boldsymbol{x} - \boldsymbol{x}^{(i)}\|) = \exp(-\|\boldsymbol{x} - \boldsymbol{x}^{(i)}\|^2). \tag{4.22}$$

MLS is essentially an adapting surrogate, and this additional flexibility can be translated in more appealing designs (especially in computer graphics applications). However, MLS is computationally more expensive than WLS, since computing the approximation for each point, $\boldsymbol{x}$, requires solving a new optimization problem.

Gaussian process regression (GPR) (Rasmussen and Williams, 2006) is another surrogate modeling technique that, like kriging, addresses the approximation problem from a stochastic point view. From this perspective, and since Gaussian processes are mathematically tractable, it is relatively easy to compute error estimations for GPR-based surrogates in the form of uncertainty distributions. Under appropriate conditions, Gaussian processes can be shown to be equivalent to large neural networks (Rasmussen and Williams, 2006). Nevertheless, Gaussian process modeling normally requires much less regression parameters than neural networks.

An interesting variation of the standard kriging interpolation (cf. Section 4.4.5) is cokriging (Forrester *et al.*, 2007; Toal and Keane, 2011). Cokriging allows the blending of information from computational models of different fidelities or even from physical experiments into one surrogate (Forrester *et al.*, 2007; Forrester, 2010; Toal and Keane, 2011). The main benefit here is that — because of exploiting knowledge embedded in the low-fidelity model — the surrogate can be created at much lower computational cost than the models exclusively based on high-fidelity data. Cokriging is a relatively recent method (first introduced in the 1980s) with relatively few applications in aerospace engineering (Laurenceau and Sagaut, 2008; Toal and Keane, 2011; Huang and Gao, 2012).

4.4.9 *Surrogate model validation*

Validation is an important step in the surrogate modeling process, necessary to estimate model quality. It should be stressed that assessing the model quality based only on its performance on the training set may be misleading (e.g. interpolative models — by definition — exhibit zero error at the training set) and should not be used. Some of the approximation techniques described above determine a surrogate model together with some estimation of the attendant approximation error (e.g. kriging or Gaussian process regression). Alternatively, there are procedures that can be used in a stand-alone manner to validate the prediction capability of a given model beyond the set of training points.

A simple and probably the most popular method for validating a model is the split-sample method (Queipo *et al.*, 2005). In this approach, the set of available data samples is divided into two subsets. The first subset is called the training subset and contains the points considered for the construction of the surrogate. The second subset is the testing subset and serves purely as a model validation objective. In general, the error estimated by a split-sample method strongly depends on how the set of data samples is partitioned. We also note that in this approach the samples available do not appear to be put to good use, since the surrogate is based only on a subset of them.

More accurate estimation of the model generalization error can be obtained using cross-validation (Geisser, 1993; Queipo *et al.*, 2005). In cross-validation the data set is divided into L subsets, and each of these subsets is sequentially used as a testing set for a surrogate constructed on the other L–1 subsets. If the number of subsets, L, is equal to the sample size, p, the approach is called leave-one-out cross-validation (Queipo *et al.*, 2005). The prediction error can be estimated with all the L error measures obtained in this process (for example, as an average value). Cross-validation provides an error estimation that is less biased than with the split-sample method (Queipo *et al.*, 2005). The disadvantage of this method is that the surrogate has to be constructed more than once. However, having multiple approximations may improve the robustness of the whole surrogate generation and validation approach, since all the data available are used with both training and testing purposes.

4.5 Physics-Based Surrogates

Here we describe the concept and selected techniques for constructing physics-based surrogate models. The main idea is exploitation of the knowledge about the system of interest embedded in an underlying low-fidelity (or coarse) model, denoted here as $c(\boldsymbol{x})$. The latter is a simplified representation of the system. It is computationally cheaper but, at the same time, less accurate than the high-fidelity model. The low-fidelity model can be obtained, for example, from the same simulator as the one used for the high-fidelity model, but using a coarse discretization (Alexandrov *et al.*, 2000). Alternatively, the coarse model can be based on simplified physics (e.g. by exploiting simplified equations (Bandler *et al.*, 2004a) or neglecting certain second-order effects) (Wu *et al.*, 2004) or on a significantly different physical description (e.g. lumped parameter versus partial differential equation-based models (Bandler *et al.*, 2004a)). In some cases, coarse models can be formulated using analytical or semi-empirical formulas (Robinson *et al.*, 2008). A discussion of low-fidelity models for aerodynamic surfaces is presented in Chapter 9.

Because of its limited accuracy, the low-fidelity model has to be corrected to create a reliable surrogate. The main advantage of physics-based models is that — because of exploiting some knowledge embedded in the low-fidelity model — a limited amount of high-fidelity data are necessary to ensure decent accuracy. For the same reason, physics-based surrogates are characterized by good generalization capability, i.e. they can provide reliable prediction of the high-fidelity model response at the designs not used in the training process. These advantages are normally translated into better efficiency (in particular, lower computational cost) when physics-based surrogates are used in the design optimization process (Koziel *et al.*, 2011b).

In general, physics-based surrogate models are:

- based on particular knowledge about the physical system of interest;
- dedicated (reuse across different designs is uncommon); and

- more expensive to evaluate and more accurate (in a global sense) than approximation-based surrogates.

It should be noted that the evaluation of a physics-based surrogate may involve, for example, the numerical solution of partial differential equations, or even actual measurements of the physical system.

The main focus of this book is surrogate-based optimization methods involving physics-based surrogates. Here, we briefly describe a few selected methods of low-fidelity model enhancement/correction. A number of other approaches will be discussed in Chapter 6, where we present specific design techniques using these types of models.

4.5.1 *Function correction*

Model correction aims at reducing misalignment between the low- and high-fidelity models. In the optimization context, it is often performed to improve the model matching at the current design, so that the surrogate model obtained in this way may be used as a reliable prediction tool that allows us to find a better design. There are three main groups of function correction techniques: compositional, additive, or multiplicative corrections. We will briefly illustrate each of these categories for correcting the low-fidelity model $c(\boldsymbol{x})$, as well as discussing if zero- and first-order consistency conditions with $f(\boldsymbol{x})$ (Alexandrov *et al.*, 2001) can be satisfied. Here, we assume that the models are scalar functions; however, generalizations for vector-valued functions are straightforward in some cases.

The compositional correction (Søndergaard, 2003)

$$s^{(i+1)}(\boldsymbol{x}) = g(c(\boldsymbol{x})) \tag{4.23}$$

is a simple scaling of the objective function. Since the mapping, g, is a real-valued function of a real variable, a compositional correction will not in general yield first-order consistency conditions. By selecting a mapping g that satisfies

$$g'(c(\boldsymbol{x}^{(i)})) = \frac{\nabla f(\boldsymbol{x}^{(i)})\nabla c(\boldsymbol{x}^{(i)})^T}{\nabla c(\boldsymbol{x}^{(i)})\nabla c(\boldsymbol{x}^{(i)})^T}, \tag{4.24}$$

the discrepancy between $\nabla f(\boldsymbol{x}^{(i)})$ and $\nabla s^{(i+1)}(\boldsymbol{x}^{(i)})$ (expressed in Euclidean norm) is minimized. It should be noted that the correction in Equation (4.24), like many transformations that ensure first-order consistency, requires a high-fidelity gradient, which may be expensive to compute. However, numerical estimates of $\nabla f(\boldsymbol{x}^{(i)})$ may yield, in practice, acceptable results.

The compositional correction also can be introduced in the parameter space (Bandler *et al.*, 2004a):

$$s^{(i+1)}(\boldsymbol{x}) = c(\boldsymbol{p}(\boldsymbol{x})). \tag{4.25}$$

If the ranges of $f(\boldsymbol{x})$ and $c(\boldsymbol{x})$ are different, then the condition $c(\boldsymbol{p}(\boldsymbol{x}^{(i)})) = f(\boldsymbol{x}^{(i)})$ is not achievable. This difficulty can be alleviated by combining both compositional corrections so that g and $\boldsymbol{p}$ take the following forms:

$$g(t) = t - c(\boldsymbol{x}^{(i)}) + f(\boldsymbol{x}^{(i)}), \tag{4.26}$$

$$\boldsymbol{p}(\boldsymbol{x}) = \boldsymbol{x}^{(i)} + \boldsymbol{J}_{\boldsymbol{p}}(\boldsymbol{x} - \boldsymbol{x}^{(i)}), \tag{4.27}$$

where $\boldsymbol{J}_{\mathbf{p}}$ is a $n \times n$ matrix for which $\boldsymbol{J}_{\boldsymbol{p}}^T \nabla c = \nabla f(\boldsymbol{x}^{(i)})$ guarantees consistency.

Additive and multiplicative corrections allow the obtaining of first-order consistency conditions. For the additive case we can generally express the correction as:

$$s^{(i+1)}(\boldsymbol{x}) = \lambda(\boldsymbol{x}) + s^{(i)}(\boldsymbol{x}). \tag{4.28}$$

The associated consistency conditions require that $\lambda(\boldsymbol{x})$ satisfies $\lambda(\boldsymbol{x}^{(i)}) = f(\boldsymbol{x}^{(i)}) - c(\boldsymbol{x}^{(i)})$ and $\nabla\lambda(\boldsymbol{x}^{(i)}) = \nabla f(\boldsymbol{x}^{(i)}) - \nabla c(\boldsymbol{x}^{(i)})$. Those requirements can be obtained by the following linear additive correction:

$$\begin{aligned} s^{(i+1)}(\boldsymbol{x}) = {} & f(\boldsymbol{x}^{(i)}) - c(\boldsymbol{x}^{(i)}) \\ & +(\nabla f(\boldsymbol{x}^{(i)}) - \nabla c(\boldsymbol{x}^{(i)}))(\boldsymbol{x} - \boldsymbol{x}^{(i)}) + c(\boldsymbol{x}). \end{aligned} \tag{4.29}$$

Multiplicative corrections (also known as the β-correlation method (Søndergaard, 2003)) can be represented generically by:

$$s^{(i+1)}(\boldsymbol{x}) = \alpha(\boldsymbol{x}) s^{(i)}(\boldsymbol{x}). \tag{4.30}$$

If $c(\boldsymbol{x}^{(i)}) \neq 0$, zero- and first-order consistency can be achieved if $\alpha(\boldsymbol{x}^{(i)}) = f(\boldsymbol{x}^{(i)})/c(\boldsymbol{x}^{(i)})$ and $\nabla\alpha(\boldsymbol{x}^{(i)}) = [\nabla f(\boldsymbol{x}^{(i)}) - f(\boldsymbol{x}^{(i)})/c(\boldsymbol{x}^{(i)})$ $\nabla c(\boldsymbol{x}^{(i)})]/c(\boldsymbol{x}^{(i)})$.

The requirement $c(\boldsymbol{x}^{(i)}) \neq 0$ is not strong in practice, since very often the range of $f(\boldsymbol{x})$ (and thus, of the surrogate $c(\boldsymbol{x})$) is known beforehand, and hence a bias can be introduced both for $f(\boldsymbol{x})$ and $c(\boldsymbol{x})$ to avoid cost function values equal to zero. In these circumstances, the multiplicative correction

$$s^{(i+1)}(\boldsymbol{x}) = \left[\frac{f(\boldsymbol{x}^{(i)})}{c(\boldsymbol{x}^{(i)})} + \frac{\nabla f(\boldsymbol{x}^{(i)})c(\boldsymbol{x}^{(i)}) - f(\boldsymbol{x}^{(i)})\nabla c(\boldsymbol{x}^{(i)})}{c(\boldsymbol{x}^{(i)})^2}(\boldsymbol{x} - \boldsymbol{x}^{(i)})\right] c(\boldsymbol{x}) \tag{4.31}$$

ensures both zero- and first-order consistency between $s^{(i+1)}$ and f.

4.5.2 *Space mapping*

SM (Bandler *et al.*, 2004a; Koziel and Bandler, 2010a) is a well-known methodology for constructing physics-based surrogate models. The idea is to enhance the low-fidelity model c and create an SM surrogate model s by means of auxiliary transformations (mappings) with parameters determined so that s matches the high-fidelity model f as much as possible at a set of training points (designs). As the low-fidelity model is assumed to be physics-based, i.e. describes the same system as the high-fidelity model, the surrogate model is likely to retain a good match with f over the entire region of interest.

The standard SM model is defined as follows (Koziel *et al.*, 2005):

$$s(\boldsymbol{x}) = \bar{s}(\boldsymbol{x}, \boldsymbol{p}). \tag{4.32}$$

Here, $\boldsymbol{p}$ is a vector of SM parameters which were obtained using the parameter extraction process (one of the trademarks of SM):

$$\boldsymbol{p} = \arg\min_{\boldsymbol{r}} \sum_{k=1}^{N} \|f(\boldsymbol{x}^k) - \bar{s}(\boldsymbol{x}^k, \boldsymbol{r})\|, \tag{4.33}$$

while $\bar{s}$ is a generic SM model, i.e. the low-fidelity model composed with suitable mappings. Vectors $\boldsymbol{x}^k, k = 1, 2, \ldots, N$ represent the training points. In many cases, it is sufficient to use a quite limited number of training points, e.g. $N = 2n + 1$, with n being the number of design variables, allocated according to so-called star-distribution (Koziel *et al.* (2005); cf. Section 4.4.2).

The model often used in practice has the form:

$$\bar{s}(\boldsymbol{x}, \boldsymbol{p}) = \bar{s}(\boldsymbol{x}, \boldsymbol{A}, \boldsymbol{B}, \boldsymbol{c}) = \boldsymbol{A} \cdot c(\boldsymbol{B} \cdot \boldsymbol{x} + \boldsymbol{c}) + \boldsymbol{d}, \tag{4.34}$$

where $\boldsymbol{A} = diag\{a_1, \ldots, a_m\}$, $\boldsymbol{B}$ is an $n \times n$ matrix, $\boldsymbol{c}$ is an $n \times 1$ vector, and $\boldsymbol{d}$ is an $m \times 1$ vector (here, m is the number of model response components).

An interesting variation of SM is implicit SM (Bandler *et al.*, 2004b). In many cases, both high- and low-fidelity models have parameters that are normally fixed and not used in the optimization process (so-called pre-assigned parameters). In implicit SM, these parameters are used as additional degrees of freedom in the low-fidelity model and adjusted in order to obtain a better match with the high-fidelity one. Let us denote the low-fidelity model exploiting the pre-assigned parameters $\boldsymbol{x}_p$ as $c(\boldsymbol{x},\boldsymbol{x}_p)$. The implicit SM surrogate takes the following form:

$$\bar{s}(\boldsymbol{x}, \boldsymbol{p}) = \bar{s}(\boldsymbol{x}, \boldsymbol{x}_p), \tag{4.35}$$

where $\boldsymbol{x}_p$ are obtained similarly as in Equation (4.33). Implicit SM can be combined with other types of SM transformations.

4.5.3 *Example: 2D airfoil model with multi-point output SM*

In this section, we illustrate the use of multi-point response correction (output SM) to enhance the low-fidelity transonic airfoil model. In combination with other corrections and trust-region-based optimization algorithms, output SM allows for a dramatic reduction of the airfoil design time (Koziel and Leifsson, 2012a).

The high-fidelity airfoil model is based on compressible Euler equations, assuming steady, inviscid, and adiabatic flow with no body forces. Numerical fluid flow simulations are performed using the computer code FLUENT (2012). The fine grid used in the simulation contains ~400,000 mesh cells. The iterative convergence of each solution is examined by monitoring the overall residual, which is the sum (over all the cells in the computational domain) of the L^2 norm of all the governing equations solved in each cell. The solution convergence criterion is the one that occurs first: either a maximum residual of 10^{-6} or a maximum number of iterations of 1,000.

The low-fidelity CFD model is constructed in the same way as the high-fidelity model, but with a coarser computational mesh and relaxed convergence criteria (see Koziel and Leifsson (2012a) for details). The average ratio of simulation times of the high- and low-fidelity models is around 80.

The model responses of interest are lift and drag coefficients, so that we have $f(\boldsymbol{x}) = [C_{l.f}(\boldsymbol{x})\ C_{d.f}(\boldsymbol{x})]^T$ and $c(\boldsymbol{x}) = [C_{l.c}(\boldsymbol{x})\ C_{d.c}(\boldsymbol{x})]^T$, where $C_{l.f}$, $C_{d.f}$, $C_{l.c}$, and $C_{d.c}$ denote the lift and drag coefficients of the high- and low-fidelity model, respectively:

$$s(\boldsymbol{x}) = \boldsymbol{A} \circ c(\boldsymbol{x}) + \boldsymbol{D} = [a_l C_{l.c}(\boldsymbol{x}) + d_l \;\; a_d C_{d.c}(\boldsymbol{x}) + d_d]^T. \tag{4.36}$$

The model parameters $\boldsymbol{A} = [a_l\, a_d]^T$ and $\boldsymbol{D} = [d_l\, d_d]^T$ are obtained as:

$$[\boldsymbol{A}, \boldsymbol{D}] = \arg \min_{[\boldsymbol{A}, \boldsymbol{D}]} \sum_{k=1}^{N} \| f(\boldsymbol{x}^{(k)}) - \boldsymbol{A} \circ c(\boldsymbol{x}^{(k)}) + \boldsymbol{D} \|^2, \tag{4.37}$$

where $\boldsymbol{x}^{(k)}$, $k = 1, \ldots, N$ are the training points with known high-fidelity model responses.

Equation (4.37) has an analytical solution, as shown in Koziel and Leifsson (2012a). In particular, the terms $\boldsymbol{A}$ and $\boldsymbol{D}$ can be obtained as:

$$\begin{bmatrix} a_l \\ d_l \end{bmatrix} = (\boldsymbol{C}_l^T \boldsymbol{C}_l)^{-1} \boldsymbol{C}_l^T \boldsymbol{F}_l, \tag{4.38}$$

$$\begin{bmatrix} a_d \\ d_d \end{bmatrix} = (\boldsymbol{C}_d^T \boldsymbol{C}_d)^{-1} \boldsymbol{C}_d^T \boldsymbol{F}_d, \tag{4.39}$$

where

$$\boldsymbol{C}_l = \begin{bmatrix} C_{l.c}\boldsymbol{x}^{(1)}) & C_{l.c}(\boldsymbol{x}^{(2)}) & \cdots & C_{l.c}(\boldsymbol{x}^{(N)}) \\ 1 & 1 & \cdots & 1 \end{bmatrix}^T, \tag{4.40}$$

$$\boldsymbol{F}_l = [C_{l.f}(\boldsymbol{x}^{(1)}) \quad C_{l.f}(\boldsymbol{x}^{(2)}) \quad \ldots \quad C_{l.f}(\boldsymbol{x}^{(N)})]^T, \tag{4.41}$$

$$\boldsymbol{C}_d = \begin{bmatrix} C_{d.c}(\boldsymbol{x}^{(1)}) & C_{d.c}(\boldsymbol{x}^{(2)}) & \cdots & C_{d.c}(\boldsymbol{x}^{(N)}) \\ 1 & 1 & \cdots & 1 \end{bmatrix}^T, \tag{4.42}$$

$$\boldsymbol{F}_d = [C_{d.f}(\boldsymbol{x}^{(1)}) \quad C_{d.f}(\boldsymbol{x}^{(2)}) \quad \ldots \quad C_{d.f}(\boldsymbol{x}^{(N)})]^T. \tag{4.43}$$

Equations (4.38) and (4.39) are least-square optimal solutions to the linear regression problems $\boldsymbol{C}_l a_l + d_l = \boldsymbol{F}_l$ and $\boldsymbol{C}_d a_d + d_d = \boldsymbol{F}_d$, equivalent to Equation (4.37). Note that the matrices $\boldsymbol{C}_l^T \boldsymbol{C}_l$ and $\boldsymbol{C}_d^T \boldsymbol{C}_d$ are non-singular for $N > 1$.

Figure 4.5 shows $c(\boldsymbol{x})$ and $f(\boldsymbol{x})$ at a few randomly selected airfoil geometries, as well as $\boldsymbol{A} \circ c(\boldsymbol{x}) + \boldsymbol{D}$, calculated using Equations (4.36) and (4.37). Note that output SM is capable of substantially reducing misalignment between the surrogate and high-fidelity model responses. In the optimization context (Koziel and Leifsson, 2012a), a supplemental additive output SM term is also locally applied to further improve surrogate model accuracy in the vicinity of the current design.

4.5.4 *Example: implicit SM modeling of a converging-diverging nozzle*

Here we illustrate the use of implicit SM modeling of a converging-diverging nozzle. We consider a problem involving axisymmetric compressible flow in a nozzle. The high-fidelity model is CFD-based and the low-fidelity model is based on analytical equations.

The model problem involves one-dimensional, steady, viscous, turbulent, non-heat-conducting compressible flow through a converging-diverging nozzle. The behavior of the flow through the nozzle is well known and a description can be found in most compressible flow textbooks (see, e.g., Anderson (2007)). A sketch of a typical nozzle is shown in Fig. 4.6. The operating parameters of the nozzle flow are the stagnation pressure, p_0, and the exit pressure, p_e. Depending on these parameters, as well as the nozzle geometry, the flow is characterized by subsonic and/or supersonic flow, or even with normal shocks occurring afterwards, behind the throat. Associated with the shock, there is a change in the flow variables (aside from the stagnation temperature), including the entropy. The objective is to improve the low-fidelity model to match the change in entropy across the normal shock of the high-fidelity model.

The high-fidelity CFD model solves the steady-state Reynolds-averaged Navier–Stokes equations with the one-equation Spalart–Allmaras turbulence model. The adiabatic no-slip boundary

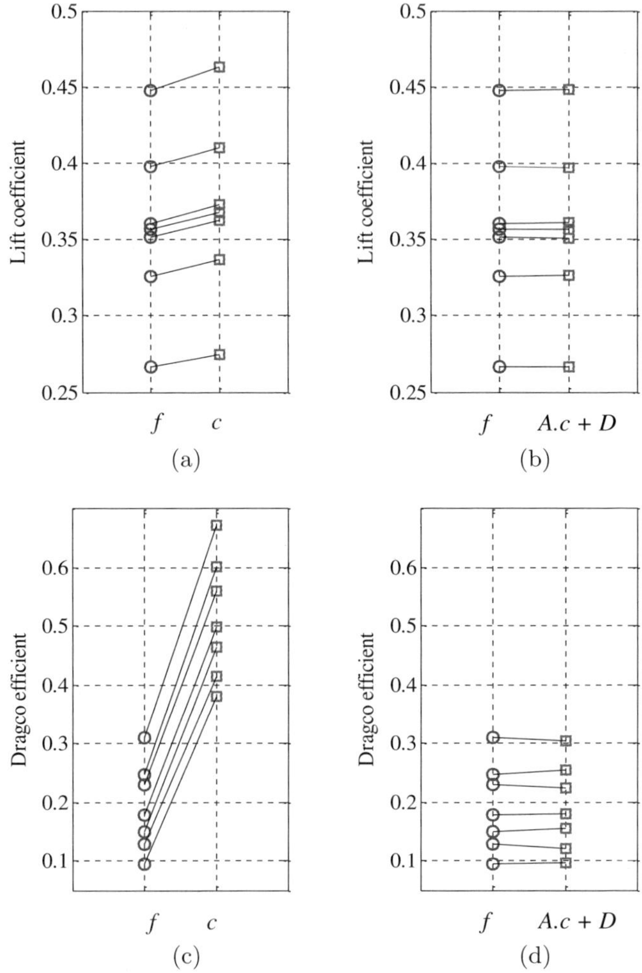

Fig. 4.5. The effect of output SM Equation (3.36) on the low- and high-fidelity model alignment. (a) High- (o) and low-fidelity (□) lift coefficient for selected airfoil geometries; (b) high- (o) and output-SM-aligned low-fidelity (□) lift coefficient for corresponding geometries; (c) the same as (a) but for drag coefficient; (d) the same as (b) but for drag coefficient. The model misalignment is greatly reduced by applying both multiplicative and additive output SM.

condition is applied to the nozzle wall. At the inlet, a constant stagnation pressure (p_0) and temperature (T_0) are specified. An axisymmetric boundary condition is applied to the centerline. At the exit, the outflow boundary is set to a constant static pressure (p_e) and

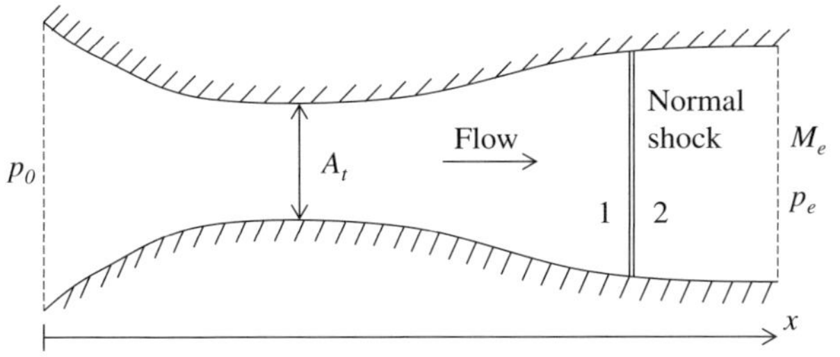

Fig. 4.6. A sketch of a convergent-divergent nozzle with the relevant inlet and outlet flow boundary conditions.

backflow stagnation temperature ($T_{0e} = T_0$). To initialize each CFD solution, inflow conditions are used. CFD calculations were performed using FLUENT (2012), with the implicit density-based flow solver using the Roe-FD flux type. The residuals, which are the sum of the L^2 norm of all governing equations in each cell, are monitored and checked for convergence. The computational grid is structured with 200 nodes in the flow direction and 120 nodes in the radial direction. The grids are generated using ICEM CFD (2013).

The low-fidelity model is an analytical one. We neglect viscous effects and assume a calorically perfect gas so that the flow in the nozzle can be described by the isentropic equations, i.e. the stagnation-to-static temperature ratio (Anderson, 2007) $T_0/T = 1+M^2(\gamma-1)/2$, where γ is the specific heat ratio and M is the Mach number, and the stagnation-to-static pressure ratio $p0/p = (1 + M^2(\gamma - 1)/2)^{\gamma/(\gamma-1)}$. If exit pressure reaches a low enough value, a normal shock standing wave will be observed inside the nozzle, after the throat, with the flow going from supersonic to subsonic. The ratio of pressures before the normal shock (point 1) and after it (point 2) is (Anderson, 2007) $p2/p1 = \alpha(2\gamma M_1^2/(\gamma + 1) - (\gamma - 1)/(\gamma + 1))$, the temperature ratio is $T_2/T_1 = (1 + M^2(\gamma - 1)/2) \cdot (2\gamma M_1^2/(\gamma - 1) - 1)$ $(0.5 \cdot M_1^2(\gamma + 1)^2/(\gamma - 1))^{-1}$, and the change in entropy is $s_2 - s_1 = c_p \ln(T_2/T_1) - R\ln(p_2/p_1)$, where c_p is the specific heat at constant pressure and R is the gas constant. The parameter α is a multiplier introduced by the implicit SM approach (cf. Section 4.5.2). $\alpha = 1$ yields the standard normal shock equations.

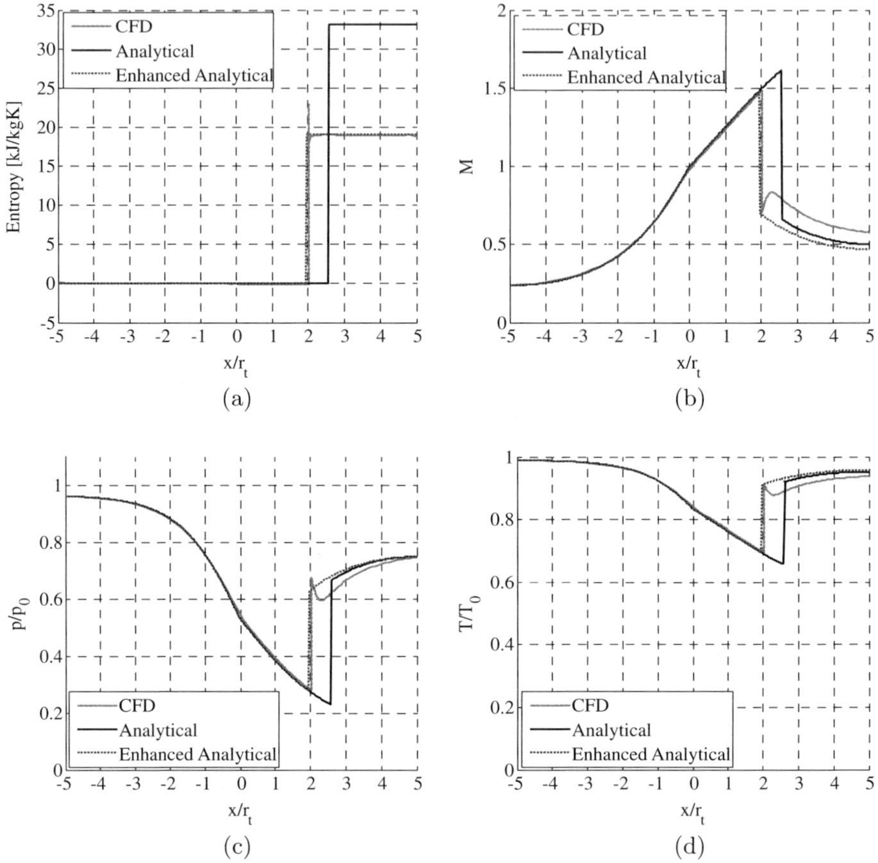

Fig. 4.7. Centerline distributions for $p_e/p_0 = 0.75$ obtained by the CFD, analytical ($\alpha = 1$), and enhanced analytical ($\alpha = 0.932$) models. (a) Entropy; (b) Mach; (c) pressure ratio; (d) temperature ratio.

Figure 4.7 shows the flow distributions obtained for the high-fidelity CFD model, the analytical model, and the implicit-SM-enhanced analytical model for the pressure ratio $p_e/p_0 = 0.75$. One can observe that the analytical model and CFD model differ significantly in terms of the shock location and strength. Therefore, there is a large difference in the entropy generation across the shock. Even though the implicit SM used here exploits only one scalar parameter, the enhanced analytical model matches the magnitude of the entropy generation across the shock. As a result, the shock strength and location match more closely the CFD model. The distributions behind the shock differ somewhat, since the CFD model captures the viscous effect of the boundary layer. Other implicit-SM-like enhancements of the low-fidelity model can be introduced, such as entropy generation due to viscous effects (the boundary layer) and the effects of distortion, in order to further reduce the misalignment between the models.

4.5.5 *Other physics-based surrogate modeling techniques*

In this chapter, only a few selected physics-based surrogate modeling techniques have been highlighted. Certain recently introduced surrogate-based techniques, such as manifold mapping (Echeverría and Hemker, 2005), shape-preserving response prediction (SPRP) (Koziel, 2010a), or adaptive response correction (Koziel *et al.*, 2009), have been specifically introduced in the context of design optimization, in which the surrogate model is normally defined locally, in the vicinity of the current iteration point considered during the optimization process (see Chapter 6 for details). However, some of these approaches can also be viewed as modeling techniques, which can be obtained by appropriate reformulation and using multiple (not a single one as in the original versions) training points. This has been done, in particular for SPRP, in Koziel and Szczepanski (2011). The details are beyond the scope of this chapter. Interested readers are referred to the literature (e.g. Koziel and Szczepanski (2011); Koziel *et al.* (2013)).

4.6 Conclusion

In this chapter, we presented an overview of surrogate-based modeling and optimization. SBO plays an important and increasing role in contemporary engineering design, particularly for problems involving expensive computer simulations. With evaluation times of high-fidelity CFD models ranging from several hours to even weeks per design, the direct use of expensive numerical models in conventional aerodynamic shape optimization procedures (such as gradient-based, even with adjoint sensitivities) is often prohibitive. Surrogate-based optimization can be a very useful approach in this context, since, apart from reducing significantly the number of high-fidelity expensive simulations in the whole design process, it helps in addressing important high-fidelity cost function issues (e.g. presence of discontinuities and/or multiple local optima). A more detailed exposition of SBO will be presented in Chapters 5 and 6, in which the surrogate-based techniques exploiting both approximation and physics-based models, respectively, will be discussed.

Chapter 5

SBO with Approximation-Based Surrogates

In this chapter, we briefly highlight a few concepts regarding optimization with approximation-based surrogate models. Although approximation models are not the main topic of this work, some exposition of related methods is necessary for the book to be self-contained. Here we explain the optimization concept and illustrate it using a simple example. We discuss response surface design methodology, exploiting low-order polynomial approximations, as well as surrogate-based optimization (SBO) techniques based on kriging interpolation. The issue of balance between exploration and exploitation of the design space is also considered in the context of global optimization. Finally, we describe the surrogate management framework as an example of a complete SBO technique, exploiting both approximation surrogates and pattern search.

5.1 Optimization with Response Surface Approximations

As discussed in Chapter 4, SBO is a paradigm in which the direct optimization of the computationally (or otherwise) expensive objective function (or a high-fidelity model) is replaced by iterative enhancement and re-optimization of its cheaper representation, a surrogate model (Forrester and Keane, 2009; Koziel *et al.*, 2011b). Another advantage of using surrogates is the possibility of handling non-smooth or noisy responses (Queipo *et al.*, 2005), performing low-cost sensitivity analysis, or even tolerance-aware design (Bandler *et al.*, 2004a).

In this chapter we focus on the methods exploiting approximation surrogates (also referred to as response surface approximations), i.e. the ones that are created from sampled high-fidelity model data. The optimization normally consists of two main stages: construction of the initial surrogate model based on the initial set of training data; and iterative prediction-correction scheme, in which the surrogate model is used to find promising locations in the design space and then updated using the high-fidelity model data evaluated at these candidate designs.

The flow of the SBO process is the following (Queipo *et al.*, 2005).

1. Set up an initial surrogate model using the set of initial training points.
2. Estimate the high-fidelity optimum design using the surrogate model.
3. Evaluate the high-fidelity model at the design found in Step 2.
4. Verify the termination condition; if satisfied, stop the algorithm.
5. Update the surrogate model using new high-fidelity data.
6. Go to 2.

Figure 5.1 illustrates the SBO process using a simple one-dimensional function minimization.

In the above scheme, infill criteria (i.e. allocating new samples) may not necessarily be based on optimizing the surrogate. Other strategies are also possible and often used in practice, in which the emphasis is put, for example, on improving the global modeling accuracy of the surrogate (Couckuyt, 2013) or on balanced exploration and exploitation of the design space (Forrester and Keane, 2009). This topic is discussed in more detail in Section 5.3. On the other hand, particularly if Step 2 is implemented as a local optimization of the surrogate, the entire procedure may be embedded in the trust-region-like framework (Forrester and Keane, 2009) for improved convergence.

It is worth noting that in some situations the SBO procedure may be reduced to one iteration only. In such one-shot approaches, the initial surrogate is constructed with a sufficient number of training samples to ensure its global accuracy. The prediction obtained by optimizing the surrogate is then taken as a final solution to the design problem. While this approach requires considerable computational effort to construct a globally accurate initial surrogate model, it might be beneficial if the model is intended for multiple use (Kabir *et al.*, 2008).

5.2 Sequential Approximate Optimization

A simple and interesting variation of SBO with approximation-based surrogates is sequential approximate optimization (SAO) (Giunta, 1997; Sobieszczanski-Sobieski and Haftka, 1997; Roux *et al.*, 1998; Giunta and Eldred, 2000; Hosder *et al.*, 2001; Pérez *et al.*, 2002; Jacobs *et al.*, 2004). In SAO, the optimization is limited to a certain (typically, rectangular) subregion of the search space, in which a local approximation surrogate (e.g. low-order polynomial) is set up based on high-fidelity model data sampled in the subregion. The new design is found by optimizing the surrogate within the subregion, which is followed by defining a new subregion according to the chosen move strategy. The SAO flow can be summarized as follows (Jacobs *et al.*, 2004).

1. Set $k = 0$ (iteration counter).
2. Evaluate the high-fidelity model f at the design $\boldsymbol{x}^{(k)}$.
3. Construct a search subregion, $r^{(k)}$, based on a move limit strategy.
4. Sample $r^{(k)}$ using a selected design of experiment plan.
5. Acquire high-fidelity data at the designs selected in Step 4.
6. Construct the surrogate model, $s^{(k)}$, using the data set of Step 5.
7. Find a new design, $\boldsymbol{x}^{(k+1)}$, by optimizing the surrogate, $s^{(k)}$, within the region, $r^{(k)}$.
8. Terminate the algorithm if the stopping criteria are met; otherwise set $k = k + 1$ and go to Step 2.

A block diagram of the SAO procedure is shown in Fig. 5.2.

One of the advantages of SAO is that, due to the problem decomposition (and, consequently, working in the usually small subset of

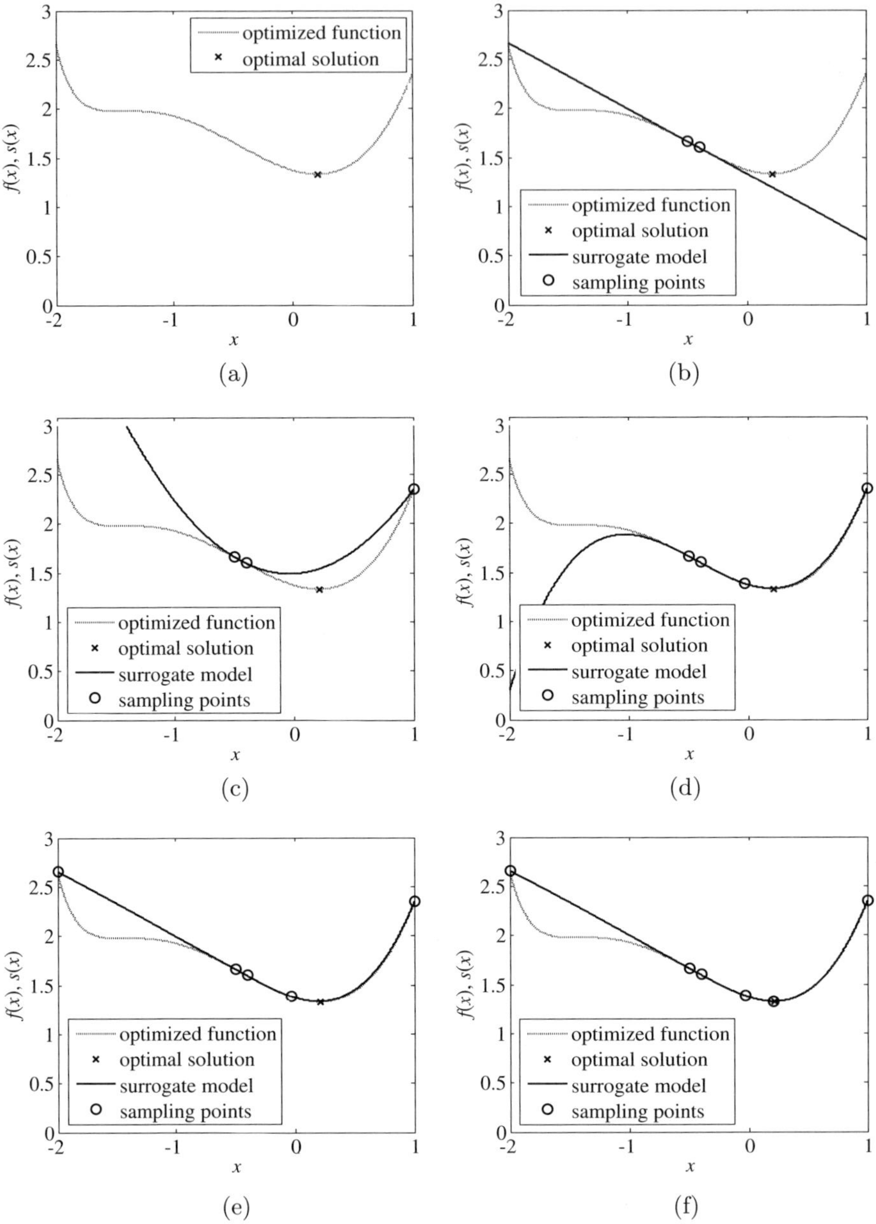
optimized function
optimal solution
surrogate model
sampling points
f(x), s(x)
x
(a)
(b)
(c)
(d)
(e)
(f)

Fig. 5.1. (figure on facing page) SBO illustration. (a) The high-fidelity model f (dotted line) is to be minimized, the minimum of f is denoted as $\times$; (b) the initial surrogate model (here cubic splines) is constructed using two data points; (c) after optimizing the surrogate, a new data point obtained by evaluating the high-fidelity model is used to update the surrogate; (d), (e), (f) after a few iterations, the surrogate model optimum adequately predicts the high-fidelity minimizer. Note that the global accuracy of the surrogate model is not of primary concern here.

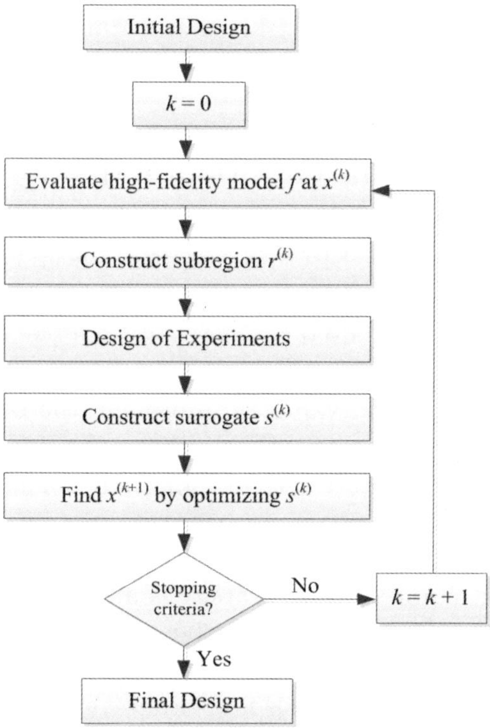

Fig. 5.2. Flow diagram of a typical SAO procedure.

the design space), the surrogate models constructed in each iteration might be simple, e.g. based on low-order polynomial approximation, and, therefore, constructed using a limited number of high-fidelity data samples.

Normally, both the location and the size of the new subregion have to be determined at the beginning of each iteration. Relocation of the search subregion is often performed in the direction of the last iteration optimum (Jacobs *et al.*, 2004), e.g. so that the last optimum becomes a center of the new region. In multi-point strategies, the last iteration optimum may become the corner point for the next search subregion (Toropov *et al.*, 1993). Adjustment of the subregion size may be based on the quality of the designs obtained by optimizing the surrogate, e.g. comparison of the actual versus predicted improvement of the high-fidelity objective function, as in a trust-region-like framework (Giunta and Eldred, 2000; Koziel and Ogurtsov, 2012).

5.3 Kriging-Based Optimization: Exploration versus Exploitation

As mentioned earlier, the SBO process starts from constructing the surrogate model based on the initial set of training samples. The surrogate is subsequently used to select a set of new points (designs), which are promising either in terms of reducing the value of the high-fidelity objective function value or improving the overall surrogate predictive power (Forrester and Keane, 2009; Couckuyt, 2013). These new points are usually referred to as infill points. The entire iterative process of adding training points based on certain infill criteria is referred to as adaptive sampling.

Probably the simplest infill strategy is allocating a single sample at the surrogate model optimum. If the surrogate optimization step is constrained using the trust region approach (Forrester and Keane, 2009), and the surrogate model is first-order consistent with the high-fidelity model (Alexandrov *et al.*, 1998), this strategy is capable of finding at least a local minimum of the high-fidelity model.

In general, infill strategy may be oriented towards global search as well as establishing a globally accurate surrogate. From this point view, the most convenient type of approximation-based surrogates is probably kriging (Jones *et al.*, 1998; Simpson *et al.*, 2001; Queipo *et al.*, 2005; Kleijnen, 2008), because, apart from the predicted response, it also gives information about the expected model error at a given point of the design space (Gorissen *et al.*, 2010).

The commonly used infill criteria include:

- maximizing expected improvement, i.e. the improvement one expects to achieve at an untried point, $\boldsymbol{x}$ (Jones *et al.*, 1998);
- minimizing predicted objective function $\hat{y}(\boldsymbol{x})$, i.e. surrogate optimization already mentioned above (here, reasonable global accuracy of the surrogate is assumed (Liu *et al.*, 2012));
- minimizing the statistical lower bound, i.e. $LB(\boldsymbol{x}) = \hat{y}(\boldsymbol{x}) - As(\boldsymbol{x})$ (Forrester and Keane, 2009), where $\hat{y}(\boldsymbol{x})$ is the surrogate model prediction, $s^2(\boldsymbol{x})$ is the variance, and A is a user-defined constant;
- maximizing the probability of improvement, i.e. finding locations that give the highest chance of improving the objective function value (Forrester and Keane, 2009); and
- maximizing the mean squared error, i.e. finding locations where the mean squared error (as predicted by the kriging model) is the highest (Liu *et al.*, 2012).

In most cases, allocating the new samples according to the infill criteria above requires the use of global optimization methods, e.g. genetic algorithms (Couckuyt, 2013).

SBO for constrained optimization problems using the above infill strategies is more involved. Interested readers can find details in the literature (Queipo *et al.*, 2005; Forrester and Keane, 2009; Liu *et al.*, 2012).

As mentioned earlier, adaptive sampling techniques may target constructing globally accurate surrogates or finding the optimum design of the high-fidelity model (either in a local or global sense). Methods focused on the exploitation of the design space are usually more efficient in the computational sense; however, they may not be able to find the best possible design. On the other hand, the exploration of the design space in most cases implies a global search. If the underlying objective function is non-convex, exploration usually boils down to performing a global sampling of the search space, for example by selecting those points that maximize some estimation of the error associated with the surrogate considered (Forrester and Keane, 2009). It should be stressed that global exploration is often impractical, especially for computationally expensive cost functions with a medium/large number of optimization variables (more than a few tens). Additionally, pure exploration may not be a good approach for updating the surrogate in an optimization context, since a great amount of computing resources may be spent in modeling parts of the search space that are either infeasible or too far from optimum.

Therefore, it appears that in optimization there should be a balance between exploitation and exploration. A good example of an infill criterion that realizes such a balanced search is the minimization of the statistical lower bound. The constant, A, in that criterion controls the exploitation/exploration balance. Here, pure exploitation, i.e. $\mathrm{LB}(\boldsymbol{x}) \rightarrow \hat{y}(\boldsymbol{x})$, is attained if $A \rightarrow 0$. On the other hand, pure exploration is attained when $A \rightarrow \infty$. It is not quite clear how to choose the constant A as it is function dependent. A possible solution is to try a number of values and position infill points where there are clusters of minima of $LB(\boldsymbol{x})$ (Forrester and Keane, 2009).

Surrogate-based methods that utilize infill criteria as a part of their formulation, i.e. balancing the need for exploration versus exploitation, are often referred to as efficient global optimization (EGO) techniques. A detailed discussion of EGO is given by Jones *et al.* (1998) and by Forrester and Keane (2009).

5.4 Cokriging

The popularity of kriging has generated a large body of research, including several extensions to kriging in order to handle different problem settings, e.g. adding gradient information in the prediction (Morris *et al.*, 1993) or by approximating stochastic simulations (Staum, 2009). Cokriging (Kennedy and O'Hagan, 2000; Forrester *et al.*, 2007) is yet another kriging-based method worth noting in the context of SBO. Cokriging is an extension of kriging that allows for the blending of data from computational models of different fidelities into one surrogate. It exploits the correlation between the low- and high-fidelity models to enhance the prediction accuracy. In practice, a relatively large number of low-fidelity data and a limited amount of high-fidelity data are used. This allows for substantial computational savings when compared to the kriging model based exclusively on high-fidelity models, without compromising model accuracy (Koziel *et al.*, 2012b). Combining variable-fidelity data through cokriging can also be used to realize computationally efficient optimization (see, for example, Huang and Gao (2012); Koziel *et al.* (2012b)).

Let us recall that kriging is an interpolation technique for deterministic noise-free data (Gorissen *et al.*, 2010). Let $X_{B.c} = \{\boldsymbol{x}_c^1, \boldsymbol{x}_c^2, \ldots, \boldsymbol{x}_c^{N.c}\}$ be the training set and $c(X_{B.c})$ the associated low-fidelity model responses. The kriging interpolant with a constant mean α is derived as:

$$s_{KR}(\boldsymbol{x}) = \alpha + r(\boldsymbol{x}) \cdot \Psi^{-1} \cdot (f(X_{B.c}) - \mathbf{1}\alpha), \tag{5.1}$$

where $\mathbf{1}$ is a column vector of ones. The coefficient vector, α, is determined by generalized least squares. $r(\boldsymbol{x})$ is an $1 \times N_{KR}$ vector of correlations between the point $\boldsymbol{x}$ and the base set $X_{B.KR}$, where the entries are $r_i(\boldsymbol{x}) = \psi(\boldsymbol{x}, \boldsymbol{x}_c^i)$, and Ψ is a $N_c \times N_c$ correlation matrix, with the entries given by $\Psi_{i,j} = \psi(\boldsymbol{x}_c^i, \boldsymbol{x}_c^j)$. One of the popular correlation functions is an exponential one, i.e. $\psi(\boldsymbol{x}, \boldsymbol{y}) = \exp(\sum_{k=1,\ldots,n} -\theta_k|\boldsymbol{x}^k - \boldsymbol{y}^k|)$, where the parameters $\theta_1, \ldots, \theta_n$ are identified by maximum likelihood estimation (MLE).

Cokriging is an extension to kriging, in which the high- and the low-fidelity data are used to make an accurate prediction. Below, we briefly describe the autoregressive cokriging model of Kennedy and O'Hagan (2000).

In a first step, a kriging model, s_{KRc}, of the low-fidelity data $(X_{B.c}, c(X_{B.c}))$ is constructed. Subsequently, in a second step, a separate kriging model, s_{KRd}, is constructed on $(X_{B.f}, d)$, where $d = f(X_{B.f}) - \rho \cdot c(X_{B.f})$. The parameter ρ of the second kriging model, s_{KRd}, is included in the MLE. Note that if the response values, $c(X_{B.f})$, are not available, they can be approximated by using the first kriging model, s_{KRc}, namely $c(X_{B.f}) \approx s_{KRc}(X_{B.f})$. The resulting cokriging interpolant is defined as:

$$s(\boldsymbol{x}) = M\alpha + r(\boldsymbol{x}) \cdot \Psi^{-1} \cdot (d - F\alpha), \tag{5.2}$$

where the block matrices M, F, $r(\boldsymbol{x})$, and Ψ can be written in function of the two kriging models, s_{KRc} and s_{KRd}:

$$r(\boldsymbol{x}) = [\rho \cdot \sigma_c^2 \cdot r_c(\boldsymbol{x}), \rho^2 \cdot \sigma_c^2 \cdot r_c(\boldsymbol{x}, X_{B.f}) + \sigma_d^2 \cdot r_d(\boldsymbol{x})],$$

$$\Psi = \begin{bmatrix} \sigma_c^2 \Psi_c & \rho \cdot \sigma_c^2 \cdot \Psi_c(X_{B.c}, X_{B.f}) \\ 0 & \rho^2 \cdot \sigma_c^2 \cdot \Psi_c(X_{B.f}, X_{B.f}) + \sigma_d^2 \cdot \Psi_d \end{bmatrix} \tag{5.3}$$

$$F = \begin{Bmatrix} \mathbf{1} & \mathbf{0} \\ \rho \cdot \mathbf{1} & \mathbf{1} \end{bmatrix}, \quad M = [\rho \quad 1],$$

where (σ_c, Ψ_c) and (σ_d, Ψ_d) are matrices obtained from the kriging models s_{KRc} and s_{KRd}, respectively. In particular, σ_c^2 and σ_d^2 are process variances, while $\Psi_c(\cdot,\cdot)$ and $\Psi_d(\cdot,\cdot)$ denote correlation matrices of two datasets with the optimized $\theta_1, \ldots, \theta_n$ parameters and correlation function of the kriging models s_{KRc} and s_{KRd}, respectively.

Figure 5.3 illustrates both kriging and cokriging applied to example analytical-function models. Using the same high-fidelity model data, cokriging is able to capture the behavior of the high-fidelity model better than kriging, which is attributed to the use of the additional low-fidelity model data.

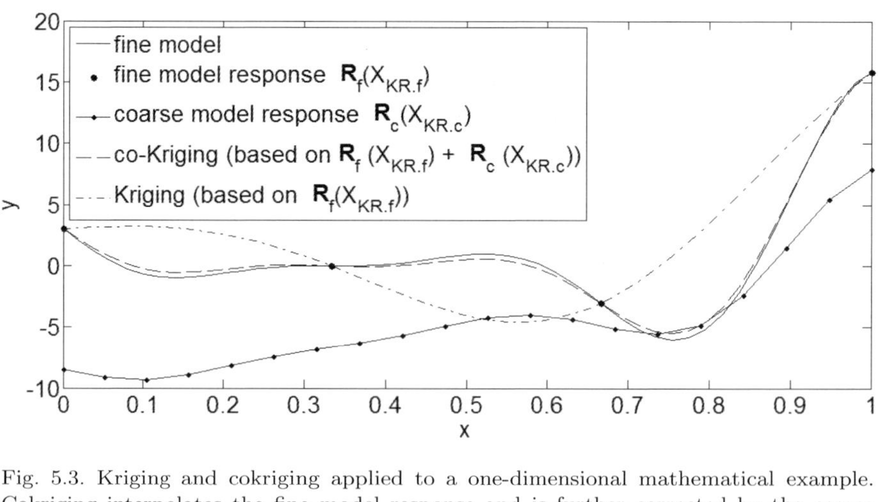

Fig. 5.3. Kriging and cokriging applied to a one-dimensional mathematical example. Cokriging interpolates the fine model response and is further corrected by the coarse model response.

5.5 Surrogate Management Framework

Surrogate management framework (SMF) (Booker *et al.*, 1999) is an optimization technique based on pattern search (Kolda *et al.*, 2003), but also exploits approximation-based surrogates (kriging). It is a general set of derivative-free optimizers that are provably convergent to first-order stationary points of the high-fidelity model. A pattern search optimization algorithm is based on exploring the search space by means of a structured set of points (pattern or stencil) that is modified along iterations.

Each iteration of the search scheme considered in Booker *et al.* (1999) consists of the two main steps: SEARCH and POLL. Each iteration starts with a pattern of size Δ, centered at $\boldsymbol{x}^{(i)}$. The SEARCH step is optional and is always performed before the POLL step. In the SEARCH step, a (small) number of points are selected from the search space (typically by means of a surrogate), and the cost function, $f(\boldsymbol{x})$, is evaluated at these points. If the cost function for some of them improves on $f(\boldsymbol{x}^{(i)})$, the SEARCH step is declared successful, the current pattern is centered at this new point, and a new SEARCH step is started. Otherwise a POLL (or a local search) step is taken. Polling requires computing $f(\boldsymbol{x})$ for neighboring points in the pattern. If one of these points is found to improve on $f(\boldsymbol{x}^{(i)})$, the POLL step is declared successful, the pattern is translated to this new point, and a new SEARCH step is performed. Otherwise the entire pattern search iteration is considered unsuccessful and the termination condition is checked. This stopping criterion is typically based on the pattern size, Δ (Booker *et al.*, 1999; Marsden *et al.*, 2004). If, after the unsuccessful pattern search iteration another iteration is needed, the pattern size, Δ, is decreased, and a new SEARCH step is taken with the pattern centered again at $\boldsymbol{x}^{(i)}$. Surrogates are incorporated in the SMF through the SEARCH step. For example, kriging (with Latin hypercube sampling) is considered in the SMF application studied in Marsden *et al.* (2004). The flow of the algorithm is shown in Fig. 5.4.

In order to guarantee convergence to the (at least local) optimum of the high-fidelity model, the set of vectors formed by each

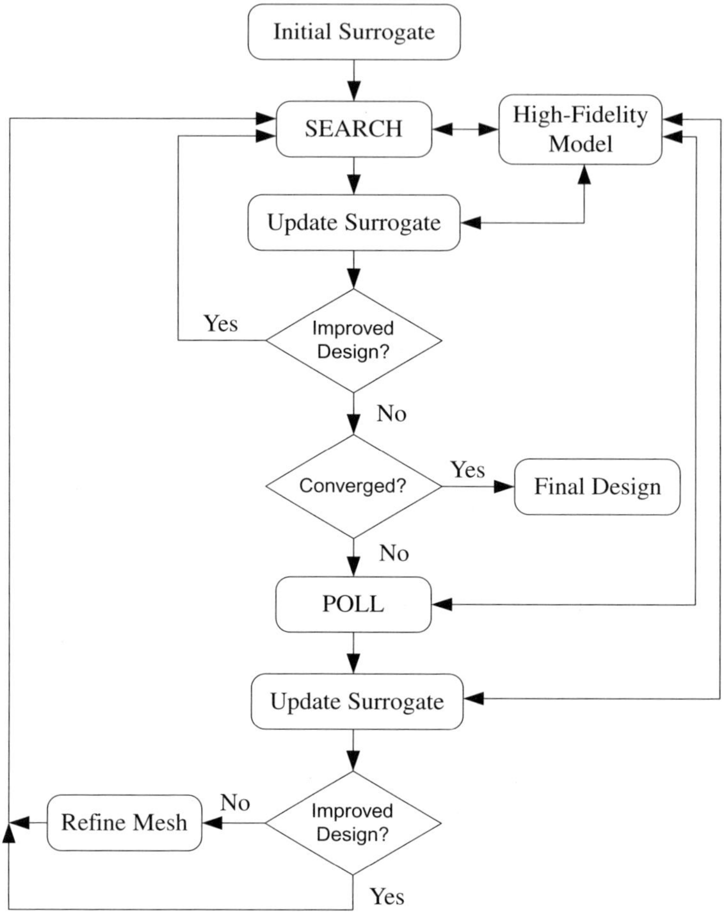

Fig. 5.4. A flowchart of the SMF.

pattern point and the pattern center should be a generating (or positive-spanning) set (Kolda *et al.*, 2003; Marsden *et al.*, 2004). A generating set for R^n consists of a set of vectors, the non-negative linear combinations of which span R^n. Generating sets are crucial in proving convergence (for smooth objective functions) due to the following property: if a generating set is centered at $\boldsymbol{x}^{(i)}$ and $\nabla f(\boldsymbol{x}^{(i)}) \neq 0$, then at least one of the vectors in the generating set defines a descent direction (Kolda *et al.*, 2003). Therefore, if $f(\boldsymbol{x})$ is smooth and $\nabla f(\boldsymbol{x}^{(i)}) \neq 0$, we can expect that for a pattern size,

Δ, small enough, some of the points in the associated stencil will improve on $f(\boldsymbol{x}^{(i)})$.

Although pattern search optimization algorithms typically require many more function evaluations than gradient-based techniques, the computations in both the SEARCH and POLL steps can be performed in a distributed fashion. On top of that, the use of surrogates, as is the case for the SMF, generally noticeably accelerates the entire optimization process.

5.6 Conclusions

In this chapter, we gave an overview of SBO techniques exploiting approximation-based models. The essence of the presented methods is in the iterative construction of the computationally cheap and smooth surrogate model and using it to find promising locations in the design space, in particular predicting the high-fidelity model optimum. One of the issues here is finding a proper balance between the exploration of the design space and the exploitation of the promising regions. While exploration usually leads to improving the global accuracy of the surrogate, it might be computationally expensive and not really necessary if the goal is to optimize the high-fidelity model. In general, approximation-based surrogates may be the right solution for problems with a small or medium number of designable parameters. For more complex problems the computational cost of creating the surrogate may be prohibitive.

It should be mentioned that approximation-based surrogates are probably the most popular option in the context of SBO for aerodynamic shape optimization today. In the next chapter, we will focus on SBO with physics-based models, which is the main emphasis of this book.

Chapter 6

SBO with Physics-Based Surrogates

Surrogate-based optimization (SBO) exploiting physics-based models is the main focus of this book. As explained in Chapter 4, physics-based surrogates offer good generalization capabilities, which is a consequence of exploiting the knowledge embedded in the underlying low-fidelity models. This, as opposed to approximation-based models, allows the construction of reliable surrogates using limited amounts of high-fidelity model data (see also Section 4.5). In some cases, a single high-fidelity model evaluation might be sufficient per optimization iteration (Koziel *et al.*, 2011b). Low-fidelity models are relaxed-accuracy but computationally cheap representations of the system/component under design. Here, an important class is variable-fidelity (or variable-resolution) models that share the same simulation tools but differ in mesh density and other parameters affecting both speed and accuracy (e.g. number of simulation iterations and termination conditions).

In this chapter, we present several physics-based SBO techniques. We discuss approximation model management optimization (AMMO) (Alexandrov and Lewis, 2001), space mapping (SM) (Bandler *et al.*, 2004a; Koziel *et al.*, 2006), shape-preserving response prediction (SPRP) (Koziel, 2010a; Koziel and Leifsson, 2011), adaptive response correction (ARC) (Koziel *et al.*, 2009), and multi-level design optimization (Koziel and Ogurtsov, 2010).

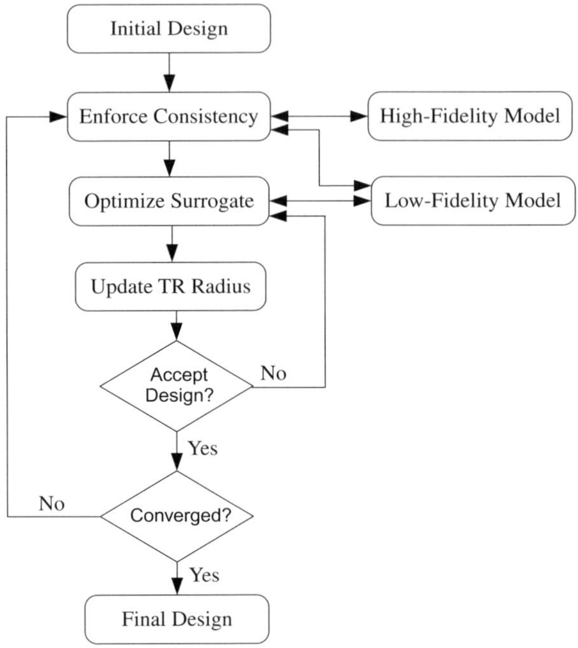

Fig. 6.1. A flowchart of the AMMO algorithm.

6.1 Approximation Model Management Optimization

The AMMO algorithm (Alexandrov and Lewis, 2001) is a general approach for controlling the use of variable-fidelity models when solving a nonlinear minimization problem, such as Equation (4.1). A flowchart of the AMMO algorithm is shown in Fig. 6.1. In the first step, the low-fidelity model is corrected to enforce consistency conditions (at least zero- and first-order) between the resulting surrogate and the high-fidelity model (Alexandrov and Lewis, 2001), i.e. agreement between the function values and the first-order derivatives at a given iteration point. The expensive high-fidelity computations are performed outside the optimization loop and serve to recalibrate the low-fidelity model in the new design obtained by optimizing the surrogate. AMMO exploits the trust-region methodology (Conn *et al.*, 2000), which is an adaptive move limit strategy for improving the global behavior of optimization algorithms based on local models. By combining the trust-region approach with the use of the low-fidelity model satisfying at least first-order consistency conditions, the convergence of AMMO to the optimum of the high-fidelity model can be guaranteed.

The correction applied to the low-fidelity model, c, to ensure first-order consistency between the surrogate and the high-fidelity model, i.e. $s^{(i)}(\boldsymbol{x}^{(i)}) = f(\boldsymbol{x}^{(i)})$ and $\nabla s^{(i)}(\boldsymbol{x}^{(i)}) = \nabla f(\boldsymbol{x}^{(i)})$, is the following. Let $\beta(\boldsymbol{x}) = f(\boldsymbol{x})/c(\boldsymbol{x})$ be the correction function. We define:

$$\beta_i(\boldsymbol{x}) = \beta(\boldsymbol{x}^{(i)}) + \nabla\beta(\boldsymbol{x}^{(i)})^T(\boldsymbol{x} - \boldsymbol{x}^{(i)}). \tag{6.1}$$

The surrogate model is then defined as:

$$s^{(i)}(\boldsymbol{x}) = \beta_i(\boldsymbol{x})c(\boldsymbol{x}), \tag{6.2}$$

and can be shown to satisfy the aforementioned consistency conditions.

First-order consistency in variable-fidelity SBO can be insufficient to achieve acceptable convergence rates, which can be similar to those achieved by first-order optimization methods, such as steepest-descent or sequential linear programming (Rao, 1996). More successful optimization methods, such as sequential quadratic programming, use at least approximate second-order information to achieve super-linear or quadratic convergence rates in the neighborhood of the minimum. Eldred *et al.* (2004) present second-order correction methods for variable-fidelity SBO algorithms. The second-order corrections enforce consistency with the high-fidelity model Hessian. Unfortunately, second-order high-fidelity sensitivity may not be available or may be too expensive to compute. In practice, various approximations based on updating formulas may be used instead (Nocedal and Wright, 2006).

6.2 Space Mapping

SM (Bandler *et al.*, 1994; Bandler *et al.*, 2004a; Koziel *et al.*, 2006) was originally developed in order to solve computationally heavy design problems in microwave engineering. It gave rise to an entire family of SBO approaches. Nowadays, its popularity is spreading across several engineering disciplines (Leary *et al.*, 2001; Redhe and Nilsson, 2004; Bandler *et al.*, 2004a; Priess *et al.*, 2011; Marheineke *et al.*, 2012; Tu *et al.*, 2013).

The initial SM optimization methodologies were based on input SM (Bandler *et al.*, 2004a), i.e. a correction of the coarse model design space. This kind of correction is well suited for many engineering problems, particularly in electrical engineering, in which the model discrepancy is mostly due to second-order effects (e.g. presence of parasitic components). In these applications the model response ranges are often similar in shape, but slightly distorted and/or shifted with respect to a sweeping parameter (e.g. signal frequency). In the case of the ranges of the low- and high-fidelity models being substantially misaligned, input-like SM is not sufficient and response correction (output SM; Bandler *et al.* (2003)) is usually necessary. Other types of SM have also been developed, as briefly outlined below (Bandler *et al.*, 2004b; Koziel *et al.*, 2006; Koziel *el al.*, 2008b).

The SM algorithm follows the SBO Equation (4.2). The SM surrogate model at iteration i is defined as follows:

$$s^{(i)}(\boldsymbol{x}) = \bar{s}(\boldsymbol{x}; \boldsymbol{p}^{(i)}), \tag{6.3}$$

where $\bar{s}$ is a generic SM surrogate, normally a composition of the low-fidelity model and suitable (usually linear) transformations. The surrogate model parameters, $\boldsymbol{p}$, are obtained using the parameter extraction process:

$$\boldsymbol{p}^{(i)} = \arg\min_{\boldsymbol{p}} \sum_{k=0}^{i} w_{i.k} \| f(\boldsymbol{x}^{(k)}) - \bar{s}(\boldsymbol{x}^{(k)}, \boldsymbol{p}) \|, \tag{6.4}$$

where $\boldsymbol{x}^{(i)}$, $i = 0, 1, \ldots$ are the previous designs generated by the SM algorithm, whereas $w_{i.k}$ are user-defined weighting factors. Two popular weight setups are: $w_{i.k} = 1$ for all k and i (equal contribution of all previous designs to the parameter extraction process), and $w_{i.k} = 1$ for $k = i$, and $w_{i.k} = 0$ otherwise (only the most recent design is used in the parameter extraction process).

SM surrogate models can be roughly divided into the four following categories (see, for example, Bandler *et al.* (2004a); Koziel *et al.* (2006)).

- Models based on a distortion of the low-fidelity model parameter space, in particular the input SM of the form $\bar{s}(\boldsymbol{x},\boldsymbol{p}) = \bar{s}(\boldsymbol{x}, \boldsymbol{B},\boldsymbol{c}) = c(\boldsymbol{B}\cdot\boldsymbol{x}+\boldsymbol{c})$ (Bandler *et al.*, 1994; Bandler *et al.*, 2004a).
- Models based on a correction of the low-fidelity model response, e.g. the output SM of the form $\bar{s}(\boldsymbol{x},\boldsymbol{p}) = \bar{s}(\boldsymbol{x},\boldsymbol{d}) = c(\boldsymbol{x})+\boldsymbol{d}$ or $\bar{s}(\boldsymbol{x},\boldsymbol{p}) = \bar{s}(\boldsymbol{x},\boldsymbol{A}) = \boldsymbol{A}\cdot c(\boldsymbol{x})$ (Bandler *et al.*, 2003; Koziel *et al.*, 2006). An example of a more involved, multi-point response correction has been discussed in Section 4.5.3.
- Models utilizing an adjustment of an additional set of parameters, which are separate from the design variables, i.e. implicit space mapping of the form $\bar{s}(\boldsymbol{x},\boldsymbol{p}) = \bar{s}(\boldsymbol{x},\boldsymbol{x}_p) = c_i(\boldsymbol{x},\boldsymbol{x}_p)$, with c_i being the low-fidelity model dependent on both the design variables, $\boldsymbol{x}$, and so-called pre-assigned parameters, $\boldsymbol{x}_p$, that are normally fixed in the high-fidelity model but can be freely altered in the low-fidelity model (Bandler *et al.* (2003); Bandler *et al.* (2004b); Koziel *et al.* (2010a); Koziel *et al.* (2010c); see also Section 5.5.4 for an aerodynamic-oriented example).
- Special models exploiting parameters specific to a given problem. A good example of such a parameter could be frequency, commonly used in microwave engineering. Frequency SM exploits a surrogate model of the form $\bar{s}(\boldsymbol{x},\boldsymbol{p}) = \bar{s}(\boldsymbol{x},\boldsymbol{F}) = c_f(\boldsymbol{x},\boldsymbol{F})$ (Bandler *et al.*, 2003), where c_f is a frequency-mapped low-fidelity model. Here, the low-fidelity model is evaluated at frequencies different from the original frequency sweep for the high-fidelity one, according to the mapping $\omega \rightarrow f_1 + f_2\omega$, with $\boldsymbol{F} = [f_1\ f_2]^T$.

The elementary SM transformations described above can be combined into more involved models (Koziel *et al.*, 2006). In general, the selection of the optimal surrogate for a given problem is not a trivial task (Koziel and Bandler, 2007a; Koziel and Bandler, 2007b; Koziel *et al.*, 2008b). In general, an SM surrogate is identified through nonlinear parameter extraction Equation (6.4); however, for output SM, the parameters can be obtained (at least in some cases) analytically. For composed models, e.g. $s^{(i)}(\boldsymbol{x}) = c(\boldsymbol{B}\cdot\boldsymbol{x}+\boldsymbol{c})+\boldsymbol{d}\ +\boldsymbol{E}\cdot(\boldsymbol{x}-\boldsymbol{x}^{(i)})$, input SM parameters are found using Equation (6.4), whereas the vector, $\boldsymbol{d}$, and the matrix, $\boldsymbol{E}$, are typically calculated based on the high-fidelity data (both response and sensitivity) at the most current design (Koziel *et al.*, 2006). In so-called corrected SM (Robinson *et al.*, 2008), all SM parameters of the above model are extracted as in Equation (6.4), which arguably improves the surrogate model generalization capability in some cases.

The performance of the SM optimization algorithm, both in terms of the robustness (likelihood of finding a high-fidelity model optimum) and the computational cost of the process, depend on the quality of the low-fidelity model but also on the proper selection of the SM transformations (Koziel and Bandler, 2007a; Koziel *et al.*, 2008a). In many applications, it has been observed that the number of points, p, needed for obtaining a satisfactory SM-based corrected surrogate is in the order of the number of optimization variables, n (Bandler *et al.*, 2004a). Though output SM can be used to obtain both zero- and first-order consistency conditions with $f(\boldsymbol{x})$, many other SM-based optimization algorithms that have been applied in practice do not satisfy those conditions, and in some occasions convergence problems have been identified (Koziel *et al.*, 2008a). Additionally, the choice of an adequate SM correction approach is not always obvious (Koziel *et al.*, 2008a). However, on multiple occasions and in several different disciplines (see, for example, Leary *et al.* (2001); Redhe and Nilsson (2004); Bandler *et al.* (2004a); Marheineke *et al.* (2012)), SM has been reported as a very efficient means for obtaining satisfactory optimal designs.

Similar to other SBO methods, SM can be embedded in the trust-region framework for improved convergence (cf. Equation (4.3)). A number of enhancements for space mapping have been suggested in the literature recently (e.g. to implement zero-order and aproximate/exact first-order consistency conditions with $f(\boldsymbol{x})$ (Koziel *et al.*, 2010b; Koziel *et al.*, 2012a) or adaptively constrained parameter extraction (Koziel *et al.*, 2010a)).

6.3 Shape-Preserving Response Prediction

SPRP is one of the recent SBO techniques, initially introduced in microwave engineering (Koziel, 2010a). SPRP is a knowledge-based method that heavily relies on the correlation between the low- and high-fidelity models. SPRP follows the generic SBO Equation (4.2), creating a surrogate model at the beginning of each iteration.

In this section, we describe the process of constructing the SPRP surrogate for the two-dimentional airfoil (Koziel and Leifsson, 2013a). The surrogate model is constructed based on the pressure distribution rather than directly on the figures of interest, such as the lift coefficient or the drag coefficient. This is due to the fact that aerodynamic forces are scalars and the size of the design variable vector is typically large, which can result in non-uniqueness of any alignment procedure that could be applied in order to match the low-fidelity model with the high-fidelity one at any given design. Now, the dimensionality of the pressure distribution can be made as large as necessary by selecting a sufficient number of control points along the airfoil chord. And since the objectives and constraints are uniquely determined by the pressure distribution (for an inviscid flow analysis), the alignment of corresponding distributions for the low- and high-fidelity models will result in the (unique) alignment of the figures of interest. The generalization of the concept presented below for other cases is straightforward (Koziel, 2010a).

The pressure distribution for the high- and low-fidelity models will be denoted as $C_{p.f}$ and $C_{p.c}$, respectively. The surrogate model is constructed assuming that the change of $C_{p.f}$ due to the adjustment of the design variables, $\boldsymbol{x}$, can be predicted using the actual changes of $C_{p.c}$. The change of $C_{p.c}$ is described by the translation vectors corresponding to a certain (finite) number of its characteristic points. These translation vectors are subsequently used to predict the change of $C_{p.f}$, whereas the actual $C_{p.f}$ at the current design, $C_{p.f}(\boldsymbol{x}^{(i)})$, is treated as a reference.

Figure 6.2(a) shows the pressure distribution, $C_{p.c}$, of the low-fidelity model at a certain reference geometry, $\boldsymbol{x}^{(i)}$ (here, corresponding to NACA 2412 airfoil, i.e. $\boldsymbol{x}^{(i)} = [m\ p\ t/c]^T = [0.02\ 0.4\ 0.12]^T$ for the following operating conditions: $M_\infty = 0.7$ and $\alpha = 1$ deg), as well as $C_{p.c}$ at a different geometry, $\boldsymbol{x}$ (specifically, $\boldsymbol{x} = [0.025\ 0.56\ 0.122]^T$). In the optimization context, $\boldsymbol{x}^{(i)}$ would denote a current design (at the ith iteration of the optimization algorithm).

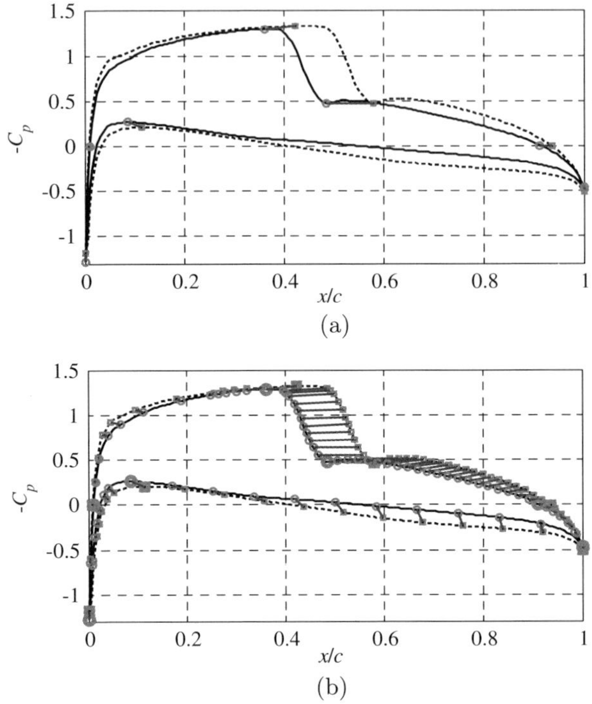

Fig. 6.2. SPRP concept. (a) Example low-fidelity model pressure distribution at the design $\boldsymbol{x}^{(i)}$, $C_{p.c}(\boldsymbol{x}^{(i)})$ (solid line), the low-fidelity model pressure distribution at other design $\boldsymbol{x}$, $C_{p.c}(\boldsymbol{x})$ (dotted line), characteristic points of $C_{p.c}(\boldsymbol{x}^{(i)})$ (circles) and $C_{p.c}(\boldsymbol{x})$ (squares), and the translation vectors (short lines); (b) low-fidelity model pressure distributions, initial characteristic points (large markers) and translation vectors from Fig. 6.2(a), as well as additional points (small markers), inserted in between the initial points either uniformly with respect to x/c (for those parts of the pressure distribution that are almost flat) or based on the relative pressure value with respect to corresponding initial points (for those parts of the pressure distribution that are "steep").

Circles denote characteristic points of $C_{p.c}(\boldsymbol{x}^{(i)})$, here representing, among others, x/c equal to 0 and 1 (leading and trailing airfoil edges, respectively), the maxima of $C_{p.c}$ for the lower and upper airfoil surfaces, and the local minimum of $C_{p.c}$ for the upper surface. The last two points are useful for locating the pressure shock. Squares denote corresponding characteristic points for $C_{p.c}(\boldsymbol{x})$, while small line segments represent the translation vectors that determine the "shift" of the characteristic points of $C_{p.c}$ when changing the design variables from $\boldsymbol{x}^{(i)}$ to $\boldsymbol{x}$.

The characteristic points shown in Fig. 6.2(a) are only to explain the SPRP concept. The actual number of points used in the process and necessary to obtain a reliable prediction is much larger. It is created by inserting additional points in between the initial ones, either uniformly with respect to x/c (for those parts of the pressure distribution that are almost flat) or based on the relative pressure value with respect to corresponding initial points (for those parts of the pressure distribution that are "steep"). Figure 6.2(b) shows the full set of characteristic points (initial points are distinguished using larger markers).

The pressure distribution of the high-fidelity model at the given design, here $\boldsymbol{x}$, can be predicted using the translation vectors applied to the corresponding characteristic points of the pressure distribution of the high-fidelity model at $\boldsymbol{x}^{(i)}$, $C_{p.f}(\boldsymbol{x}^{(i)})$. This is illustrated in Fig. 6.3(a), where only initial characteristic points and translation vectors are shown for clarity. Figure 6.3(b) shows the predicted pressure distribution of the high-fidelity model at $\boldsymbol{x}$ as well as the actual $C_{p.f}(\boldsymbol{x})$. The agreement between both curves is very good. The SPRP modeling flow showing the relationships between the pressure distributions of the low- and high-fidelity models is shown in Fig. 6.4.

The description above outlines the concept of SPRP. The technique can be rigorously formulated as follows. Let $C_{p.f}(\boldsymbol{x}) = [c_{p.f}(\boldsymbol{x}, y_1) \ldots c_{p.f}(\boldsymbol{x}, y_m)]^T$ and $C_{p.c}(\boldsymbol{x}) = [c_{p.c}(\boldsymbol{x}, y_1) \ldots\ c_{p.c}(\boldsymbol{x}, y_m)]^T$, where y_j, $j = 1, \ldots, m$ are control points on the x/c axis (we assume that $y_{j+1} > y_j$ and $0 \leq y_j \leq 1$ for all j). To simplify the notation we assume that $C_{p.f}$ ($C_{p.c}$) is the pressure distribution for the upper surface only. Formulation for the lower surface is identical. Let $p_j^f = [y_j^f\ r_j^f]^T$, $p_j^{c0} = [y_j^{c0}\ r_j^{c0}]^T$, and $p_j^c = [y_j^c\ r_j^c]^T$, $j = 1, \ldots, K$ denote the sets of characteristic points of $C_{p.f}(\boldsymbol{x}^{(i)})$, $C_{p.c}(\boldsymbol{x}^{(i)})$, and $C_{p.f}(\mathbf{x})$, respectively. Here, y and r denote the x/c and magnitude components of the respective point. The translation vectors of the low-fidelity model pressure distribution are defined as $T_i = [y_j^t\ r_j^t]^T$, $j = 1, \ldots, K$, where $y_j^t = y_j^c - y_j^{c0}$ and $r_j^t = r_j^c - r_j^{c0}$.

The SPRP surrogate model is defined as follows:

$$C_{p.s}^{(i)} = [c_{p.s}^{(i)}(\boldsymbol{x}, y_1)\ \ldots\ c_{p.s}^{(i)}(\boldsymbol{x}, y_m)]^T, \tag{6.5}$$

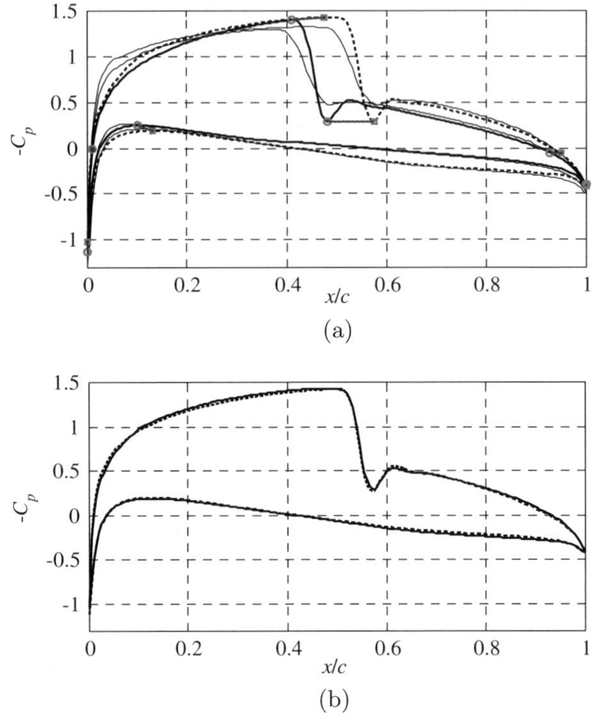

Fig. 6.3. SPRP concept. (a) High-fidelity model pressure distribution at $\boldsymbol{x}^{(i)}$, $C_{p.f}(\boldsymbol{x}^{(i)})$ (solid line) and the predicted high-fidelity model C_p at $\boldsymbol{x}$ (dotted line) obtained using SPRP based on characteristic points of Fig. 6.2(b). Characteristic points of $C_{p.f}(\boldsymbol{x}^{(i)})$ (circles) and the translation vectors (short lines) were used to find the characteristic points (squares) of the predicted high-fidelity model pressure distribution (only initial points are shown for clarity). Low-fidelity model distributions $C_{p.c}(\boldsymbol{x}^{(i)})$ and $C_{p.c}(\boldsymbol{x})$ are plotted using thin solid and dotted lines, respectively; (b) high-fidelity model pressure distribution at $\boldsymbol{x}$, $C_{p.f}(\boldsymbol{x})$ (solid line) and the predicted high-fidelity model pressure distribution at $\boldsymbol{x}$ obtained using SPRP (dotted line).

where:

$$c_{p.s}^{(i)}(\boldsymbol{x}, y_j) = \bar{c}_{p.f}(\boldsymbol{x}^{(i)}, F(y_j, \{-y_k^t\}_{k=1}^{K})) + r(y_j, \{r_k^t\}_{k=1}^{K}) \tag{6.6}$$

for $j = 1, \ldots, m$. $\bar{c}_{p.f}(\boldsymbol{x}, y)$ is an interpolation of $\{c_{p.f}(\boldsymbol{x}, y_1), \ldots, c_{p.f}(\boldsymbol{x}, y_m)\}$ onto the interval $[0, 1]$. The scaling function F interpolates the data pairs $\{y_1, y_1\}$, $\{y_1^f, y_1^f - y_1^t\}, \ldots, \{y_K^f, y_K^f - y_K^t\}$, $\{y_m, y_m\}$ onto the interval $[0, 1]$. The function r does a similar interpolation for data pairs $\{y_1, r_1\}, \{y_1^f, r_1^f - r_1^t\}, \ldots, \{y_K^f, r_K^f - r_K^t\}$, $\{y_m, r_m\}$; here

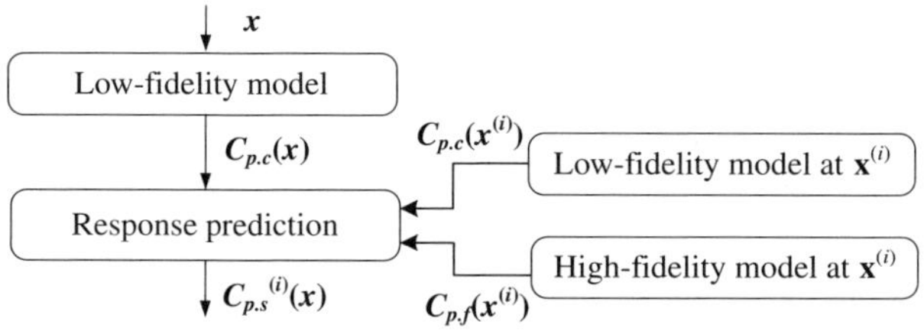

Fig. 6.4. Surrogate model construction at design iteration i using shape-preserving response prediction. The high-fidelity pressure distribution at a design $\boldsymbol{x}$, the surrogate $C_{p.s}(\boldsymbol{x})$, is predicted using the high-fidelity model data at a reference design $\boldsymbol{x}^{(i)}$, $C_{p.f}(\boldsymbol{x}^{(i)})$, and the change of the low-fidelity pressure distribution from $\boldsymbol{x}^{(i)}$ to $\boldsymbol{x}$, $C_{p.c}(\boldsymbol{x}^{(i)})$ and $C_{p.c}(\boldsymbol{x})$, respectively.

$r_1 = c_{p.c}(\boldsymbol{x}, y_1) - c_{p.c}(\boldsymbol{x}^r, y_1)$ and $r_m = c_{p.c}(\boldsymbol{x}, y_m) - c_{p.c}(\boldsymbol{x}^r, y_m)$. Note that $C_{p.s}^{(i)}(\boldsymbol{x}^{(i)}) = C_{p.f}^{(i)}(\boldsymbol{x}^{(i)})$ as all translation vectors are zero at $\boldsymbol{x} = \boldsymbol{x}^{(i)}$.

It should be emphasized that SPRP comes with some limitations. In particular, its fundamental assumption is that the high- and low-fidelity model pressure distributions have corresponding sets of characteristic points, and this assumption is valid in all relevant regions of the design space. While this is usually the case for the practical ranges of design variables in which the overall shape of the distributions are similar for both models, it is not automatically guaranteed. In the case of a lack of correspondence, original definitions of characteristic points are replaced by their closest counterparts. The typical example would be the non-existence of the local minimum of the pressure distribution for the upper surface for the high- and/or low-fidelity model at certain designs. In this case, the original point (local minimum) is replaced by the points characterized by the largest curvature.

6.4 Adaptive Response Correction

Similar to SPRP, ARC (Koziel *et al.*, 2009) was initially introduced in microwave engineering. ARC has recently been applied to airfoil design (Koziel and Leifsson, 2012b). It also exploits the knowledge embedded in the low-fidelity model; however, as opposed to SPRP, it does not rely on one-to-one correspondence between characteristic points, which was one of the fundamental assumptions of SPRP (cf. Section 6.3); The ARC optimization algorithm uses the generic SBO Equation (4.2).

Here we explain the construction of the ARC surrogate for airfoils based on the pressure distribution. However, ARC works for any types of responses that are functions of certain free parameters (such as the chord length in case of an airfoil). Having established the model of the pressure distribution (and/or other relevant quantities), other figures of interest can be uniquely determined from it. The pressure distribution for the high- and low-fidelity models will be denoted as $C_{p.f}$ and $C_{p.c}$, respectively. For simplicity, the ARC technique is explained for the pressure distribution on the upper surface of the airfoil. The concept and formulation for the lower surface is analogous.

Adaptive response correction is a generalization of an additive correction, in which the surrogate model, $C_{p.s}^{(i)}$, at iteration i of the SBO algorithm is constructed as follows:

$$C_{p.s}^{(i)}(\boldsymbol{x}) = C_{p.c}(\boldsymbol{x}) + \Delta C_p^{(i)}, \tag{6.7}$$

where

$$\Delta C_p^{(i)} = C_{p.f}(\boldsymbol{x}^{(i)}) - C_{p.c}(\boldsymbol{x}^{(i)}). \tag{6.8}$$

This definition ensures a zero-order consistency condition (Alexandrov and Lewis, 2001) between the high-fidelity model and the surrogate at $\boldsymbol{x}^{(i)}$, i.e. $C_{p.s}^{(i)}(\boldsymbol{x}^{(i)}) = C_{p.f}^{(i)}(\boldsymbol{x}^{(i)})$. This type of correction is often referred to as output SM (Bandler *et al.*, 2004a). In many cases, it can be used as an efficient auxiliary correction technique (Koziel *et al.*, 2006).

Unfortunately, this simple model does not ensure good generalization capability if the high- or low-fidelity model response contains sharp parts, such as pressure rise due to a shock. Figure 6.5(a) shows the high- and low-fidelity model pressure distributions, $C_{p.f}(\boldsymbol{x}^{(i)})$ and $C_{p.c}$ $(\boldsymbol{x}^{(i)})$, respectively, for the reference design $\boldsymbol{x}^{(i)}$. Figure 6.5(b) shows the additive correction term $\Delta C_p^{(i)}$ obtained using Equation (6.8).

As illustrated in Fig. 6.5(c), the surrogate model response at some other design $\boldsymbol{x}$, $C_{p.s}^{(i)}(\boldsymbol{x})$, is not an accurate representation of the

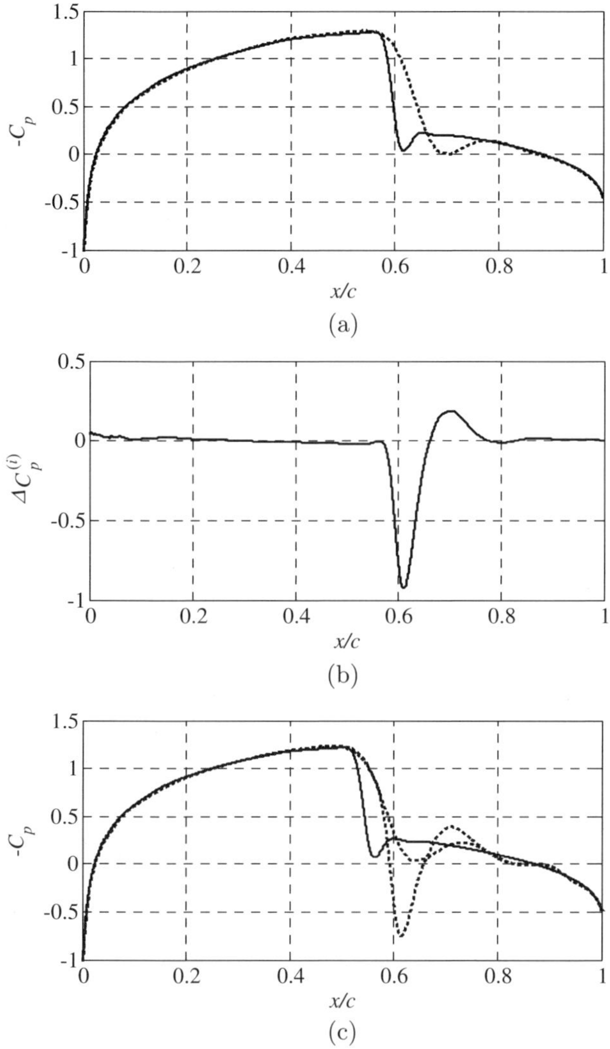

Fig. 6.5. ARC concept. (a) High-fidelity (solid line) and low-fidelity (dashed line) model pressure distributions at $\boldsymbol{x}^{(i)}$, $C_{p.c}^{(i)}(\boldsymbol{x}^{(i)})$, and $C_{p.c}^{(i)}(\boldsymbol{x}^{(i)})$; (b) additive correction term $\Delta C_p^{(i)} = C_{p.f}(\boldsymbol{x}^{(i)}) - C_{p.c}(\boldsymbol{x}^{(i)})$; (c) low-fidelity model response at other design $\boldsymbol{x}$ (dotted line), $C_{p.c}(\boldsymbol{x})$, high-fidelity model at $\boldsymbol{x}$, $C_{p.f}(\boldsymbol{x})$ (solid line), and the surrogate model response at $\boldsymbol{x}$, $C_{p.s}^{(i)}(\boldsymbol{x}^{(i)})$ (dashed line). Note a severe discrepancy between $C_{p.s}^{(i)}(\boldsymbol{x}^{(i)})$ and $C_{p.f}(\boldsymbol{x})$ around the shock.

high-fidelity model response at $\boldsymbol{x}$, $C_{p.f}(\boldsymbol{x})$, around the shock on the upper surface, where the pressure distribution changes sharply. The shock region, on the other hand, is critically important to accurately calculate the aerodynamic forces.

Adaptive response correction is a generalization of the additive correction technique that aims to alleviate the issues highlighted in Fig. 6.5. The ARC surrogate model, $C_{p.ARC}^{(i)}$, is defined as follows:

$$C_{p.ARC}^{(i)}(\boldsymbol{x}) = C_{p.c}(\boldsymbol{x}) + \Delta C_{p.ARC}(\boldsymbol{x}, \boldsymbol{x}^{(i)}), \tag{6.9}$$

where $\Delta C_{p.ARC}(\boldsymbol{x},\ \boldsymbol{x}^{(i)})$ is the response correction term dependent on the design variables, $\boldsymbol{x}$. As we still want to maintain a perfect match between the high-fidelity model and the surrogate at $\boldsymbol{x}^{(i)}$, $\Delta C_{p.ARC}(\boldsymbol{x}, \boldsymbol{x}^{(i)})$ must satisfy:

$$\Delta C_{p.ARC}(\boldsymbol{x}^{(i)}, \boldsymbol{x}^{(i)}) = \Delta C_p^{(i)} = C_{p.f}(\boldsymbol{x}^{(i)}) - C_{p.c}(\boldsymbol{x}^{(i)}). \tag{6.10}$$

The idea behind ARC is to account for the difference between the low-fidelity model responses at $\boldsymbol{x}$ and $\boldsymbol{x}^{(i)}$ and to modify the correction term accordingly, with the initial correction $\Delta C_p^{(i)}$ taken as a reference. In particular, we want to modify the initial correction term $\Delta C_p^{(i)}$ so that this modification reflects changes to $C_{p.c}$ during the process of surrogate model optimization. If the response of $C_{p.c}$ shifts or changes its shape with respect to the coordinate value x/c, the response correction term should track these changes.

As an illustration, consider Fig. 6.6(a), which shows the high- and low-fidelity model responses at two different designs. It is seen that the general relation between the low- and high-fidelity model responses is preserved (e.g. the pressure shock shifts in a similar way for both models). Figure 6.6(b) shows the additive correction terms, ΔC_p, corresponding to the responses shown in Fig. 6.6(a). As we can observe, the relation between these terms as functions of x/c is similar to the relation between the low-fidelity model responses, so that proper tracking of low-fidelity model changes helps determine necessary changes to the correction terms.

In our implementation, the tracking of the low-fidelity model response changes is realized through a proper mapping that describes the shape relations between two low-fidelity model responses and

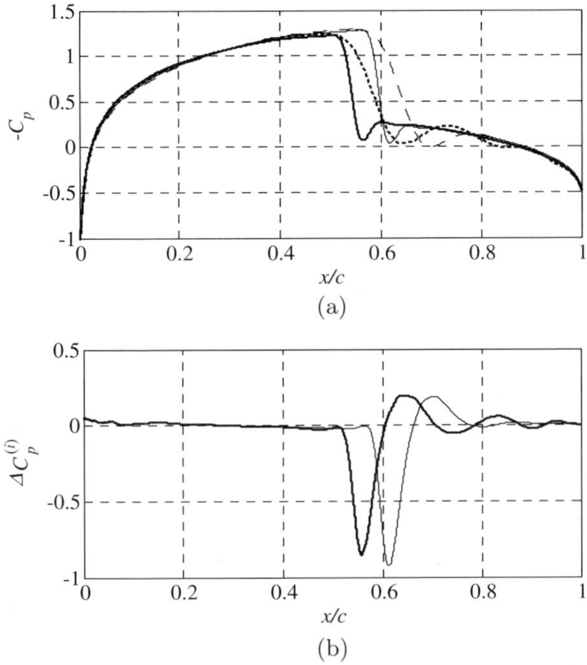

Fig. 6.6. ARC concept. (a) High-fidelity (thin solid line) and low-fidelity model (thin dashed line) responses at a certain design, as well as high-fidelity (thick solid line) and low-fidelity model (thick dashed line) responses at a different design; (b) additive correction terms corresponding to responses shown in Fig. 6.6(a).

updates the response correction term so that the initial and current correction terms correspond to the same (x/c-based) relationship as the initial and current coarse model responses. A rigorous description and technical details of this process are given below.

For the sake of simplicity, we will use the symbol w to denote the coordinate x/c. For the purpose of ARC formulation, we can consider the pressure distribution as a function of two variables, $\boldsymbol{x}$ and w, so that $C_{p.c}(\boldsymbol{x},w)$ is the value of $C_{p.c}(\boldsymbol{x})$ at coordinate $w = x/c$.

The core of the adaptive response correction is a mapping $F^{(i)}$: $X \times \Omega \to \Omega$, where X stands for the low- and high- fidelity model domain, whereas $\Omega = [0, 1]$ (the range of normalized coordinate $w = x/c$). $F^{(i)}$ is established at iteration i in such a way that the difference between $C_{p.c}(\boldsymbol{x}, w)$, i.e. the low-fidelity model response at

$\boldsymbol{x}$, and $C_{p.c}(\boldsymbol{x}^{(i)}, F^{(i)}(\boldsymbol{x}, w))$, i.e. $C_{p.c}(\boldsymbol{x}^{(i)})$ evaluated at $F^{(i)}(\boldsymbol{x}, w)$, is minimized in a given sense, e.g. $\int_w [(C_{p.c}(\boldsymbol{x}^{(i)}, F^{(i)}(\boldsymbol{x}, w)) - C_{p.c}(\boldsymbol{x}, w)]^2 dw$ to realize L-square minimization.

The function $C_{p.c}(\boldsymbol{x}^{(i)}, F^{(i)}(\boldsymbol{x}, w))$ is nothing but $C_{p.c}(\boldsymbol{x}^{(i)}, w)$ scaled with respect to frequency in order to be as similar to $C_{p.c}(\boldsymbol{x}, w)$ as possible in a given sense. In other words, mapping $F^{(i)}$ is supposed to be defined in such a way that mapped coordinate $F^{(i)}(\boldsymbol{x}, w)$ reflects the shape change of the low-fidelity model response at $\boldsymbol{x}$ with respect to its original shape at $\boldsymbol{x}^{(i)}$. For the sake of consistency with Equation (6.10), $F^{(i)}(\boldsymbol{x}^{(i)}, w)$ should be equal to w, i.e. the mapped coordinate should be the same as the original coordinate at $\boldsymbol{x}^{(i)}$.

Having the mapping $F^{(i)}$, we can define the response correction term as follows (here the dependence of $\Delta C_{p.ARC}(\boldsymbol{x}, \boldsymbol{x}^{(i)})$ on w is shown explicitly):

$$\Delta C_{p.ARC}(\boldsymbol{x}, \boldsymbol{x}^{(i)}; w) = C_{p.f}(\boldsymbol{x}^{(i)}, F^{(i)}(\boldsymbol{x}, w)) - C_{p.c}(\boldsymbol{x}^{(i)}, F^{(i)}(\boldsymbol{x}, w)). \tag{6.11}$$

In the following, we describe one realization of the mapping $F^{(i)}$. Other possible realizations can be found in Koziel *et al.* (2009). The function $F^{(i)}(\boldsymbol{x}, w)$ is defined as:

$$F^{(i)}(\boldsymbol{x}, w) = [\boldsymbol{\lambda}^{(i)}(\boldsymbol{x})]^T \boldsymbol{v}(w), \tag{6.12}$$

where $\lambda^{(i)}(x) = [\lambda_1^{(i)}(x)\ \lambda_2^{(i)}(x) \ldots \lambda_p^{(i)}(x)]^T$ are scaling coefficients determined as:

$$\boldsymbol{\lambda}^{(i)}(\boldsymbol{x}) = \arg\min_{\boldsymbol{\lambda}} \int_0^1 [C_{p.c}(\boldsymbol{x}^{(i)}, \boldsymbol{\lambda}^T \boldsymbol{v}(w)) - C_{p.c}(\boldsymbol{x}, w)]^2 dw, \tag{6.13}$$

while $\boldsymbol{v}(\omega) = [v_1(\omega)\ v_2(\omega) \ldots v_p(\omega)]^T$ are basis functions. Equation (6.13), which is readily solved, has the following constraints: $d([\lambda^{(i)}(\boldsymbol{x})]^T \boldsymbol{v}(w))/dw > 0$, i.e. scaling function F is monotonic; $[\lambda^{(i)}(\boldsymbol{x})]^T \boldsymbol{v}(0) = 0$; and $[\lambda^{(i)}(\boldsymbol{x})]^T \boldsymbol{v}(1) = 1$, i.e. $F^{(i)}$ maps 0 into 0 and 1 into 1. Here, we use a third-order polynomial scaling, i.e. we have $p = 4$ and $\boldsymbol{v}(w) = [1\ w\ w^2\ w^3]^T$.

Figures 6.7 and 6.8 illustrate the operation of the adaptive response correction technique. The responses of the low-fidelity

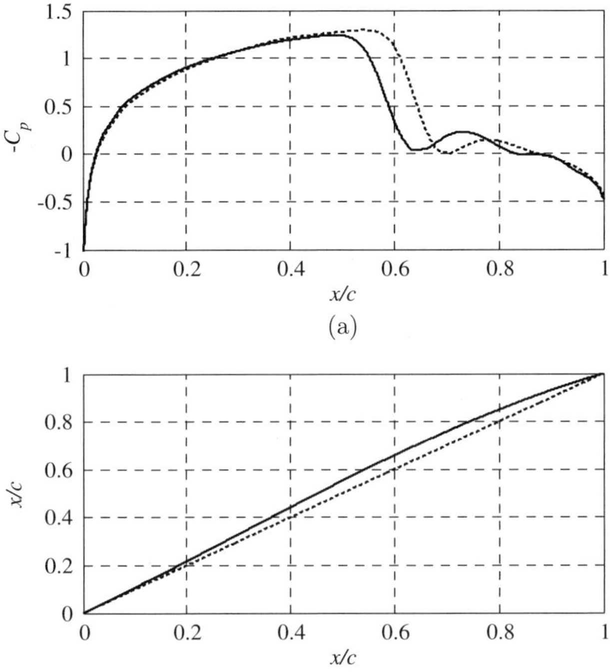

(a)

(b)

Fig. 6.7. ACR concept. (a) Low-fidelity model response at two designs, $\boldsymbol{x}^{(i)}$ and $\boldsymbol{x}$, $C_{p.c}(\boldsymbol{x}^{(i)})$ (dashed line), and $C_{p.c}(\boldsymbol{x})$ (solid line); (b) mapping function $F^{(i)}(\boldsymbol{x})$ (solid line) obtained using Equations (6.12) and (6.13). The identity function (which equals $F^{(i)}(\boldsymbol{x}^{(i)})$) is marked using dashed line.

model at two designs, $\boldsymbol{x}^{(i)}$ and $\boldsymbol{x}$, are shown in Fig. 6.7(a). Figure 6.7(b) shows the mapping function, $F^{(i)}$, obtained using Equations (6.12) and (6.13). In Fig. 6.8(a), we can observe the ARC correction term, $\Delta C_{p.ARC}$, at both $\boldsymbol{x}^{(i)}$, i.e. $\Delta C_{p.ARC}(\boldsymbol{x}^{(i)}, \boldsymbol{x}^{(i)})$, and at $\boldsymbol{x}$, i.e. $\Delta C_{p.ARC}(\boldsymbol{x}, \boldsymbol{x}^{(i)})$. Recall that $\Delta C_{p.ARC}(\boldsymbol{x}^{(i)}, \boldsymbol{x}^{(i)}) = \Delta C_p^{(i)}$.

Finally, Fig. 6.8(b) shows the predicted high-fidelity model response at $\boldsymbol{x}^{(i)}$ obtained using ARC, i.e. $C_{p.ARC}^{(i)}(\boldsymbol{x}) = C_{p.c}(\boldsymbol{x}) + \Delta C_{p.ARC}(\boldsymbol{x}, \boldsymbol{x}^{(i)})$ (cf. Equation (6.9)), which, as opposed to the prediction produced by the simple additive correction (Equations (6.7) and (6.8)), is in a good agreement with the actual high-fidelity model response $C_{p.f}(\boldsymbol{x})$.

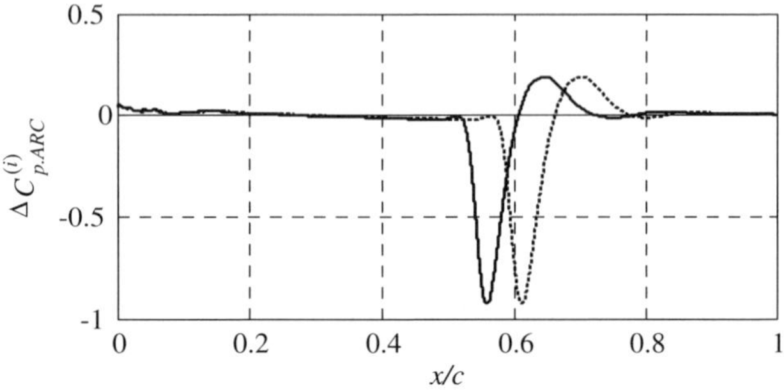

(a)

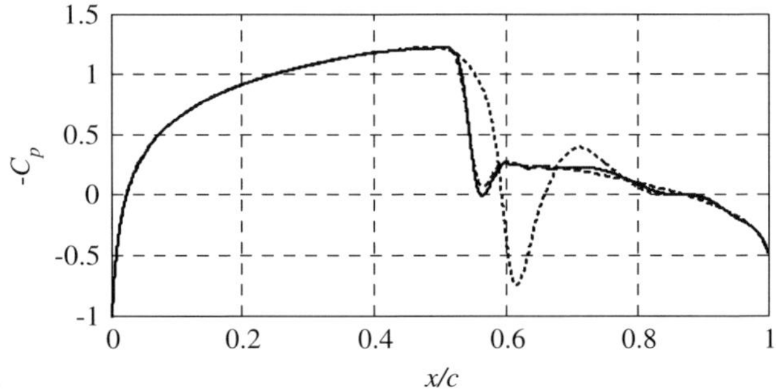

(b)

Fig. 6.8. ACR concept. (a) ARC correction terms $\Delta C_{p.ARC}(\boldsymbol{x}^{(i)}, \boldsymbol{x}^{(i)})$ (dashed line) and $\Delta C_{p.ARC}(\boldsymbol{x}, \boldsymbol{x}^{(i)})$ (solid line); (b) predicted high-fidelity model response at $\boldsymbol{x}^{(i)}$ obtained using ARC, i.e. $C^{(i)}_{p.ARC}(\boldsymbol{x}) = C_{p.c}(\boldsymbol{x}) + \Delta C_{p.ARC}(\boldsymbol{x}, \boldsymbol{x}^{(i)})$ (solid line), actual high-fidelity model response $C_{p.f}(\boldsymbol{x})$ (dashed line), and predicted high fidelity model response $C^{(i)}_{p.s}(\boldsymbol{x}) = C_{p.c}(\boldsymbol{x}) + \Delta C^{(i)}_{p}$ (dotted line).

It should be mentioned that the ARC correction term (Equation (6.9)) modifies the original correction (Equations (6.7) and (6.8)) in a "horizontal" manner (i.e. distorts it with respect to the x/c axis), but it does not change its range. In many cases it is sufficient, particularly that the construction and optimization of the ARC is an iterative process (cf. Equation (4.2)), which allows us to account for possible "vertical" response changes. In Koziel and Leifsson (2012c) an adaptive response prediction technique has been proposed that allows for both "horizontal" and "vertical" modifications of the response correction term, which, in some cases, results in better performance of the surrogate-based optimization process. Interested readers are referred to Koziel and Leifsson (2012c) for more details.

6.5 Multi-Level Design Optimization

The last technique described in this chapter is multi-level optimization (MLO) (Koziel and Ogurtsov, 2010; Koziel and Leifsson, 2013b). On the one hand, MLO differs from the methods discussed before as it utilizes several models of different fidelity. On the other hand, it is conceptually simpler because — until the final stage of the optimization process — it does not require any modification of the low-fidelity model. Below we provide a brief description of the method. Its applications for aerodynamic shape optimization will be discussed in Chapter 10.

MLO exploits a family of low-fidelity models denoted as $\{c_j\}$, $j = 1, \ldots, K$, all evaluated by the same computational fluid dynamics solver as the one used for the high-fidelity model f. Discretization of the model c_{j+1} is finer than that of the model c_j, which results in a better accuracy but also a longer evaluation time. In practice, $K = 2$ or 3. The discretization density may be controlled by solver-dependent parameters (e.g. the meshing parameters).

The MLO workflow is the following. Starting from the initial design, $\boldsymbol{x}^{(0)}$, the lowest-fidelity (coarsest) model, c_1, is optimized to produce a first approximation of the high-fidelity model optimum, $\boldsymbol{x}^{(1)}$. The vector, $\boldsymbol{x}^{(1)}$, is used as a starting point to find the next approximation of the high-fidelity model optimum, $\boldsymbol{x}^{(2)}$, which is obtained by optimizing the next model, c_2. The process continues until the optimum, $\boldsymbol{x}^{(K)}$, of the last low-fidelity model, c_K.

Having $\boldsymbol{x}^{(K)}$, we evaluate the model c_K at all perturbed designs around $\boldsymbol{x}^{(K)}$, i.e. at $\boldsymbol{x}_k^{(K)} = [\boldsymbol{x}_1^{(K)} \ldots \boldsymbol{x}_k^{(K)} + \text{sign}(k) \cdot d_k \ldots \boldsymbol{x}_n^{(K)}]^T$, $k = -n, -n+1, \ldots, n-1, n$. We use the notation $c^{(k)} = c_K(\boldsymbol{x}_k^{(K)})$. These data are used to refine the final design without directly optimizing the high-fidelity model f. More specifically, we set up an approximation model involving $c^{(k)}$ and optimize it in the neighborhood of $\boldsymbol{x}^{(K)}$, defined as $[\boldsymbol{x}^{(K)} - d, \ \boldsymbol{x}^{(K)} + d]$, where $d = [d_1 \ d_2 \ldots d_n]^T$. The size of the neighborhood can be selected based on sensitivity analysis of

c_1 (the cheapest of the low-fidelity models); usually d equals a few percent of $\boldsymbol{x}^{(K)}$.

Here, approximation is performed using a reduced quadratic model, $q(\boldsymbol{x}) = [q_1\ q_2 \ldots q_m]^T$, defined as:

$$\begin{aligned} q_j(\boldsymbol{x}) &= q_j([x_1 \ldots x_n]^T) \\ &= \lambda_{j.0} + \lambda_{j.1}x_1 + \cdots + \lambda_{j.n}x_n + \lambda_{j.n+1}x_1^2 + \cdots + \lambda_{j.2n}x_n^2. \end{aligned} \tag{6.14}$$

The coefficients $\lambda_{j.r}$, $j = 1, \ldots, m, r = 0, 1, \ldots, 2n$ are uniquely obtained by solving the linear regression problems:

$$\begin{bmatrix} 1 & x_{-n.1}^{(K)} & \cdots & x_{-n.n}^{(K)} & (x_{-n.1}^{(K)})^2 & \cdots & (x_{-n.n}^{(K)})^2 \\ \vdots & \vdots & & \vdots & & \vdots & \vdots \\ 1 & x_{0.1}^{(K)} & \cdots & x_{0.n}^{(K)} & (x_{0.1}^{(K)})^2 & \cdots & (x_{-n.n}^{(K)})^2 \\ \vdots & \vdots & & \vdots & & \vdots & \vdots \\ 1 & x_{n.1}^{(K)} & \cdots & x_{n.n}^{(K)} & (x_{n.1}^{(K)})^2 & \cdots & (x_{-n.n}^{(K)})^2 \end{bmatrix} \cdot \begin{bmatrix} \lambda_{j.0} \\ \lambda_{j.1} \\ \vdots \\ \lambda_{j.2n} \end{bmatrix} = \begin{bmatrix} c_j^{(-n)} \\ \vdots \\ c_j^{(0)} \\ \vdots \\ c_j^{(n)} \end{bmatrix}, \tag{6.15}$$

where $\boldsymbol{x}_{k.j}^{(K)}$ is a jth component of the vector $\boldsymbol{x}_k^{(K)}$, and $c_j^{(k)}$ is a jth component of the vector $c^{(k)}$. In the case of an airfoil, the components of the response vector may consist of the lift and drag coefficients, as well as the cross-section area.

In order to account for unavoidable misalignment between c_K and f, instead of optimizing the quadratic model q, it is recommended to optimize a corrected model $q(\boldsymbol{x})+[f(\boldsymbol{x}^{(K)})-c_K(\boldsymbol{x}^{(K)})]$, which ensures a zero-order consistency (Alexandrov and Lewis, 2001) between c_K and f. The refined design can be then found as:

$$\boldsymbol{x}^* = \arg \min_{\boldsymbol{x}^{(K)}-d \le \boldsymbol{x} \le \boldsymbol{x}^{(K)}+d} H(q(\boldsymbol{x}) + [f(\boldsymbol{x}^{(K)}) - c_K(\boldsymbol{x}^{(K)})]). \tag{6.16}$$

This kind of correction is also known as output SM. If necessary, Equation (6.16) can be performed a few times, starting from a refined design, i.e. $\boldsymbol{x}^* = \text{argmin}\{\boldsymbol{x}^{(K)} - d \leq \boldsymbol{x} \leq \boldsymbol{x}^{(K)} + d\colon H(q(\boldsymbol{x}) + [f(\boldsymbol{x}^*) - c_K(\boldsymbol{x}^*)])\}$. It should be noted that the high-fidelity model is not evaluated until executing the refinement step (Equation (6.16)). Also, each refinement iteration requires only a single evaluation of f.

The optimization procedure can be summarized as follows (where K is the number of models).

1. Set $j = 1$.
2. Select the initial design $\boldsymbol{x}^{(0)}$.
3. Starting from $\boldsymbol{x}^{(j-1)}$, find $\boldsymbol{x}^{(j)} = \arg\min\{\boldsymbol{x}\colon H(c_j(\boldsymbol{x}))\}$.
4. Set $j = j + 1$; if $j < K$ go to Step 3.
5. Obtain a refined design according to Equation (6.16).

The main benefit of using models of different fidelity is that starting from a less accurate but fast model allows us to quickly find an approximate location of the optimum design. Switching to finer models at the later stages allows us to locate the optimum more accurately without excessive computational effort because each algorithm iteration starts from an already reasonable approximation of the optimum. Another benefit of this procedure is that, aside from the refinement stage, no enhancement/correction of the low-fidelity models is necessary, which is in contrast to most of other SBO techniques, as discussed before. Therefore, the MLO approach is less dependent on the low-fidelity model quality.

An important step in the described procedure is the selection of the low-fidelity models. Currently, this step has to be done manually by the user. However, a few guidelines that should facilitate the process can be formulated. The simplest setup uses two low-fidelity models and a high-fidelity one: c_1, c_2, and f. The high-fidelity model f should have sufficiently fine mesh in order be an accurate representation of the geometry under consideration. On the other hand, the coarsest model, c_1, should be as fast as possible, i.e. it should have a quite coarse mesh; however, its response should still capture all relevant features of f. This can be done by visual inspection of both model responses at the initial design, particularly while observing the pressure distribution (e.g. the presence and the location of the pressure shock). The setup for the intermediate model, c_2, should be done by ensuring that the time evaluation ratio for f and c_2 is comparable to that of c_2 and c_1. With such a setup, the computational burden will be more or less split uniformly between all the models involved in the design process.

The algorithm presented above is the basic version of MLO. Several variations have been described in the literature, including the computational-budget-driven MLO algorithm (Koziel and Ogurtsov, 2013a), as well as MLO with performance-driven low-fidelity model switching (Koziel and Ogurtsov, 2013b).

6.6 Conclusions

In this chapter, we outlined several SBO techniques exploiting physics-based surrogate models. The major advantage of this type of method is computational efficiency: because of exploiting the knowledge about the system of interest embedded in the low-fidelity model, a reliable surrogate can be created with the limited amount of high-fidelity model data. Also, the number of SBO algorithm iterations (cf. Equation (4.2)) is usually relatively small.

One of the important issues in physics-based methods is the proper selection of the low-fidelity model, c. On the one hand, it should be as fast as possible; on the other hand, it should be reasonably accurate, so that it captures the essential features of the high-fidelity model (such as a pressure shock in case of an airfoil). Setting up the low-fidelity model is a problem in itself. Finding the trade-off between the low-fidelity model speed and accuracy is crucial for the performance of the SBO algorithms, both in terms of the computational cost of the design process and in terms of the quality of the final solution generated by the optimization procedure. A more extensive discussion of this issue will be provided in Chapter 9.

Apart from the methods presented in this chapter, a few other techniques can be found in the literature, including aggressive space mapping (Bandler *et al.*, 1995), manifold mapping (Echeverría and Hemker, 2005), multi-point response correction (Toropov *et al.*, 1993), or adaptively adjusted design specifications approach (Koziel, 2010b). Interested readers are referred to the original articles to find more details about these methods.

Part III
Aerodynamics Modeling

Chapter 7

Geometry Parameterization

Describing an aerodynamic surface in an efficient and robust manner is a critical part of the design process. Airfoils and wings belong to the most complex shapes encountered in engineering design. For example, in terms of performance for transonic designs, the position and strength of the compression shock is very important, demanding accurate control of the surface. Allowing too much freedom of aerodynamic surface shape leads to a large design space and a corresponding increase in the computational cost of the design process. On the other hand, insufficient parameterization flexibility may not allow for rendering a satisfactory design. This chapter provides an overview of several widely used shape parameterization methods for airfoil design. A detailed formulation of each method is given, along with example fits to various well-known airfoil shapes.

7.1 Bump Functions

Hicks and Henne (1978) were the first to use bump functions for airfoil design. The basic idea is to take a baseline airfoil shape, $z(x)_{baseline}$, and deform it to yield a new airfoil shape, $z(x)$. The new airfoil shape can be written as:

$$z(x) = z(x)_{baseline} + \Delta z(x), \tag{7.1}$$

where

$$\Delta z(x) = \sum_{n=1}^{N} \delta_n f_n(x) \tag{7.2}$$

is the total deformation, with

$$f_n(x) = \sin^3(\pi\, x^{e_n}) \tag{7.3}$$

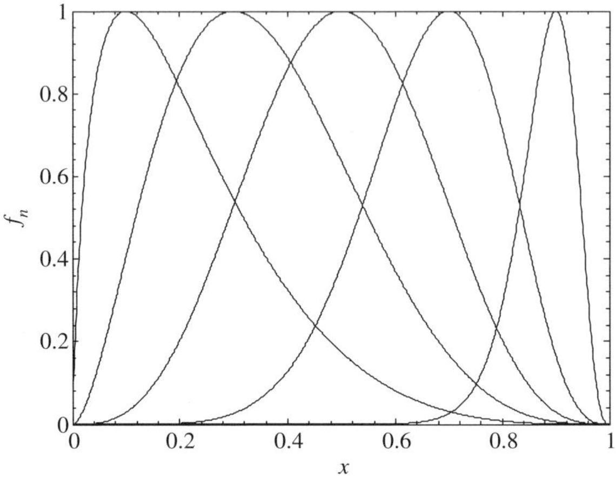

Fig. 7.1. Hicks–Henne bump functions with maxima at $x_n = 0.1, 0.3, 0.5, 0.7$, and 0.9, $b = 3.0$, and $\delta = 1.0$.

being the Hicks–Henne bump functions, δ_n is the deformation amplitude, N is the number of deformations, and

$$e_n = \frac{\log_{10}(0.5)}{\log_{10}(x_n)}, \tag{7.4}$$

where $x_n \in [0, 1]$ is the location of the function maximum. Examples of a few Hicks–Henne bump function shapes are shown in Fig. 7.1.

In airfoil shape optimization, the number of bumps, N, and their locations of maxima, x_n, are typically fixed. The designable parameters are then the amplitudes, δ_n, with the design variable vector written as $\boldsymbol{x} = [\delta_1 \delta_2 \cdots \delta_N]^T$.

To illustrate how the approach works, let us consider the RAE 2822 airfoil (see, for example, Cook *et al.* (1979)) as a target shape and the ONERA M6 airfoil (see, for example, Schmitt and Charpin (1979)) as the baseline shape. In other words, we want find the bump shapes required to deform the ONERA M6 airfoils to obtain the RAE 2822 airfoil shape.

Figure 7.2 shows the baseline, target, and deformed shapes for four and 20 bump functions, for each surface, distributed uniformly between $x = 0.01$ and 0.99. Clearly, using more bumps yields a better approximation to the target shape. The function shapes for the case

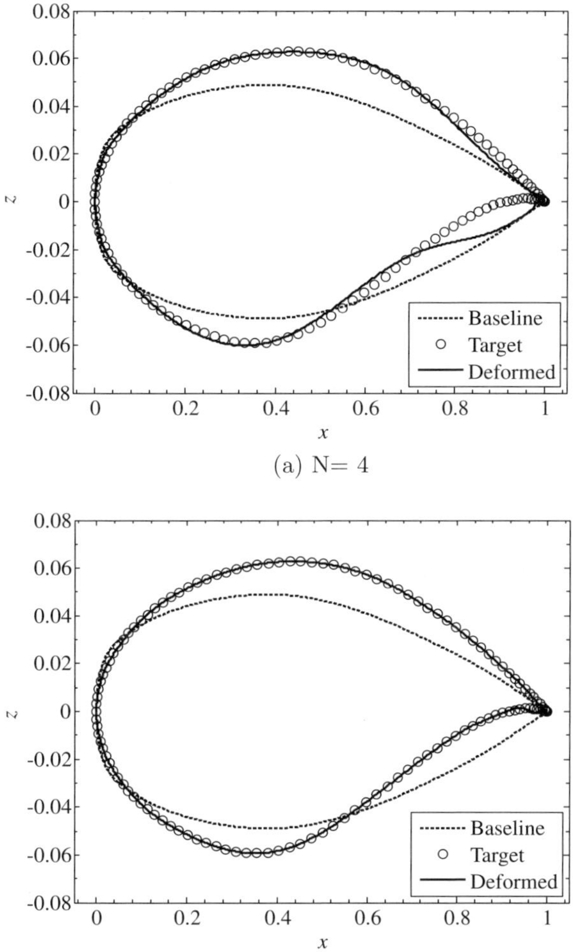

(a) N= 4

(b) N= 20

Fig. 7.2. Deformation of the ONERA M6 airfoil (baseline) to the RAE 2822 airfoil (target) using Hicks–Henne bump functions.

with four bumps per surface are shown in Fig. 7.3, and the numerical values are given in Table 7.1.

7.2 PARSEC

The PARSEC airfoil parameterization was developed by Sobieczky (1998). Unlike the bump function approach described in the last

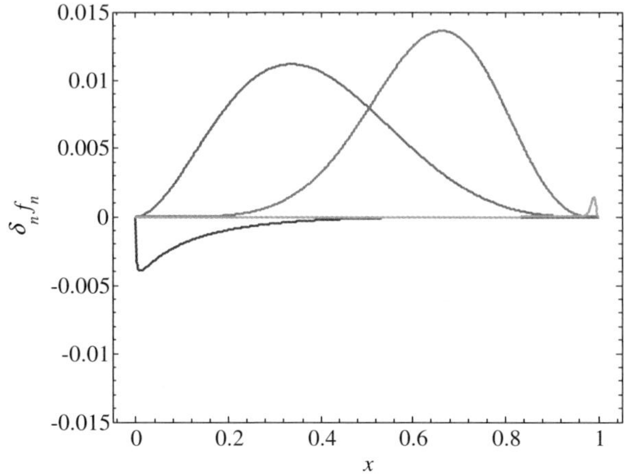

(a) Upper surface

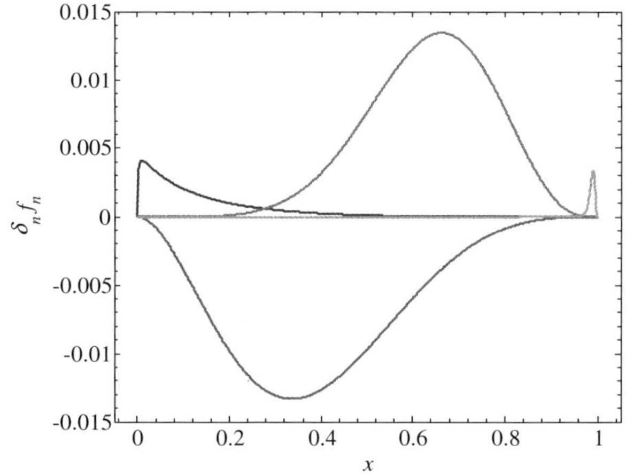

(b) Lower surface

Fig. 7.3. Bump function shapes needed to deform the ONERA M6 airfoil to match, approximately, the RAE 2822 airfoil (the results are shown in Fig. 7.2(a)).

section, the PARSEC method uses well-defined geometric parameters, such as the leading-edge (LE) radius and airfoil thickness. The parameters, therefore, have a specific meaning and may appeal more to the airfoil designer. Moreover, a baseline airfoil shape is not needed.

Table 7.1 Bump functions needed to deform the ONERA M6 airfoil to match the RAE 2822.

Bump	$x_{n,u}$	$\delta_{n,u}$
1	0.0100	−3.9397e-3
2	0.3367	1.1157e-2
3	0.6633	1.3631e-2
4	0.9900	1.4178e-3
Bump	$x_{n,l}$	$\delta_{n,l}$
1	0.0100	4.0784e-3
2	0.3367	−1.3324e-2
3	0.6633	1.3443e-2
4	0.9900	3.3683e-3

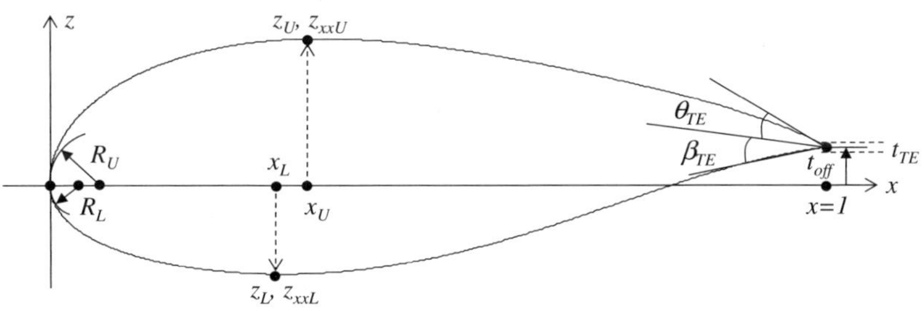

Fig. 7.4. PARSEC airfoil geometry parameters.

In the PARSEC method, the upper surface of the airfoil is described by a sixth order polynomial:

$$z_u(x) = \sum_{i=1}^{6} a_i x^{i-(1/2)}, \tag{7.5}$$

where a_i, $i = 1, \ldots, 6$, are the polynomial coefficients, and x is the nondimensional chordwise location, $0 \leq x \leq 1$. The coefficients, shown in Fig. 7.4, are determined by six geometrical constraints.

1. Maximum upper surface z-coordinate z_U at x_U, i.e.:

$$z_u(x_U) = z_U. \tag{7.6}$$

2. First order derivative is zero at x_U, i.e.:

$$\left.\frac{dz_u}{dx}\right|_{x=x_U} = 0. \tag{7.7}$$

3. Prescribed second order derivative z_{xxU} at x_U, i.e.:

$$\left.\frac{d^2 z_u}{dx^2}\right|_{x=x_U} = z_{xxU}. \tag{7.8}$$

4. Trailing edge (TE) off-set, t_{off}, and thickness, t_{TE}, i.e.:

$$z_u(x = 1) = t_{off} + \frac{1}{2}t_{TE}. \tag{7.9}$$

5. TE direction angle, θ_{TE}, and wedge angle, β_{TE}, i.e.:

$$\left.\frac{dz_u}{dx}\right|_{x=1} = \tan\left(\theta_{TE} - \frac{1}{2}\beta_{TE}\right). \tag{7.10}$$

6. Prescribed LE radius of upper surface, R_U. A circle at $(x_c,\ z_c)$ with a radius R_U, and is defined as:

$$(x - x_c)^2 + (z - z_c)^2 = R_U. \tag{7.11}$$

The LE of the airfoil is at the origin, so $(x_c,\ z_c) = (0, 0)$. We can write:

$$z = (2R_U - x)^{1/2} \cdot x^{1/2}. \tag{7.12}$$

The upper surface polynomial can be written as:

$$z_u = a_1 \cdot x^{1/2} + \sum_{i=2}^{6} a_i x^{i-(1/2)}. \tag{7.13}$$

Comparing Equations (7.12) and (7.13) for small x, we obtain:

$$a_1 \cdot x^{1/2} \approx (2R_U)^{1/2}, \tag{7.14}$$

which is the condition that has been met at the LE.

The geometrical constraints (Equations (7.6)–(7.10) and (7.14)) form a linear system of equations:

$$\boldsymbol{P}\boldsymbol{a} = \boldsymbol{q}, \tag{7.15}$$

where:

$$\boldsymbol{P} = \begin{bmatrix} x_U^{1/2} & x_U^{3/2} & x_U^{5/2} & x_U^{7/2} & x_U^{9/2} & x_U^{11/2} \\ 1/2x_U^{-1/2} & 3/2x_U^{1/2} & 5/2x_U^{3/2} & 7/2x_U^{5/2} & 9/2x_U^{7/2} & 11/2x_U^{9/2} \\ -1/4x_U^{-3/2} & 3/4x_U^{-1/2} & 15/4x_U^{1/2} & 35/4x_U^{3/2} & 63/4x_U^{5/2} & 99/4x_U^{7/2} \\ 1 & 1 & 1 & 1 & 1 & 1 \\ 1/2 & 3/2 & 5/2 & 7/2 & 9/2 & 11/2 \\ 1 & 0 & 0 & 0 & 0 & 0 \end{bmatrix}, \tag{7.16}$$

$$\boldsymbol{q} = \begin{bmatrix} z_U \\ 0 \\ z_{xxU} \\ t_{\mathit{off}} + 1/2t_{TE} \\ \tan(\theta_{TE} - 1/2\beta_{TE}) \\ (2R_U)^{1/2} \end{bmatrix}, \tag{7.17}$$

$$\boldsymbol{a} = [a_1 \quad \cdots \quad a_6]^T. \tag{7.18}$$

The linear system has a unique solution given by:

$$\boldsymbol{a} = \boldsymbol{P}^{-1}\boldsymbol{q}. \tag{7.19}$$

The lower surface is configured in a similar fashion. We have:

$$z_l(x) = \sum_{i=1}^{6} b_i x^{i-(1/2)}, \tag{7.20}$$

with the following constraints:

$$z_l(x_L) = z_L, \tag{7.21}$$

$$\left.\frac{d^2 z_l}{dx^2}\right|_{x=x_L} = z_{xxL}, \tag{7.22}$$

$$z_l(x=1) = t_{\mathit{off}} - \frac{1}{2}t_{TE}, \tag{7.23}$$

$$\left.\frac{dz_l}{dx}\right|_{x=1} = \tan\left(\theta_{TE} + \frac{1}{2}\beta_{TE}\right), \tag{7.24}$$

$$b_1 \cdot x^{1/2} \approx -(2R_L)^{1/2}. \tag{7.25}$$

The linear system of equations is:

$$\boldsymbol{E}\boldsymbol{b} = \boldsymbol{v}, \tag{7.26}$$

where E = P and

$$\boldsymbol{v} = \begin{bmatrix} z_L \\ 0 \\ z_{xxL} \\ t_{off} - 1/2t_{TE} \\ \tan(\theta_{TE} + 1/2\beta_{TE}) \\ -(2R_L)^{1/2} \end{bmatrix}, \tag{7.27}$$

$$\boldsymbol{b} = [b_1 \quad \cdots \quad b_6]^T. \tag{7.28}$$

The solution to Equation (7.28) is:

$$\boldsymbol{b} = \boldsymbol{E}^{-1}\boldsymbol{v}. \tag{7.29}$$

Altogether, there are 12 parameters in the PARSEC formulation. Some of the parameters can be fixed during an optimization. For example, setting $t_{TE} = 0$ will yield a sharp closed TE. Also, it is possible to work only on the upper or the lower surface.

An example PARSEC fit to an airfoil profile is given in Fig. 7.5, along with the corresponding parameter values in Table 7.2.

7.3 Bezier Curves

A Bezier curve of order n is defined as (Samareh, 2001):

$$p(u) = \sum_{i=0}^{n} P_i J_{n,i}(u), \tag{7.30}$$

where P_i are the control points (which may or may not lie on the curve), $u \in [0,1]$, and $J_{n,i}$ are the Bernstein basis polynomials,

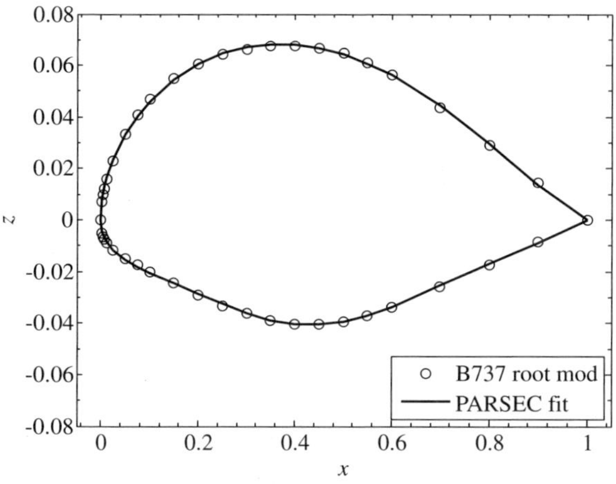

Fig. 7.5. A PARSEC fit to a modified Boeing 737 root airfoil profile (coordinates from Selig (2014)).

Table 7.2 Values of a PARSEC fit to a modified Boeing 737 root airfoil profile (see Selig (2014) for the coordinates).

Parameter	Value
x_U	0.3656
z_U	0.0681
z_{xxU}	−0.4779
R_U	0.0112
x_L	0.4348
z_L	−0.0406
z_{xxL}	0.5343
R_L	0.0033
t_{off}	0.0000
t_{TE}	0.0000
θ_{TE}	0.2990
β_{TE}	12.8392

defined as:

$$J_{n,i}(u) = C_{n,i}u^i(1-u)^{n-1}, \tag{7.31}$$

with $C_{n,i}$ being the binomial coefficient:

$$C_{n,i} = \binom{n}{i} = \frac{n!}{i!(n-i)!}. \tag{7.32}$$

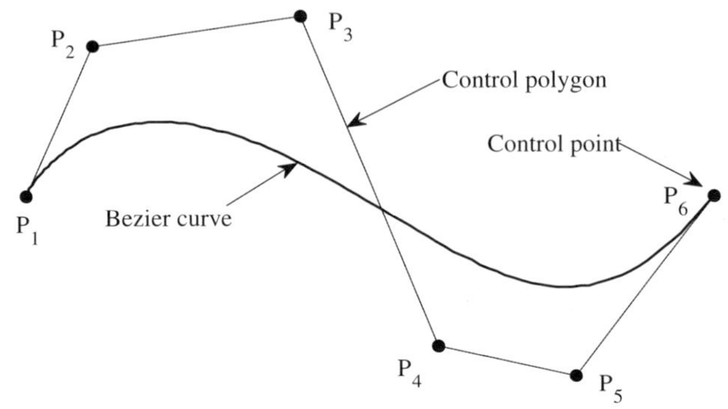

Fig. 7.6. An example Bezier curve with six control points.

An example Bezier curve is given in Fig. 7.6. Any number of control points can be used to represent a curve.

To obtain a typical airfoil shape, a few control points need to be set at specific locations:

- a control point is fixed at the LE, typically set at $(x, z) = (0, 0)$;
- a control point is fixed at the TE, i.e. at $x = 1$. If the TE is fixed on the x-axis, then it is set at $(x, z) = (0, 1)$; and
- to ensure a rounded LE, two control points are allocated on the line $x = 0$.

With these control points in place, the upper and lower surfaces are determined by the remaining control points. In general, the remaining control points need not be bounded. However, often in practice, e.g. within an optimization run, the control points are set at fixed x-locations and allowed to move in the z-direction. This is done to prevent unrealistic airfoil shapes or shapes which cannot be handled by the fluid flow solver.

Figure 7.7 shows examples of two Bezier curve approximations to the Whitcomb supercritical airfoil (see coordinates in Selig (2014)). The airfoil shapes are found by minimizing the norm of the difference between the target airfoil shape and the curve shapes generated by Equation (7.30). Tables 7.3 and 7.4 give the coordinates of the control points for the upper and lower surfaces.

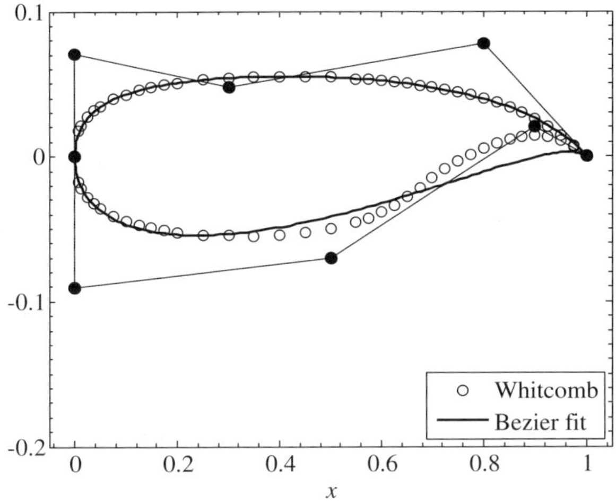

(a) Six free control points

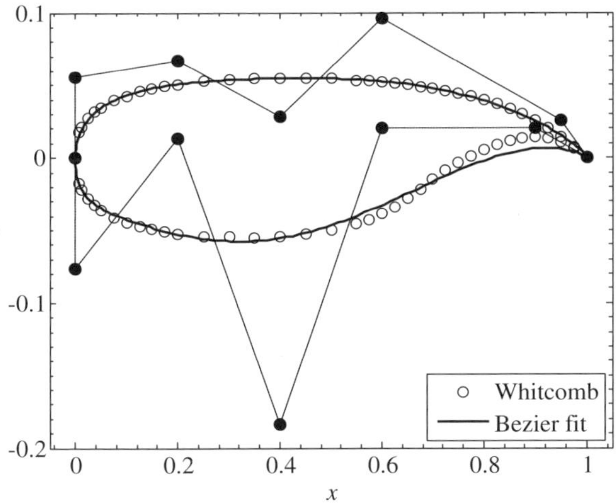

(b) 10 free control points

Fig. 7.7. Bezier curve fits to the Whitcomb airfoil (see coordinates in Selig (2014)). The free control points can only move vertically.

Adding more control points clearly yields a better approximation to the given airfoil shape. This will also translate to an improved control of the airfoil shape during optimization. However, with the additional control points comes additional computational cost.

Table 7.3 Bezier fit to the Whitcomb airfoil with six free control points.

Control point	$x_{p,u}$	$z_{p,u}$
1	0.0	0.0
2	0.0	7.0283e-2
3	0.3	4.7832e-2
4	0.8	7.7703e-2
5	1.0	0.0
Control point	$x_{p,l}$	$z_{p,l}$
1	0.0	0.0
2	0.0	−9.0046e-2
3	0.5	−7.0031e-2
4	0.9	2.0000e-2
5	1.0	0.0

Table 7.4 Bezier fit to the Whitcomb airfoil with ten free control points.

Control point	$x_{p,u}$	$z_{p,u}$
1	0.0	0.0
2	0.0	5.5948e-2
3	0.2	6.7424e-2
4	0.4	2.8235e-2
5	0.6	9.6065e-2
6	0.95	2.5247e-2
7	1.0	0.0
Control point	$x_{p,l}$	$z_{p,l}$
1	0.0	0.0
2	0.0	−7.6658e-2
3	0.2	1.3315e-2
4	0.4	−1.8337e-2
5	0.6	2.0000e-2
6	0.9	2.0000e-2
7	1.0	0.0

7.4 Bezier–PARSEC

The Bezier–PARSEC (BP) parameterization, developed by Derksen and Rogalsky (2010), utilizes the advantages of both the Bezier and the PARSEC parameterizations. In the BP parameterization, the PARSEC variables are used as parameters to define four separate

Bezier curves governing the airfoil shape. The curves define the leading and trailing portions of the camber line, and the leading and trailing portions of the thickness distribution. Each curve is in effect a Bezier curve, but described by the PARSEC parameters, which are more aerodynamically oriented than the parameters used in the Bezier formulation.

There are two different formulations of the BP parameterization. One uses third-degree Bezier curves for all four curves; the so-called BP 3333 parameterization. The other formulation, called BP 3434, uses third-degree Bezier curves for both LE portions and fourth-degree Bezier curves for both TE portions. The BP 3333 involves 12 PARSEC parameters, whereas the BP 3434 includes an additional five Bezier parameters.

7.4.1 *BP 3333*

The curves defining the BP 3333 thickness distribution and camber line are shown in Figs. 7.8 and 7.9, respectively. The BP 3333 relies exclusively on the following 12 parameters: LE radius (r_{le}), trailing camber line angle (α_{te}), trailing wedge angle (β_{te}), TE vertical displacement (z_{te}), LE direction (γ_{te}), location of the camber crest (x_c, z_c), curvature of the camber crest (κ_c), location of the thickness crest (x_t, z_t), curvature of the thickness crest (κ_t), and the half thickness of the TE (dz_{te}).

The notation 3333 refers to the fact that the LE and TE portions are described by a third-degree Bezier curve, given parametrically by:

$$x(u) = x_0(1-u)^3 + 3x_1u(1-u)^2 + 3x_2u^2(1-u) + x_3u^3 \quad (7.33)$$

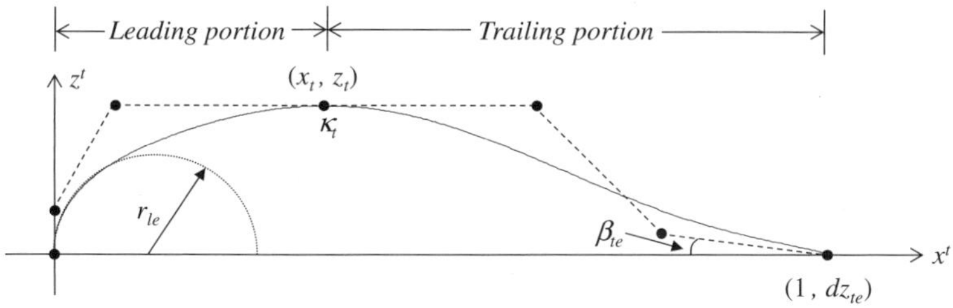

Fig. 7.8. The Bezier–PARSEC 3333 thickness distribution parameters.

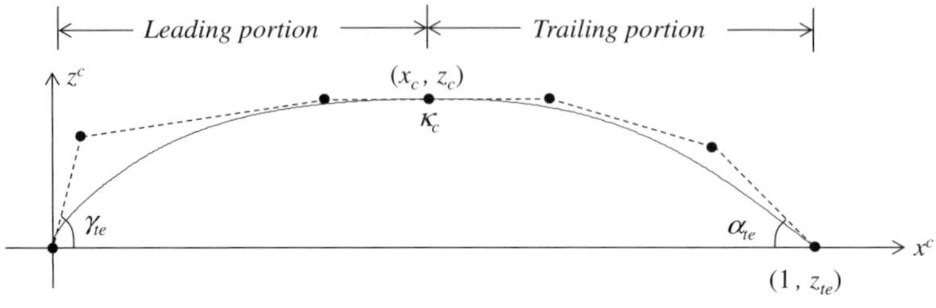

Fig. 7.9. The Bezier–PARSEC 3333 camber line parameters.

and

$$z(u) = z_0(1-u)^3 + 3z_1u(1-u)^2 + 3z_2u^2(1-u) + z_3u^3, \quad (7.34)$$

where u is the parameter that runs from 0 at the LE to 1 at the TE, and x_i, $z_i, i = 1, \ldots, 6$ are the control points.

The LE thickness curve control points are given by:

$$\begin{array}{ll} x_0 = 0 & z_0 = 0 \\ x_1 = 0 & z_1 = 3\kappa_t(x_t - r_t)^2/2 + z_t \\ x_2 = r_t & z_2 = z_t \\ x_3 = x_t & z_3 = z_t \end{array}. \quad (7.35)$$

The parameter r_t is the smallest root of:

$$\begin{aligned} &27\kappa_t^2 r_t^4/4 - 27\kappa_t^2 x_t r_t^3 + (9\kappa_t z_t + 81\kappa_t^2 x_t^2/2) r_t^2 \\ &\quad + (2r_{le} - 18\kappa_t x_t z_t - 27\kappa_t^2 r_t^3) r_t \\ &\quad + (3z_t^2 + 9\kappa_t x_t^2 z_t + 27\kappa_t^2 x_t^4/4) = 0, \end{aligned} \quad (7.36)$$

within the bounds given by:

$$\max(0, x_t - \sqrt{-2z_t/3\kappa_t}) < r_t < x_t. \quad (7.37)$$

The TE thickness curve control points are given by:

$$\begin{array}{ll} x_0 = x_t & z_0 = z_t \\ x_1 = 2x_t - r_t & z_1 = z_t \\ x_2 = 1 + [dz_{te} - (3\kappa_t(x_t - r_t)^2/2 + z_t)]\cot(\beta_{te}) & z_2 = 3\kappa_t(x_t - r_t)^2/. \\ & \quad 2 + z_t \\ x_3 = 1 & z_3 = dz_{te} \end{array} \quad (7.38)$$

The LE camber curve control points are given by:

$$\begin{array}{ll} x_0 = 0 & z_0 = 0 \\ x_1 = r_c \cot(\gamma_{te}) & z_1 = r_c \\ x_2 = x_c - \sqrt{2(r_c - z_c)/3\kappa_c} & z_2 = z_c \\ x_3 = x_c & z_3 = z_c \end{array}. \tag{7.39}$$

The parameter r_c is computed from:

$$r_c = \frac{16 + 3\kappa_c(\cot\gamma_{le} + \cot\alpha_{te})(1 + z_{te}\cot\alpha_{te})}{3\kappa_c(\cot\gamma_{le} + \cot\alpha_{te})} \pm 4\sqrt{\begin{array}{l} 16 + 6\kappa_c(\cot\gamma_{le} + \cot\alpha_{te})(1 - y_c(\cot\gamma_{le} + \cot\alpha_{te}) \\ \quad + z_{te}\cot\alpha_{te}) \end{array}}, \tag{7.40}$$

and must be within the following bounds:

$$0 < r_c < z_c. \tag{7.41}$$

The TE camber curve control points are given by:

$$\begin{array}{ll} x_0 = x_c & z_0 = z_c \\ x_1 = x_c + \sqrt{2(r_c - z_c)/3\kappa_c} & z_1 = z_c \\ x_2 = 1 + (z_{te} - r_c)\cot\alpha_{te} & z_2 = r_c \\ x_3 = 1 & z_3 = z_{te} \end{array}. \tag{7.42}$$

7.4.2 *BP 3434*

The BP 3434 thickness distribution and camber line are shown in Figs. 7.10 and 7.11, respectively. The BP 3434 parameterization relies on the parameters r_{le}, α_{te}, β_{te}, z_{te}, γ_{te}, x_c, z_c, x_t, z_t, and dz_{te}, as well as five free Bezier points: b_0, b_2, b_8, b_{15}, and b_{17}.

The leading portions are described by third-degree Bezier curves, whereas the trailing portions are described by a fourth-degree Bezier

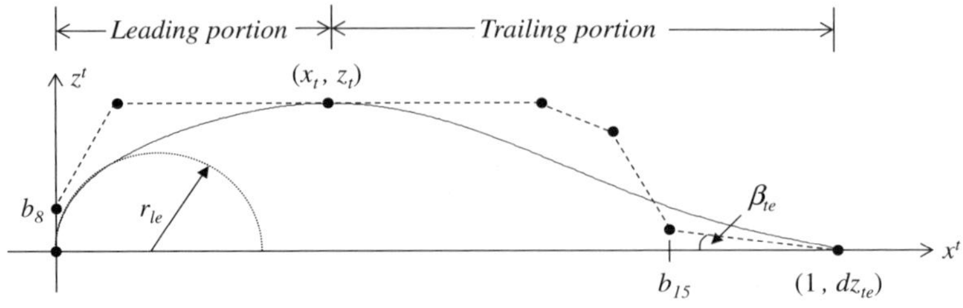

Fig. 7.10. The Bezier–PARSEC 3434 thickness distribution parameters.

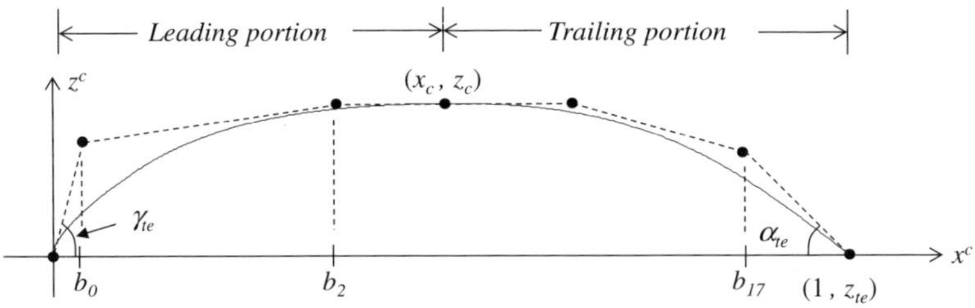

Fig. 7.11. The Bezier–PARSEC 3434 camber line parameters.

curve, given parametrically by:

$$\begin{aligned} x(u) &= x_0(1-u)^4 + 4x_1u(1-u)^3 + 6x_2u^2(1-u)^2 \\ &\quad + 4x_3u^3(1-u) + x_4u^4 \end{aligned} \tag{7.43}$$

$$\begin{aligned} z(u) &= z_0(1-u)^4 + 4z_1u(1-u)^3 + 6z_2u^2(1-u)^2 \\ &\quad + 4z_3u^3(1-u) + z_4u^4. \end{aligned} \tag{7.44}$$

The LE thickness curve control points are given by:

$$\begin{array}{ll} x_0 = 0 & z_0 = 0 \\ x_1 = 0 & z_1 = b_8 \\ x_2 = -3b_8^2/2r_{le} & z_2 = z_t \\ x_3 = x_t & z_3 = z_t \end{array}. \tag{7.45}$$

The parameter b_8 is subject to the following restriction:

$$0 < b_8 < \min(y_t, \sqrt{-2r_{le}x_t/3}). \tag{7.46}$$

Table 7.5 Numerical values of the Bezier–PARSEC 3434 parameters for a fit to the Eppler 266 airfoil profile (see coordinates in Selig (2014)).

Parameter	Value	Parameter	Value	Parameter	Value
r_{le}	−0.0139	z_{te}	0.0000	b_0	−0.0444
x_t	0.4059	dz_{te}	0.0000	b_2	0.1908
y_t	0.0989	γ_{le}	−0.0493	b_8	0.0453
x_c	0.3221	α_{te}	0.1000	b_{15}	0.8665
y_c	0.0389	β_{te}	0.0854	b_{17}	0.8289

The TE thickness curve control points are given by:

$$\begin{array}{ll} x_0 = x_t & z_0 = z_t \\ x_1 = \left(7x_t + 9b_8^2/2r_{le}\right)/4 & z_1 = z_t \\ x_2 = 3x_t + 15b_8^2/4r_{le} & z_2 = (z_t + b_8)/2 \\ x_3 = b_{15} & z_3 = dz_{te} + (1 - b_{15})\tan(\beta_{te}) \\ x_4 = 1 & z_4 = dz_{te} \end{array} . \tag{7.47}$$

The LE camber curve control points are given by:

$$\begin{array}{ll} x_0 = 0 & z_0 = 0 \\ x_1 = b_0 & z_1 = -b_0\tan(\gamma_{le}) \\ x_2 = b_2 & z_2 = z_c \\ x_3 = x_c & z_3 = z_c \end{array} . \tag{7.48}$$

The TE camber curve control points are given by:

$$\begin{array}{ll} x_0 = x_c & z_0 = z_c \\ x_1 = (7x_c - 3b_2)/4 & z_1 = z_c \\ x_2 = (b_0 - 5b_2 + 7x_c)/2 & z_2 = (z_c + b_0\tan(\gamma_{le}))/2 \\ x_3 = b_{17} & z_3 = z_{te} + (1 - b_{17})\tan(\alpha_{le}) \\ x_4 = 1 & z_4 = z_{te} \end{array} . \tag{7.49}$$

A fit to the Eppler 266 airfoil profile is given in Table 7.5 and Fig. 7.12, showing both the airfoil profile and the thickness and camber line distributions.

7.5 B-Splines

A B-spline is a piecewise polynomial function of the order k in a variable t defined over the range $t_0 \le t \le t_m$, $m = k + 1$, where the points $t = t_j$ are known as knots (or break-points). A spline

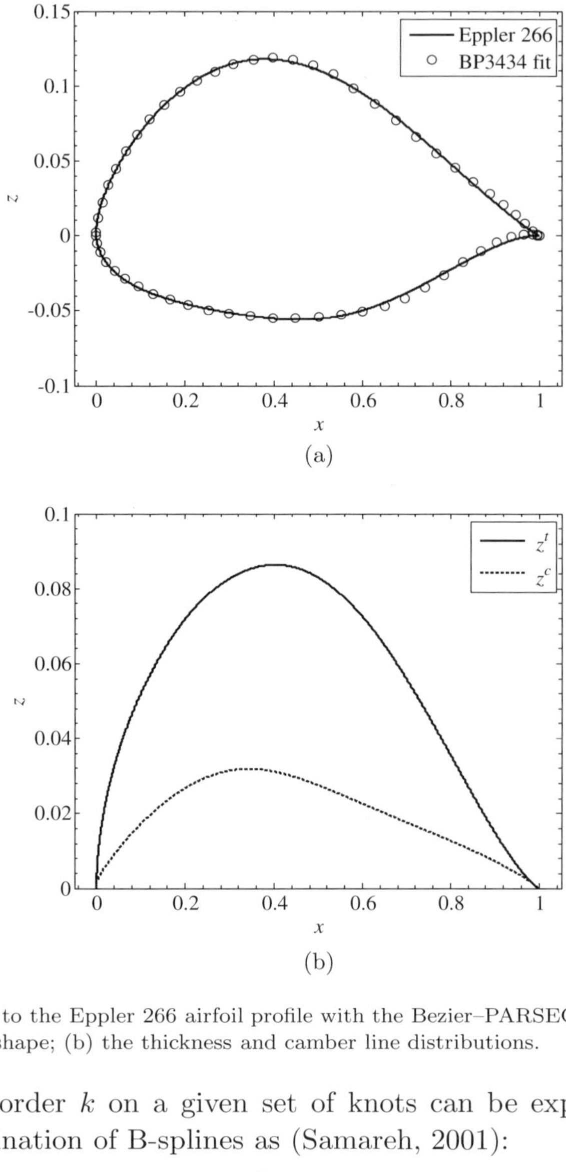

Fig. 7.12. A fit to the Eppler 266 airfoil profile with the Bezier–PARSEC 3434 method. (a) The airfoil shape; (b) the thickness and camber line distributions.

function of order k on a given set of knots can be expressed as a linear combination of B-splines as (Samareh, 2001):

$$p(t) = \sum_{i=0}^{n} N_{i,k}(t) P_i, \tag{7.50}$$

where $\mathrm{N}_{i,k}$ are the B-spline blending functions, also referred to as the basis functions, and P_i, $i = 0, 1, \ldots, n$ are control points. The basis

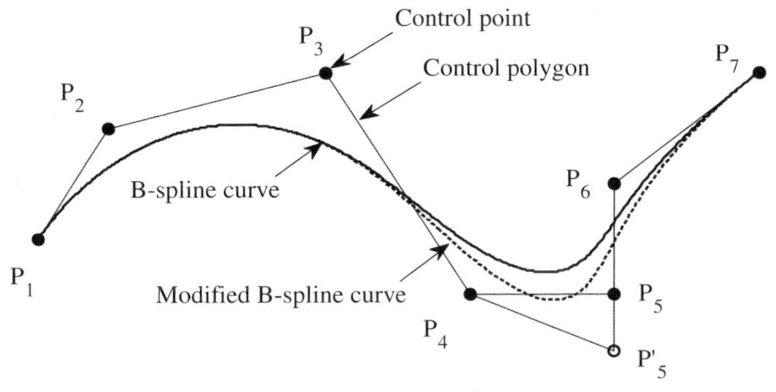

Fig. 7.13. Example B-spline curves with one different control point.

functions are calculated from the Cox–DeBoor recursion relation as:

$$N_{i,1} = \begin{cases} 1 & \text{if } t_i \le t \le t_{i+1} \\ 0 & \text{elsewhere} \end{cases}, \tag{7.51}$$

and

$$N_{i,k}(t) = \frac{t - t_i}{t_{i+k-1} - t_i} N_{i,k-1}(t) + \frac{t_{i+k} - t}{t_{i+k} - t_{i+1}} N_{i+1,k-1}(t), \tag{7.52}$$

for $k = 2, 3, \ldots, K$, as well as for all needed values of i. The basis functions $N_{i,k}$ can be polynomials of the order 1, 2, and up.

Figure 7.13 shows an example of two B-spline curves of the order 3. We can observe that the curves have the same control points, aside from control point 5. The two curves, however, are very similar, except locally, near the perturbed control point. This highlights the main difference between the B-spline curves and the Bezier curves (see Section 7.3). B-spline curves offer more local control than the Bezier curves. Figure 7.14 shows two Bezier curves with the same control points as in Fig. 7.13. We can see that the change in control point 5 affects all of the Bezier curve.

Airfoil shapes can be configured in the same manner as Bezier curves (see Section 7.3). Figure 7.15 shows B-spline fits to the Whitcomb airfoil with six and ten design variables (three and five free control points for each surface). Tables 7.6 and 7.7 give the numerical data.

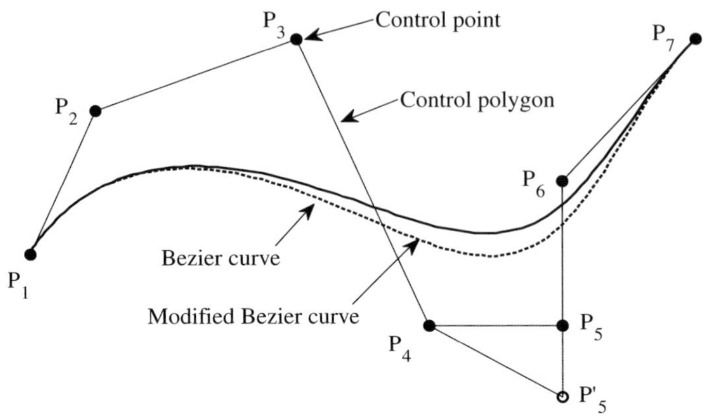

Fig. 7.14. Example Bezier curves with one different control point.

7.6 NACA Airfoil Shapes

The NACA (National Advisory Committee for Aeronautics) airfoils are constructed by combining a thickness function, $z_t(x)$, with a mean camber line function, $z_c(x)$. The x-coordinates are (Abbott and Von Doenhoff, 1959):

$$x_{u,l} = x \mp z_t \sin\theta, \tag{7.53}$$

and the z-coordinates are:

$$z_{u,l} = z_c \pm z_t \cos\theta, \tag{7.54}$$

where u and l refer to the upper and lower surfaces, respectively, $z_t(x)$ is the thickness function, $z_c(x)$ is the mean camber line function, and

$$\theta = \tan^{-1}\left(\frac{dz_c}{dx}\right) \tag{7.55}$$

is the mean camber line slope. The NACA four-digit thickness distribution is given by:

$$z_t = t(a_0 x^{1/2} - a_1 x - a_2 x^2 + a_3 x^3 - a_4 x^4), \tag{7.56}$$

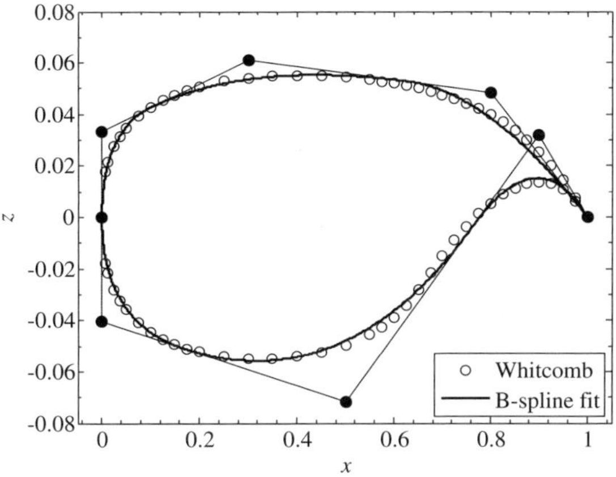

(a) Six free control points

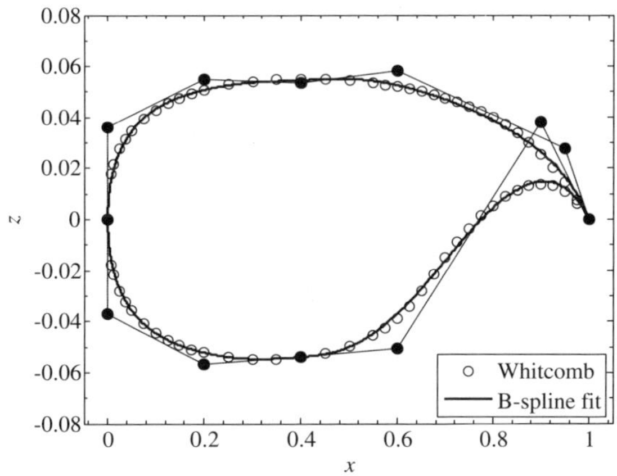

(b) 10 free control points

Fig. 7.15. B-spline curve fits to the Whitcomb airfoil (see coordinates in Selig (2014)). The free control points can only move vertically.

where $a_0 = 1.4845$, $a_1 = 0.6300$, $a_2 = 1.7580$, $a_3 = 1.4215$, $a_4 = 0.5075$, and t is the maximum thickness. The mean camber line is given by:

$$z_c = \begin{cases} \frac{m}{p^2}(2px - x^2), & x < p \\ \frac{m}{(1-p)^2}(1 - 2p + 2px - x^2), & x \geq p \end{cases}, \tag{7.57}$$

Table 7.6 B-spline fit to the Whitcomb airfoil with six free control points.

Control point	$x_{p,u}$	$z_{p,u}$
1	0.0	0.0
2	0.0	3.3436e-2
3	0.3	6.1023e-2
4	0.8	4.8300e-2
5	1.0	0.0
Control point	$x_{p,l}$	$z_{p,l}$
1	0.0	0.0
2	0.0	−4.0170e-2
3	0.5	−7.1993e-2
4	0.9	3.1785e-2
5	1.0	0.0

Table 7.7 B-spline fit to the Whitcomb airfoil with ten free control points.

Control point	$x_{p,u}$	$z_{p,u}$
1	0.0	0.0
2	0.0	3.6289e-2
3	0.2	5.4688e-2
4	0.4	5.3525e-2
5	0.6	5.8278e-2
6	0.95	2.7490e-2
7	1.0	0.0
Control point	$x_{p,l}$	$z_{p,l}$
1	0.0	0.0
2	0.0	−3.7246e-2
3	0.2	−5.6637e-2
4	0.4	−5.4126e-2
5	0.6	−5.0682e-2
6	0.9	3.7873e-2
7	1.0	0.0

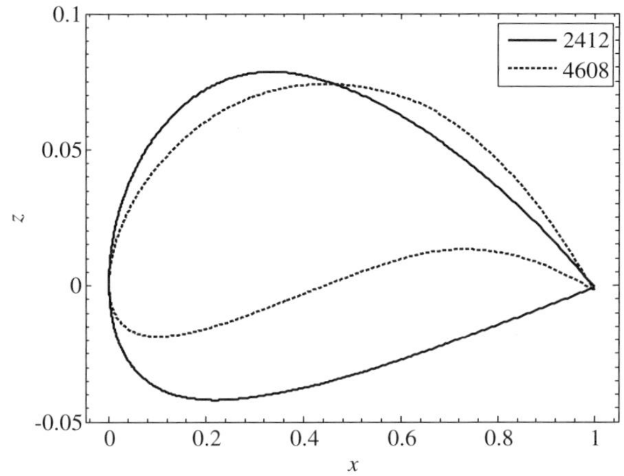

Fig. 7.16. Two different NACA four-digit airfoil sections are shown: NACA 2412 ($m = 0.02$, $p = 0.4$, $t/c = 0.12$) and NACA 4608 ($m = 0.04$, $p = 0.6$, $t/c = 0.08$).

where m is the maximum ordinate of the camber line and p is the location of the maximum ordinate. Example NACA four-digit airfoils are shown in Fig. 7.16.

7.7 Conclusion

The parameterization methods presented in this chapter can be broadly divided into three categories.

In the first category, the airfoil contours are found by choosing a baseline shape and adding suitable perturbation shapes (e.g. the bump function approach). The design variables are the perturbation shapes, and the quality of the results depend crucially on their choice.

Previous works have used simple mathematical bumps (typically sine, exponential, or polynomial shapes), which have no physical basis.

The second category involves curves controlled with several control points, such as the Bezier and B-spline methods. The control points are typically allowed to move within a specified range, either in both the horizontal and vertical directions or just along the vertical one. These methods offer the highest level of flexibility in attainable shapes. However, unrealistic shapes can also be obtained, which could present problems for the aerodynamic analysis. Moreover, the control points do not have any physical meaning.

Other methods that fit into the second category include the nonuniform rational splines (NURBS) and the freeform deformation (FFD) (see Samareh (2001) for the theory). NURBS are an extension of B-splines, using fractions of the same interpolation functions. Those fractions are weighted by a coefficient assigned to each control point. The family of curves that can be represented by NURBS is much wider than those with B-splines, as well as Bezier curves. FFD is a popular parameterization approach in aerodynamic design. The basic idea is to deform an object. While other commonly used parameterization approaches directly manipulate an object, FFD deforms a lattice built around the object and, consequently, manipulates the entire space in which the object is embedded. The lattice has the topology of a cube in the case of three-dimensional objects and a rectangle in the two-dimensional case.

In the third category, the parameters controlling the airfoil shape are physical geometric features, such as the LE radius and the airfoil maximum thickness. The NACA, PARSEC, and BP fit in this category (although BP also has a few free nongeometric control points). Even though these approaches do not offer the same degree of flexibility as the methods in the second category, the physical meaning of the design variables has great value for the airfoil designer.

There are other methods available, not presented in this chapter, that do not fit into any of the abovementioned categories. An example of such a method is the one developed by Aidala *et al.* (1983). Their approach is based on design variables with specific aerodynamic origins; the so-called aerofunction shapes. An aerofunction shape is a wing section ordinate perturbation that has been defined by a two-dimensional inverse code for a requested pressure change. The inverse code translates the aerodynamic input (a pressure change) into a geometric shape perturbation (a geometry change). Example pressure modifications include: LE expansion control; smooth upper surface shock; airfoil thickness control with lower surface modification; upper surface supersonic plateau control; move upper surface shock forward; and move upper surface shock aft. When used in optimization, the aerofunction shapes would be added to a baseline geometry by a scaling factor, which would be the design variable controlled by the optimizer. Although this approach is similar to the methods in the first category, the shapes have a sound physical basis and are therefore more intuitive for the design engineer.

This chapter does not give an exhaustive description of available geometry parameterization methods. The interested reader is directed to Samareh (2001) for a good survey of available methods and approaches.

Chapter 8

High-Fidelity Aerodynamic Models

This chapter describes the elements of modeling for high-fidelity aerodynamic analysis of airfoils and wings using computational fluid dynamics (CFD). The chapter is not intended to give an exhaustive description of the available methods; rather, the aim is to give a general overview of typical modeling approaches. The chapter presents several CFD models for airfoil and wing analysis in inviscid and viscous flows. Development of low-fidelity models — essential for design approaches considered in this book — is covered in the next chapter. These models are subsequently used in the following chapters on optimal shape design using variable-fidelity optimization algorithms.

8.1 Elements of Computational Fluid Dynamics Analysis

In general, a single CFD analysis is composed of four steps: geometry modeling, grid generation, flow solution, and calculation of the measure of merit (see Fig. 8.1). The inputs to the CFD model are the geometry design parameters, $\boldsymbol{x}$, as well as any model parameters, $\boldsymbol{p}$, such as physical properties and grid discretization. The CFD model delivers the performance characteristics of the aerodynamic surface, $f(\boldsymbol{x},\boldsymbol{p})$.

Typically, each step is executed with separate computer codes, which could be, for example, programmed by the designer himself, an in-house code, or executed by commercial software. The entire CFD process, including the flow of data between codes, needs to be fully scripted and automated so that it can be used in an optimization algorithm.

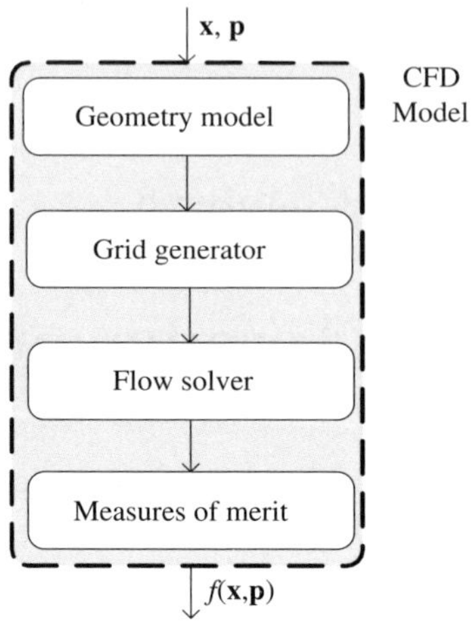

Fig. 8.1. The main elements of a CFD model.

Fig. 8.2. Three-dimensional rendering of a typical transport wing, where the surface is lofted between the sections with a linear variation in chord, thickness, and twist.

8.1.1 *Geometry model and parameterization*

The geometry model takes the design parameters, $\mathbf{x}$, as an input and generates the shape, which is subsequently picked up by the grid generator.

Several airfoil parameterization methods are described in Chapter 7. A clean wing shape can be described by means of airfoil sections at given spanwise locations and assuming a linear variation in the geometry properties between the stations. Figure 8.2 shows an example of such a wing. The wing geometry is then described by the parameters of the airfoil shapes, as well as the planform shape properties (see Chapter 2).

8.1.2 *Solution domain and grid generation*

The flow past the aerodynamic surfaces is described by nonlinear equations (provided in the next section), which cannot generally be solved analytically but can be solved by various approximate methods, including finite difference, finite volume, and finite element methods. To use these methods, it is necessary to discretize the solution domain using a grid.

A grid is a set of points distributed over a calculation field for a numerical solution of a set of partial differential equations (PDE). This set may be structured, e.g. formed by the intersections of curvilinear coordinate surfaces, or unstructured, i.e. with no relation to coordinate directions. In the former, the grid generally consists of quadrilateral cells in two dimensions, or hexahedral cells in three dimensions. The unstructured grid may be made of hexahedra or elements of any shape in general. In its most basic form, unstructured grids consist of triangles and tetrahedra in two and three dimensions, respectively.

The grid needs to resolve the entire solution domain, as well as any detailed features of the aerodynamic geometry. Furthermore, the grid needs to be sufficiently fine in order to capture the flow physics accurately. For example, a fine grid resolution is necessary near the airfoil surface, especially near the leading edge (LE), where the flow gradients are large. Also, if viscous effects are included, the grid needs to be fine near the entire airfoil surface (and any other wall surface in the solution domain). However, the grid cannot be too dense because the solution may be impractical to obtain. The grid can be much coarser several chord lengths away from the aerodynamic surface. For a detailed discussion on grid generation methods the reader is referred to Hirsch (2007) and Tannehill *et al.* (1997).

For illustration purposes, a typical structured grid for an airfoil flow simulation, generated using the commercial computer program ICEM CFD (2013), is shown in Fig. 8.3. This is a structured

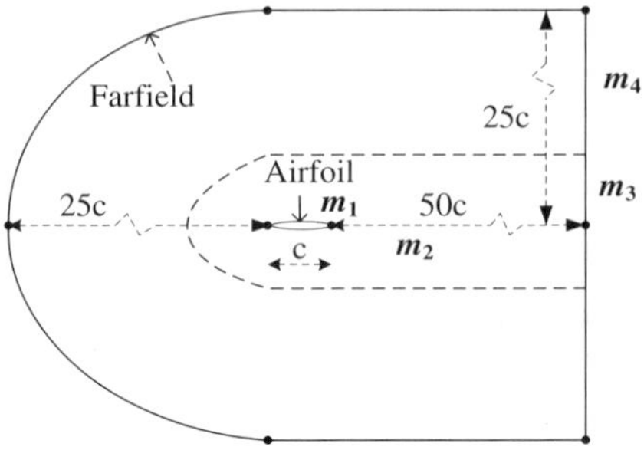

(a) Solution domain

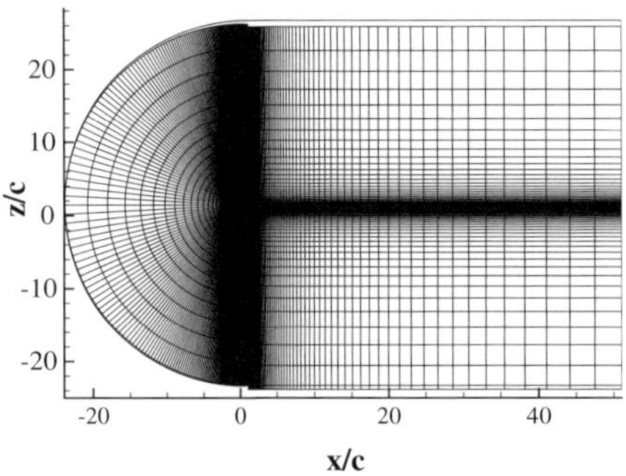

(b) Farfield

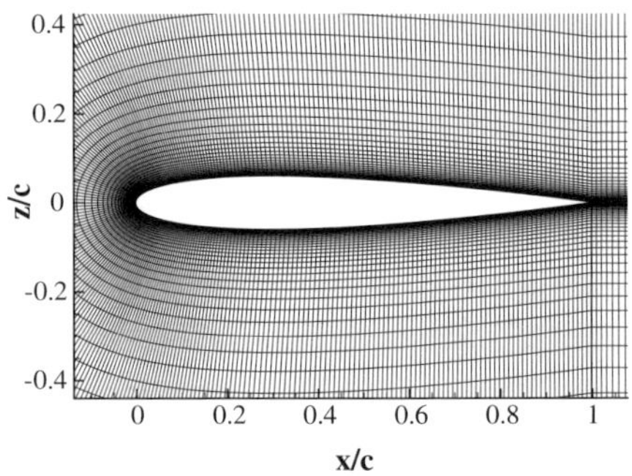

(c) Close to the airfoil surface

Fig. 8.3. Structured curvilinear body-fitted C-topology grid for transonic flow simulation of airfoils.

curvilinear body-fitted grid of C-topology (a topology that can be associated to the letter C, i.e. at the inlet the grid surrounds the LE of the airfoil but is open at the other end).

One way of generating structured grids is to solve a set of hyperbolic PDE equations (Hirch, 2007). This approach is popular for CFD analysis of aerodynamic surfaces. The advantage of this approach is that the grid can be constructed to be orthogonal and smooth. Numerical noise can be small in these types of grids if sufficiently fine. An example of this type of grid is the hyperbolic C-grid of Kinsey and Barth (1984), shown in Fig. 8.4 for two-dimensional flow past airfoil shapes. Notice the orthogonal grid lines on the airfoil surface. This grid can also be extended to three-dimensional flow past wing shapes, as shown in Fig. 8.5.

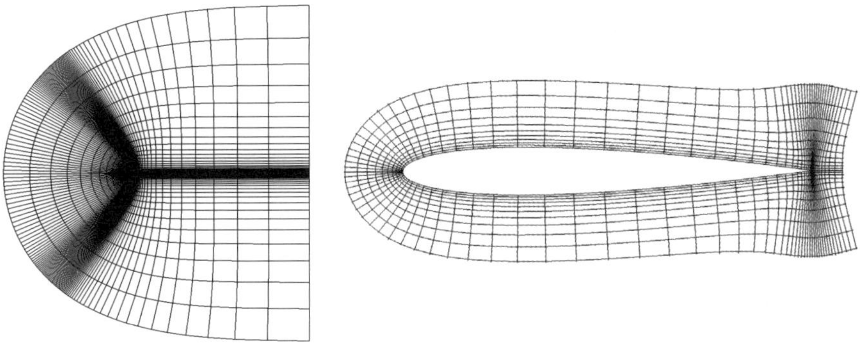

(a) Farfield (b) Close to the airfoil surface

Fig. 8.4. A structured curvilinear body-fitted hyperbolic C-topology grid for transonic flow simulation of airfoils.

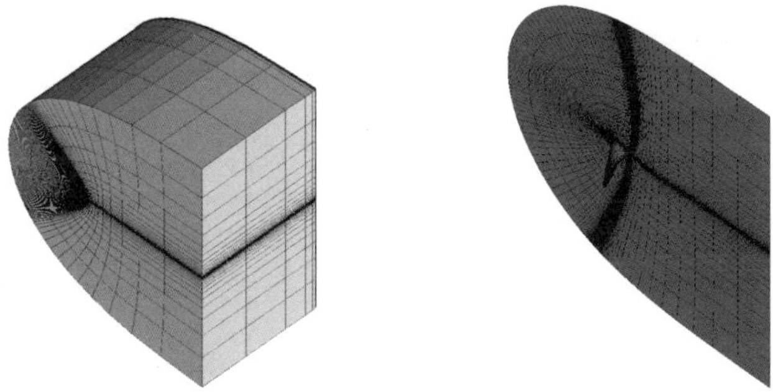

(a) Farfield (b) Symmetry plane and wing surface

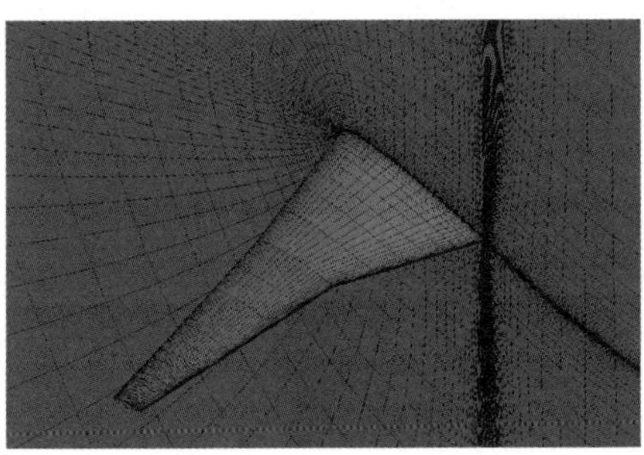

(c) Close up to the wing surface

Fig. 8.5. A structured curvilinear body-fitted hyperbolic C-topology grid for transonic flow simulation of wings.

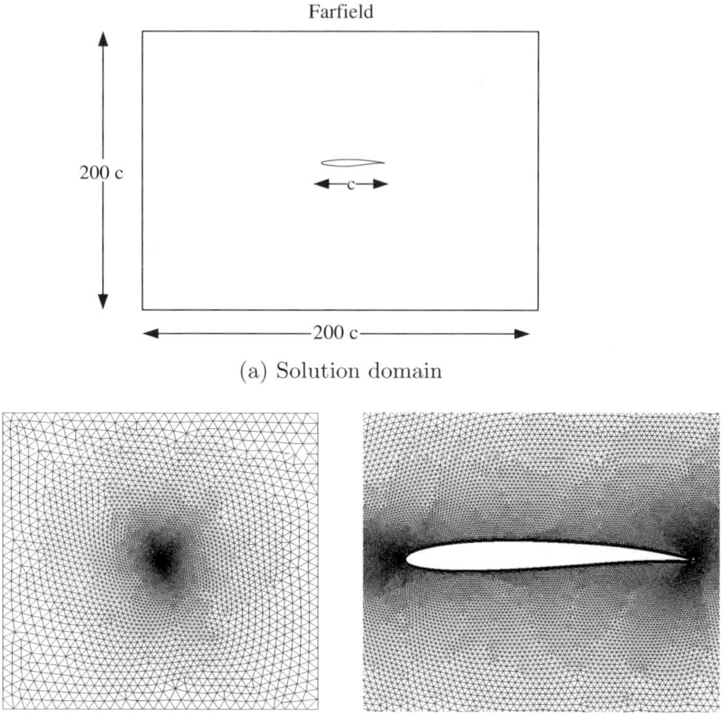

(a) Solution domain

(b) Farfield

(c) Close to the airfoil surface

Fig. 8.6. Unstructured grid with a box-shaped farfield for transonic airfoil flow simulations.

Unstructured grids can be applied to complicated geometries in two and three dimensions. This is one of the most attractive features of unstructured grids and is useful in aerodynamic analysis when handling multiple surfaces. An example of a two-dimensional unstructured grid with a box-shaped farfield is shown in Fig. 8.6. Near the airfoil surface, there are quadrilateral elements to capture the boundary layer effects. In the outer field, there are triangular elements. The three-dimensional version is shown in Fig. 8.7.

8.1.3 *Governing equations and flow solution*

The fluid flow past an aerodynamic surface is governed by the Navier–Stokes equations. For a Newtonian fluid, compressible viscous flows in three dimensions without body forces, mass diffusion, finite-rate chemical reactions, heat conduction, or external heat addition, the Navier–Stokes equations can be written (in Cartesian coordinates) as (Tannehill *et al.*, 1997):

$$\frac{\partial \boldsymbol{U}}{\partial t} + \frac{\partial \boldsymbol{E}}{\partial x} + \frac{\partial \boldsymbol{F}}{\partial y} + \frac{\partial \boldsymbol{G}}{\partial z} = 0, \tag{8.1}$$

where **U**, **E**, **F**, and **G** are vectors given by:

$$\boldsymbol{U} = \begin{bmatrix} \rho \\ \rho u \\ \rho v \\ \rho w \\ E_t \end{bmatrix}, \tag{8.2}$$

$$\boldsymbol{E} = \begin{bmatrix} \rho u \\ \rho u^2 + p - \tau_{xx} \\ \rho u v - \tau_{xy} \\ \rho u w - \tau_{xz} \\ (E_t + p)u - u\tau_{xx} - v\tau_{xy} - w\tau_{xz} \end{bmatrix}, \tag{8.3}$$

$$\boldsymbol{F} = \begin{bmatrix} \rho v \\ \rho u v - \tau_{xy} \\ \rho v^2 + p - \tau_{yy} \\ \rho v w - \tau_{yz} \\ (E_t + p)v - u\tau_{xy} - v\tau_{yy} - w\tau_{yz} \end{bmatrix}, \tag{8.4}$$

$$\boldsymbol{G} = \begin{bmatrix} \rho w \\ \rho u w - \tau_{xz} \\ \rho v w - \tau_{yz} \\ \rho w^2 + p - \tau_{zz} \\ (E_t + p)w - u\tau_{xz} - v\tau_{yz} - w\tau_{zz} \end{bmatrix}. \tag{8.5}$$

Here, ρ is the fluid density, u, v, and w are the x, y, and z velocity components, respectively, p is the static pressure, $E_t = \rho(e + V^2/2)$ is the total energy per unit volume, e is the internal energy per unit mass, $V^2/2$ is the kinetic energy, and τ is the viscous shear stress

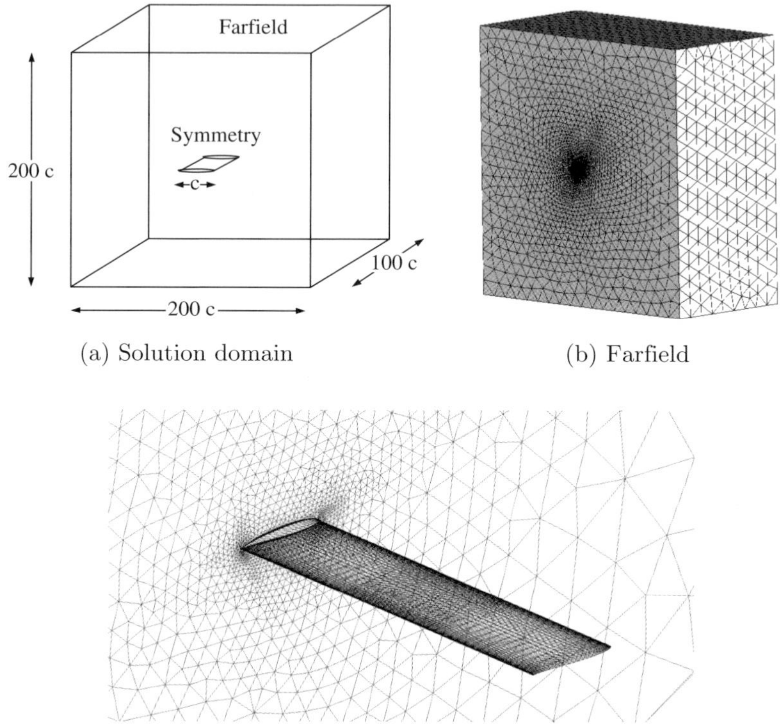

(a) Solution domain

(b) Farfield

(c) Close to the wing surface

Fig. 8.7. Unstructured grid with a box-shaped farfield for transonic wing flow simulations.

tensor given by:

$$\tau_{xx} = \frac{2}{3}\mu\left(2\frac{\partial u}{\partial x} - \frac{\partial v}{\partial y} - \frac{\partial w}{\partial z}\right), \quad \tau_{yy} = \frac{2}{3}\mu\left(2\frac{\partial v}{\partial y} - \frac{\partial u}{\partial x} - \frac{\partial w}{\partial z}\right)$$

$$\tau_{zz} = \frac{2}{3}\mu\left(2\frac{\partial w}{\partial z} - \frac{\partial u}{\partial x} - \frac{\partial v}{\partial y}\right), \quad \tau_{xy} = \mu\left(\frac{\partial u}{\partial y} + \frac{\partial v}{\partial x}\right) = \tau_{yx}$$

$$\tau_{xz} = \mu\left(\frac{\partial w}{\partial x} + \frac{\partial u}{\partial z}\right) = \tau_{zx}, \quad \tau_{yz} = \mu\left(\frac{\partial v}{\partial z} + \frac{\partial w}{\partial y}\right) = \tau_{zy},$$

where μ is the dynamic viscosity of the fluid. The first row of Equation (8.1) corresponds to the continuity equation, the second, third, and fourth rows are the momentum equations, and the fifth row is the energy equation. These five scalar equations contain six unknowns, namely ρ, p, e, u, v, and w. An equation of state is needed to close the system of equations. For most problems in gas dynamics, it is possible to assume a perfect gas, which is defined as a gas in which intermolecular forces are negligible. A perfect gas obeys the perfect gas equation of state, i.e. $p = \rho RT$, where R is the gas constant.

The governing equations are a set of coupled, highly nonlinear partial differential equations. The numerical solution of these equations is quite challenging. What complicates things even further is that all flows will become turbulent above a critical value of the Reynolds number $Re = VL/\upsilon$, where V and L are representative values of velocity and length scales and υ is the kinematical viscosity. Turbulent flows are characterized by the appearance of statistical fluctuations of all the variables (ρ, p, e, u, v, w) around mean values.

By making appropriate assumptions about the fluid flow, the governing equations can be simplified and their numerical solution becomes computationally less expensive. In general, there are two approaches that differ in either neglecting the effects of viscosity or including them into the analysis. The hierarchy of the governing flow equations depending on the assumptions made about the fluid flow situation is shown in Fig. 8.8.

Direct numerical simulation (DNS) has an objective to simulate the entire range of the turbulent statistical fluctuations at all relevant physical scales. This is a formidable challenge, which grows with increasing Reynolds number, as the total computational effort for DNS simulations is proportional to Re^3 for homogeneous turbulence (Hirsch, 2007). Due to the limitations of computational capabilities, DNS is not available for typical engineering flows, such as those encountered in airfoil design for typical aircraft and turbomachinery, i.e. with Reynolds numbers from 10^5 to 10^7.

Large-eddy simulation (LES) is of the same category as DNS, in that it computes directly the turbulent fluctuations in space and time, but only above a certain length scale. Below that scale, the

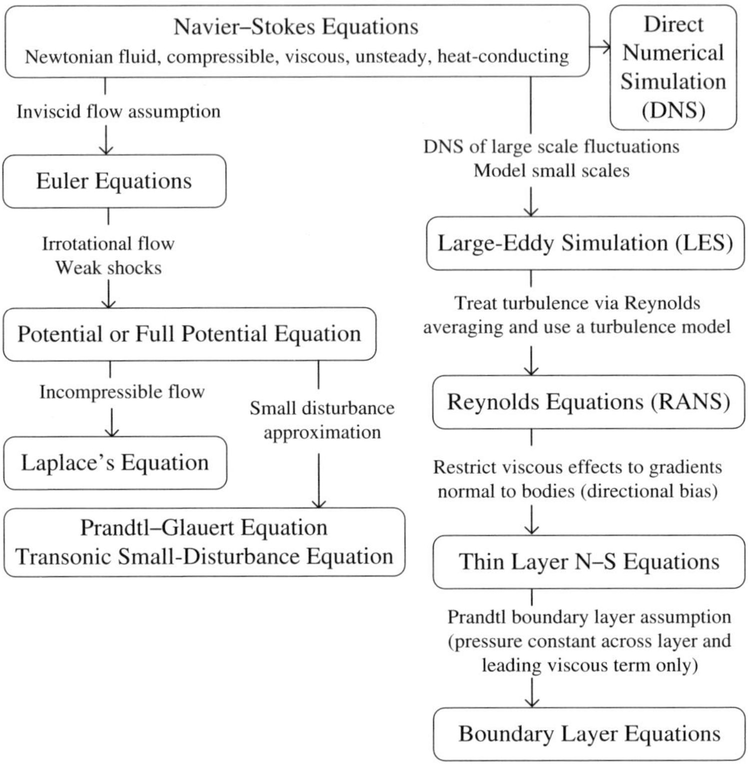

Fig. 8.8. A hierarchy of the governing fluid flow equations with the associated assumptions and approximations.

turbulence is modeled by semi-empirical laws. The total computational effort for LES simulations is proportional to $Re^{9/4}$, which is significantly lower than for DNS (Hirch, 2007). However, it is still excessively high for large Reynolds number applications.

The Reynolds equations (also called the Reynolds-averaged Navier–Stokes (RANS) equations) are obtained by time-averaging of a turbulent quantity into their mean and fluctuating components. This means that turbulence is treated through turbulence models. As a result, a loss in accuracy is introduced, as the available turbulence models are not universal. A widely used turbulence model for simulation of the flow past airfoils and wings is the Spalart–Allmaras one-equation turbulence model (Spalart and Allmaras, 1979). The model was developed for aerospace applications and is considered accurate for attached wall-bounded flows and flows with mild separation and recirculation.

However, the RANS approach retains the viscous effects in the fluid flow and, at the same time, significantly reduces the computational effort, since there is no need to resolve all the turbulent scales (as it is done in DNS and partially in LES). This approach is currently the most widely applied approximation in the CFD practice and can be applied to both low-speed, such as takeoff and landing conditions of an aircraft, and high-speed design.

The inviscid flow assumption will lead to the Euler equations. These equations hold, in the absence of separation and other strong viscous effects, for any shape of the body, thick or thin, and at any angle of attack (Anderson, 2007). Shock waves appear in transonic flow in which the flow goes from being supersonic to subsonic. Across the shock, there is almost a discontinuous increase in pressure, temperature, density, and entropy, but a decrease in Mach number (from supersonic to subsonic). The shock is termed weak if the change in pressure is small, and strong if the change in pressure is large. The entropy change is of third order in terms of shock strength. If the shocks are weak, the entropy change across shocks is small, and the flow can be assumed to be isentropic. This, in turn, allows for the assumption of irrotational flow. Then the Euler equations cascade to a single nonlinear partial differential equation, called the full potential equation (FPE). In the case of a slender body at a small angle of attack, we can make the assumption that there is a small disturbance. Then the FPE becomes the transonic small-disturbance equation (TSDE). These three different sets of equations, i.e. the Euler equations, FPE, and TSDE, represent a hierarchy of models for the analysis of inviscid, transonic flow past airfoils (Anderson, 2007). The Euler equations are exact, while FPE is an approximation (weak shocks) to those equations, and TSDE is a further approximation (thin airfoils at small angles of attack). These approaches can be applied effectively for high-speed design, such as the cruise design of transport aircraft wings and the design of turbomachinery blades.

There are numerous airfoil and wing models that are not typical CFD models, but they are nevertheless widely used in aerodynamic

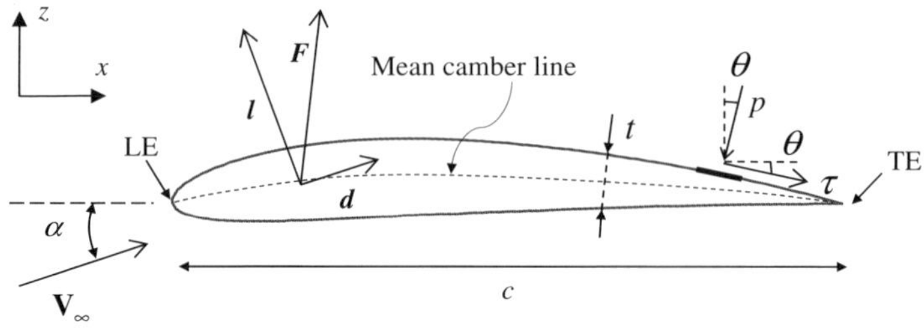

Fig. 8.9. Nomenclature of airfoil forces.

design. Examples of such methods include thin airfoil theory, lifting line theory (unswept wings), vortex lattice methods (wings), and panel methods (airfoils and wings). These methods are out of the scope of this chapter, but the interested reader is directed to Katz and Plotkin (2001) and Anderson (2007) for details.

8.1.4 *Measure of merit*

A single-element airfoil section of chord length c and thickness $t(x)$ is shown in Fig. 8.9. The free-stream velocity, $\boldsymbol{V}_\infty$, is at an angle of attack, α, relative to the x-axis. $\boldsymbol{F}$ is the force acting on the airfoil due to the airflow, where the component perpendicular to $\boldsymbol{V}_\infty$ is called the lift force, $\boldsymbol{l}$, and the component parallel to $\boldsymbol{V}_\infty$ is called the drag force, $\boldsymbol{d}$. p is the pressure acting normal to a surface element of length ds, and τ is the viscous wall shear stress acting parallel to the surface element. θ is the angle that p and τ make relative to the z- and x-axis, respectively, positive clockwise.

The function of the airfoil is to generate the lift force, $\boldsymbol{l}$, at a range of operating conditions (Mach number, M_∞, Reynolds number, angle of attack, α). Normally, the drag force, $\boldsymbol{d}$, is to be minimized. These forces are due to the pressure distribution, p (acting normal to the surface), and the shear stress distribution, τ (acting parallel to the surface), over the surface of the airfoil.

The non-dimensional form of the lift is the lift coefficient, defined as:

$$C_l \equiv \frac{l}{q_\infty S}, \tag{8.6}$$

where l is the magnitude of the lift force, $q_\infty \equiv (1/2)\rho_\infty V_\infty^2$ is the dynamic pressure, ρ_∞ is the air density, V_∞ is the free-stream velocity, and S is a reference surface. For a two-dimensional airfoil, the reference area is taken to be the chord length multiplied by a unit depth, i.e. $S = c$. Similarly, the drag coefficient is defined as:

$$C_d \equiv \frac{d}{q_\infty S}, \tag{8.7}$$

where d is the magnitude of the drag force.

The aerodynamic forces can be calculated by integrating the pressure (p) and the viscous wall shear stress (τ), as defined in Figure 8.9, over the surface of the airfoil. The pressure coefficient is defined as:

$$C_p \equiv \frac{p - p_\infty}{q_\infty}, \tag{8.8}$$

where p_∞ is the free-stream pressure. Similarly, the shear stress coefficient is defined as:

$$C_f \equiv \frac{\tau}{q_\infty}. \tag{8.9}$$

The normal force coefficient (parallel to the z-axis) acting on the airfoil is (Anderson, 2007):

$$C_n = -\int (C_{p_u} \cos\theta + C_{f_u} \sin\theta) ds_u + \int (C_{p_l} \cos\theta + C_{f_l} \sin\theta) ds_l, \tag{8.10}$$

where ds is the length of a surface element. The subscripts u and l refer to the upper and lower airfoil surfaces, respectively. The horizontal force coefficient (parallel to the x-axis) acting on the airfoil is:

$$C_a = \int (-C_{p_u} \sin\theta + C_{f_u} \cos\theta) ds_u + \int (C_{p_l} \sin\theta + C_{f_l} \cos\theta) ds_l. \tag{8.11}$$

The lift force coefficient is calculated as:

$$C_l = C_n \cos\alpha - C_a \sin\alpha, \tag{8.12}$$

and the drag force coefficient is calculated as:

$$C_d = C_n \sin\alpha + C_a \cos\alpha. \tag{8.13}$$

For a three-dimensional wing surface the calculation is performed in the same way, but an extra dimension along the wing surface is added.

8.2 Inviscid Airfoil CFD Model

A CFD model for the two-dimensional inviscid flow analysis past an airfoil section is described.

8.2.1 *Governing equations*

The flow is assumed to be steady, inviscid, and adiabatic, with no body forces. The compressible Euler equations are taken to be the governing fluid flow equations and can be written in Cartesian coordinates in vector form in three dimensions as (Tannehill *et al.*, 1997):

$$\frac{\partial \boldsymbol{U}}{\partial t} + \frac{\partial \boldsymbol{E}}{\partial x} + \frac{\partial \boldsymbol{F}}{\partial y} + \frac{\partial \boldsymbol{G}}{\partial z} = 0, \tag{8.14}$$

where **U**, **E**, **F**, and **G** are vectors given by:

$$\boldsymbol{U} = \begin{bmatrix} \rho \\ \rho u \\ \rho v \\ \rho w \\ E_t \end{bmatrix}, \tag{8.15}$$

$$\boldsymbol{E} = \begin{bmatrix} \rho u \\ \rho u^2 + p \\ \rho u v \\ \rho u w \\ (E_t + p)u \end{bmatrix}, \tag{8.16}$$

$$\boldsymbol{F} = \begin{bmatrix} \rho v \\ \rho u v \\ \rho v^2 + p \\ \rho v w \\ (E_t + p)v \end{bmatrix}, \tag{8.17}$$

$$\boldsymbol{G} = \begin{bmatrix} \rho w \\ \rho u w \\ \rho v w \\ \rho w^2 + p \\ (E_t + p)w \end{bmatrix}. \tag{8.18}$$

Here, ρ is the fluid density, u, v, and w are the x, y, and z velocity components, respectively, p is the static pressure, $E_t = \rho(e + V^2/2)$ is the total energy per unit volume, e is the internal energy per unit mass, and $V^2/2$ is the kinetic energy.

8.2.2 *Flow solver*

The flow solver FLUENT (2012) is used to solve the two-dimensional inviscid compressible flow. FLUENT is a commercially available CFD flow solver and is based on the finite volume method (FVM). In the FVM, the solution domain is subdivided into a finite number of small control volumes (cells) by a grid. The grid defines the boundaries of the control volumes, while the computational node lies at the center of the control volume. Integral conservation of mass, momentum, and energy are satisfied exactly over each control volume. The result is a set of linear algebraic equations, one for each control volume. The set of equations are then solved iteratively or simultaneously. Iterative solving is usually performed with relaxation to suppress numerical oscillations in the flow solution that result from numerical errors. The iterative process is repeated until the change in the flow variables in two subsequent iterations becomes smaller than the prescribed convergence threshold. Further information on FVM and solution procedures can be found in Hirch (2007) and Tannehill *et al.* (1997).

The particular formulation used in this model is based on an implicit density-based formulation. The inviscid fluxes are calculated by an upwind-biased, second-order, spatially accurate Roe flux scheme. Asymptotic convergence to a steady state solution is obtained in each case. The flow solution convergence criterion for the high-fidelity model is the one that occurs first of either a reduction in all the residuals by six orders or a maximum number of iterations of 1,000.

8.2.3 *Solution domain and grid generation*

A C-type solution boundary, as shown in Fig. 8.3(a), is used in this model. The boundaries are placed at 25 chord lengths in front of the airfoil, 50 chord lengths behind it, and 25 chord lengths above and below it. The elements cluster around the airfoil and grow in size with distance from the airfoil surface. The grid distribution is parameterized in each direction, as well as on the airfoil surface (see the m_i parameters in Fig. 8.3(a)). The computer code ICEM CFD (2013) is used for the mesh generation.

A grid convergence study for the NACA (National Advisory Committee for Aeronautics) 0012 airfoil at Mach number $M_\infty = 0.75$ and angle of attack $\alpha = 1$ degree is shown in Fig. 8.10. First the fine mesh (marked as f on the graph) was constructed. Then the mesh density was doubled in each direction to produce the extra fine mesh (xf). The lift and drag coefficients changed by less than one lift count ($\Delta C_l = 10^{-2}$) and one drag count ($\Delta C_d = 10^{-4}$), respectively. The remaining meshes (c_i, $i = 1, 2, 3, 4$) were generated by halving the mesh density in each direction. It is concluded that the fine mesh (f) is adequate, and roughly 408,000 cells are needed for mesh convergence.

The overall simulation time for the case considered with the fine mesh is around 33 minutes (Fig. 8.10(b)). The solver reached a fully converged solution (maximum residual was less than 10^{-6}) in this case after 217 iterations (Fig. 8.11). The solver was able to reach a fully converged solution in all the cases.

Figure 8.12 shows the Mach contours of the flow past the NACA 0012 airfoil obtained by the fine mesh and Fig. 8.13 shows the corresponding pressure distribution. One can observe a strong shock on the upper surface.

8.3 Inviscid Airfoil CFD Model with a Hyperbolic C-Grid

The CFD model in Section 8.2 is modified by replacing the grid with the hyperbolic C-grid of Kinsey and Barth (1984), shown in Fig. 8.4. Other components of the CFD model are the same.

8.3.1 *Grid setup*

The farfield is set 100 chord lengths away from the airfoil surface. The grid points were clustered at the trailing edge (TE) and the LE of the airfoil to give a minimum streamwise spacing of 0.001×

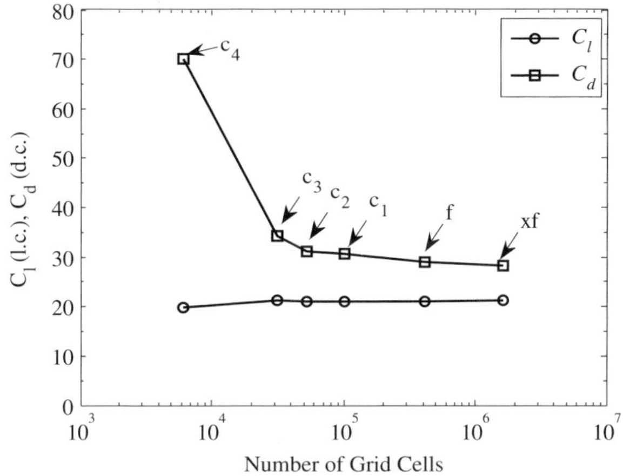

(a) Lift and drag coefficient convergence

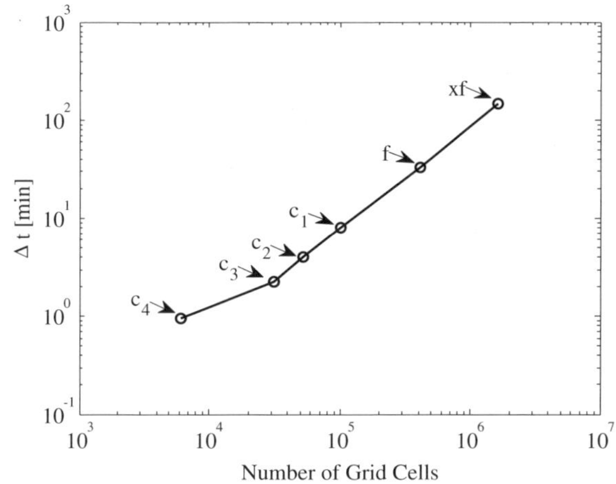

(b) Variation of simulation time

Fig. 8.10. Grid convergence study using the NACA 0012 airfoil at Mach number $M_\infty = 0.75$ and angle of attack $\alpha = 1$ degree.

chord length. The distance from the airfoil surface to the first node is $5 \cdot 10^{-4} \times$ chord length.

8.3.2 *Grid convergence*

The results of the grid convergence study of NACA 0012 at $M_\infty = 0.75$ and $\alpha = 1$ degree, given in Fig. 8.14 and Table 8.1, show that

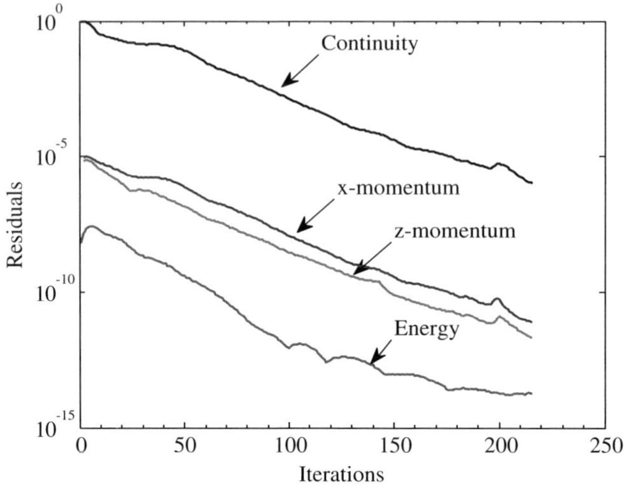

Fig. 8.11. Flow solver convergence history for the fine grid.

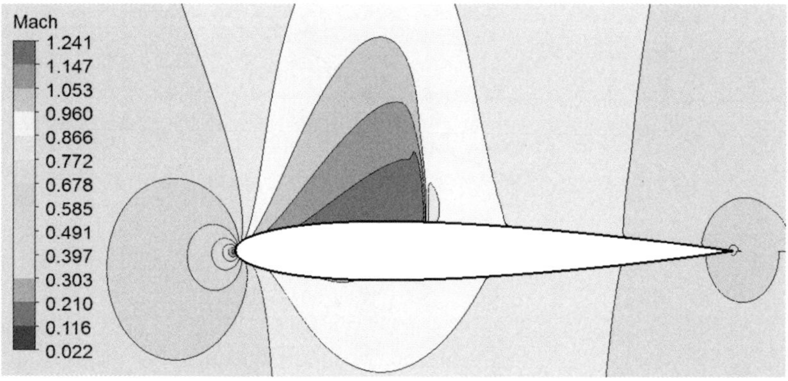

Fig. 8.12. Contours of Mach number of the NACA 0012 at $M_\infty = 0.75$ and $\alpha = 1$ degree.

the grid converges in terms of lift and drag to less than 0.1 drag counts with around 163,000 cells. The grids have 270 points in the streamwise direction and 180 points in the direction normal to the airfoil surface. The region behind the airfoil to the farfield contains 180 points. The overall evaluation time is about 15 minutes.

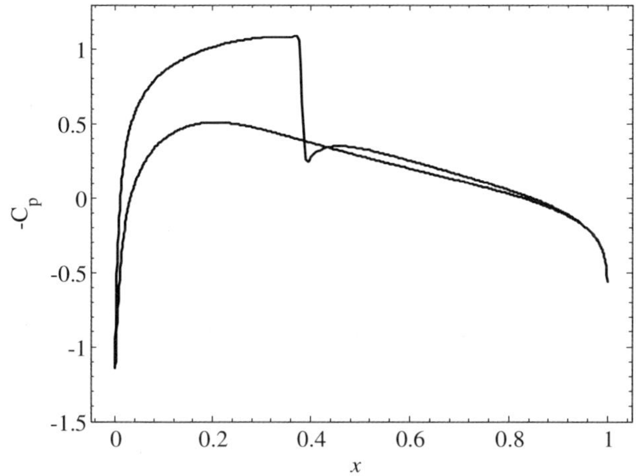

Fig. 8.13. Pressure distribution of the NACA 0012 at $M_\infty = 0.75$ and $\alpha = 1$ degree. The lift and drag coefficients are $C_l = 28.9$ l.c. and $C_d = 21.1$ d.c., respectively.

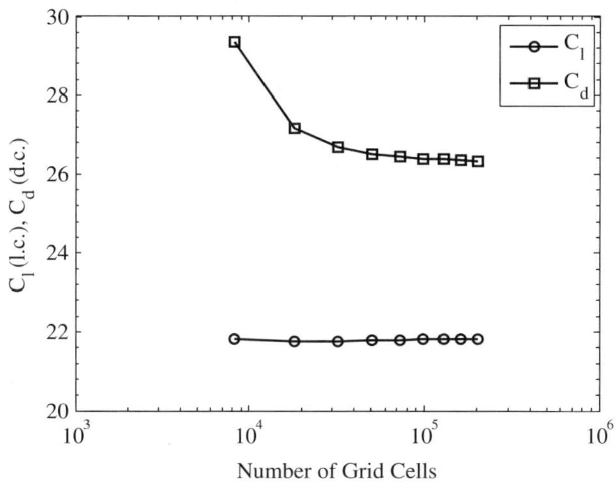

Fig. 8.14. Grid convergence study with the hyperbolic C-grid using the NACA 0012 airfoil at Mach number $M_\infty = 0.75$ and angle of attack $\alpha = 1$ degree.

8.4 Viscous Airfoil CFD Model

A CFD model for the two-dimensional viscous flow analysis past an airfoil section is described.

Table 8.1 Grid convergence study of NACA 0012 at $M_\infty = 0.75$ and $\alpha = 1$ degree.

Number of cells	C_l(l.c.)	C_d(d.c.)
34,000	21.6	26.8
73,000	21.7	26.6
163,000	21.8	26.3
393,000	21.8	26.2

8.4.1 *Governing equations*

The flow is assumed to be steady, viscous, and compressible. The RANS equations are taken as the governing flow equations with the Spalart–Allmaras turbulence model.

The Reynolds-averaged continuity equation can be written as (in Cartesian tensor form) (Tannehill *et al.*, 1997):

$$\frac{\partial \bar{\rho}}{\partial t} + \frac{\partial}{\partial x_j}(\bar{\rho}\bar{u}_j + \overline{\rho' u'_j}) = 0, \tag{8.19}$$

where $\bar{u}_i$ is the time-averaged velocity component $(i, j = 1, 2, 3)$, $\bar{\rho}$ is the time-averaged fluid density. Here, time averaging of a quantity f over a time interval t_0 to $t_0 + \Delta t$ is defined as:

$$\bar{f} \equiv \frac{1}{\Delta t}\int_{t_0}^{t_0+\Delta t} f\, dt. \tag{8.20}$$

The flow variables have the components:

$$f = \bar{f} + f', \tag{8.21}$$

where f' is the randomly changing variable.

The momentum equations are written as:

$$\frac{\partial}{\partial t}(\bar{\rho}\,\bar{u}_i + \overline{\rho' u'_i}) + \frac{\partial}{\partial x_j}(\bar{\rho}\,\bar{u}_i\,\bar{u}_j + \bar{u}_i\overline{\rho' u'_j})$$
$$= -\frac{\partial \bar{p}}{\partial x_i} + \frac{\partial}{\partial x_j}(\bar{\tau}_{ij} - \bar{u}_j\overline{\rho' u'_i} - \bar{\rho}\,\overline{u'_i u'_j} - \overline{\rho' u'_i u'_j}), \tag{8.22}$$

where $\bar{p}$ is the time-averaged pressure.

The energy equation is:

$$\frac{\partial}{\partial t}(\bar{\rho}\,\bar{H} + \overline{\rho' H'})$$
$$+\frac{\partial}{\partial x_j}\left(\bar{\rho}\,\bar{u}_j\bar{H} + \bar{\rho}\overline{u'_j H'} + \overline{\rho' u'_j}\bar{H}' + \overline{\rho' u'_j H'} + \bar{u}_j\overline{\rho' H'} - k\frac{\partial \bar{T}}{\partial x_j}\right)$$
$$=\frac{\partial \bar{p}}{\partial t} + \frac{\partial}{\partial x_j}\left[\begin{array}{c}\bar{u}_i\left(-\frac{2}{3}\mu\delta_{ij}\frac{\partial \bar{u}_k}{\partial x_k}\right) + \mu\bar{u}_i\left(\frac{\partial \bar{u}_j}{\partial x_i} + \frac{\partial \bar{u}_i}{\partial x_j}\right)\\ -\frac{2}{3}\mu\delta_{ij}\overline{u'_i\frac{\partial u'_k}{\partial x_k}} + \mu\left(\overline{u'_i\frac{\partial u'_j}{\partial x_i}} + \overline{u'_i\frac{\partial u'_i}{\partial x_j}}\right)\end{array}\right], \tag{8.23}$$

where

$$\bar{\tau}_{ij} = \mu\left[\left(\frac{\partial \bar{u}_i}{\partial x_j} + \frac{\partial \bar{u}_j}{\partial x_i}\right) - \frac{2}{3}\delta_{ij}\frac{\partial \bar{u}_k}{\partial x_k}\right] \tag{8.24}$$

is the time-averaged viscous stress tensor, and δ_{ij} is the kronecker delta.

The last component of Equation (8.22), $-\overline{\rho' u'_i u'_j}$, commonly called the Reynolds stress, represents the effects of turbulence and needs to be modeled in order to close this equation.

The Spalart–Allmaras turbulence model employs the Boussinesq approach, which provides an equation for the Reynolds stress as:

$$-\rho\,\overline{u'_i u'_j} = \mu_T\left(\frac{\partial u_i}{\partial x_j} + \frac{\partial u_j}{\partial x_i}\right) - \frac{2}{3}\delta_{ij}\left(\mu_T\frac{\partial u_k}{\partial x_k} + \rho\bar{k}\right), \tag{8.25}$$

where δ_{ij} is Kronecker delta, $\bar{k}$ is the kinetic energy of turbulence, $\bar{k} = \overline{u'_i u'_i}/2$, and μ_T is the turbulent viscosity. The turbulent viscosity is defined by the Spalart–Allmaras model as:

$$\mu_T = \rho\tilde{\nu}f_{\nu 1}, \tag{8.26}$$

where the parameter $\tilde{\nu}$ is obtained from the solution of the transport equation:

$$u_j\frac{\partial \tilde{\nu}}{\partial x_j} = \frac{1}{\sigma}\frac{\partial}{\partial x_k}\left[(\nu + \tilde{\nu})\frac{\partial \tilde{\nu}}{\partial x_k}\right] + c_{b1}(1 - f_{\nu 2})\tilde{S}\tilde{\nu}$$
$$- c_{w1}f_w\left(\frac{\tilde{\nu}}{d}\right)^2 + \frac{c_{b2}}{\sigma}\frac{\partial \tilde{\nu}}{\partial x_k}\frac{\partial \tilde{\nu}}{\partial x_k}, \tag{8.27}$$

where the closure coefficients and functions are given by:

$$c_{b1} = 0.1355, \quad c_{b2} = 0.622, \quad c_{\nu 1} = 7.1, \quad \sigma = 2/3,$$

$$c_{w1} = \frac{c_{b1}}{\kappa^2} + \frac{(1 + c_{b2})}{\sigma}, \quad c_{w2} = 0.3, \quad c_{w3} = 2, \quad \kappa = 0.41,$$

$$f_{\nu 1} = \frac{\chi^3}{\chi^3 + c_{\nu 1}^3}, \quad f_{\nu 2} = 1 - \frac{\chi}{1 + \chi f_{\nu 1}},$$

$$f_w = g\left(\frac{1 + c_{w3}^6}{g^6 + c_{w3}^6}\right), \quad \chi = \frac{\tilde{\nu}}{\nu},$$

$$g = r + c_{w2}(r^6 - r), \quad r = \frac{\tilde{\nu}}{\tilde{S}\kappa^2 d^2}, \quad \tilde{S} = S + \frac{\tilde{\nu}}{\kappa^2 d^2} f_{\nu 2},$$

$$S = \sqrt{2\Omega_{ij}\Omega_{ij}}, \quad \Omega_{ij} = \frac{1}{2}\left(\frac{\partial u_i}{\partial x_j} - \frac{\partial u_j}{\partial x_i}\right),$$

and d is the distance from the closest surface.

8.4.2 *Flow solver and grid setup*

The FLUENT (2012) flow solver described in Section 8.2.1 is used in this model with the same convergence criteria. The hyperbolic C-grid by Kinsey and Barth (1984), shown in Fig. 8.4, is used here. The farfield is set 100 chord lengths away from the airfoil surface. The grid points are clustered at the TE and the LE of the airfoil to give a minimum streamwise spacing of $0.001 \times$ chord length. The distance from the airfoil surface to the first node is $5 \cdot 10^{-6} \times$ chord length.

8.4.3 *Grid convergence*

A grid convergence study of RAE 2822 at $M_\infty = 0.734$ and $\alpha = 2.63$ degrees, given in Table 8.2 and Fig. 8.15, shows that the grid converges in terms of lift and drag to less than 0.1 drag counts with around 393,000 cells. The grids have 420 points in the streamwise direction, 280 points in the direction normal to the airfoil surface, and 280 points in the region behind the airfoil to the farfield. The evaluation time is around 40 minutes. The non-dimensional wall distance $y^+ < 1.0$ over the entire airfoil surface.

Table 8.2 Grid convergence study of RAE 2822 at $M_\infty = 0.734$ and $\alpha = 2.63$ degrees.

Number of cells	C_l(l.c.)	C_d(d.c.)
163,000	77.2	164.0
243,000	77.4	164.6
393,000	77.6	165.1
452,000	77.7	165.2

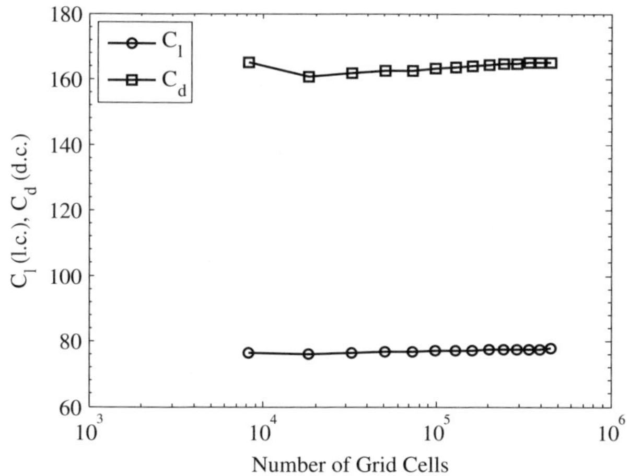

Fig. 8.15. Grid convergence study with the hyperbolic C-grid using the RAE 2822 airfoil at Mach number $M_\infty = 0.734$ and angle of attack $\alpha = 2.63$ degrees in viscous flow.

Mach contours of the flow past the NACA 0012 airfoil obtained by the fine mesh are given in Fig. 8.16, and the corresponding pressure distribution is shown in Fig. 8.17.

8.5 Inviscid Wing CFD Model

A CFD model for the three-dimensional inviscid flow analysis past a wing is described.

8.5.1 *Model setup*

The flow is governed by the steady compressible Euler equations given in Section 8.2.1. FLUENT (2012) is used for the flow solution, as described in Section 8.2.2. The computational grid is a

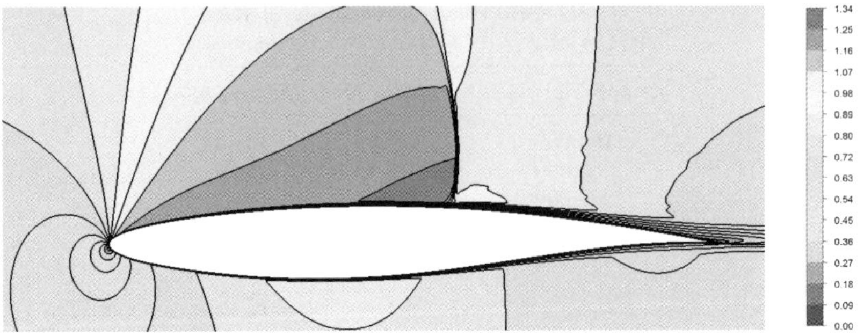

Fig. 8.16. Contours of Mach number of the RAE 2822 at $M_\infty = 0.734$ and $\alpha = 2.63$ degrees.

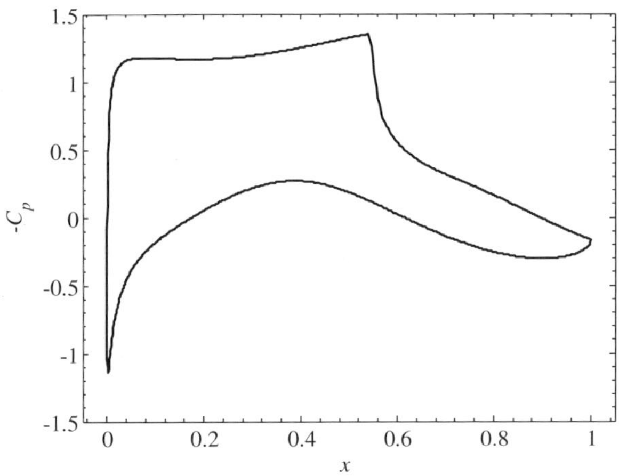

Fig. 8.17. Pressure distribution of the RAE 2822 at $M_\infty = 0.734$ and $\alpha = 2.63$ degrees. The lift and drag coefficients are $C_l = 77.6$ l.c. and $C_d = 165.1$ d.c., respectively.

three-dimensional extension of the hyperbolic C-grid by Kinsey and Barth (1984), as shown in Fig. 8.5. The farfield is set 100 chord lengths away from the wing surface (both in the symmetry plane as well as from the wing tip). The distance from the wing surface to the first node is $5 \cdot 10^{-4} \times$ chord length.

8.5.2 *Grid convergence*

A grid convergence study of the ONERA M6 wing is performed. The ONERA M6 wing is a trapezoidal swept wing with symmetric airfoil

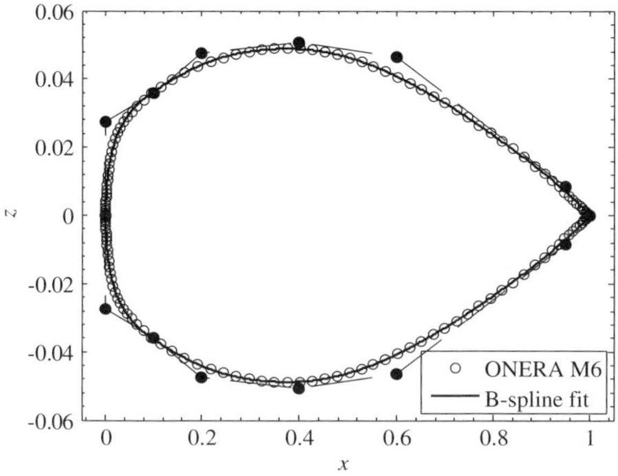

Fig. 8.18. The ONERA M6 airfoil profile.

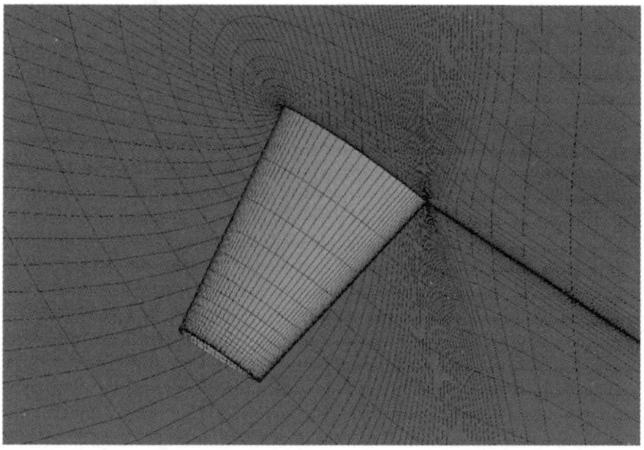

Fig. 8.19. Surface grid of the ONERA M6 wing.

sections at the root and the tip (shown in Fig. 8.18). The wing has a semi-span b/2 = 1193.6 mm, root chord $c_r = 805.9$ mm, LE sweep $\Lambda_{LE} = 30$ degrees, taper ratio $\lambda = 0.56$, and aspect ratio $AR = 3.8$. The wing surface grid and symmetry plane are shown in Fig. 8.19.

The grid study is carried out at $M_\infty = 0.8395$ and $\alpha = 3.06$ degrees. The results are given in Fig. 8.20. The grid converges within one lift and drag count on the last two grids. Therefore, roughly

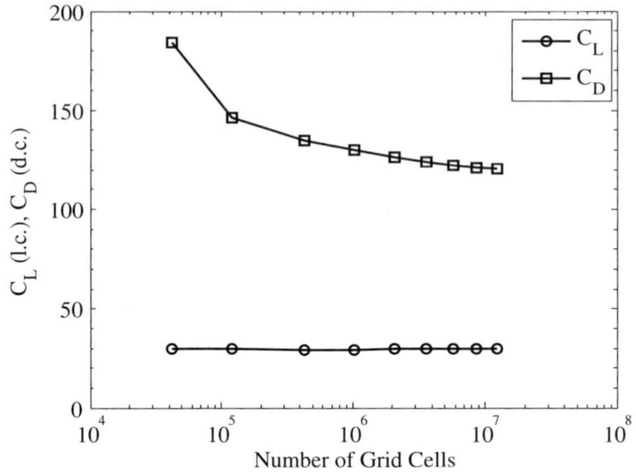

(a) Lift and drag coefficient convergence

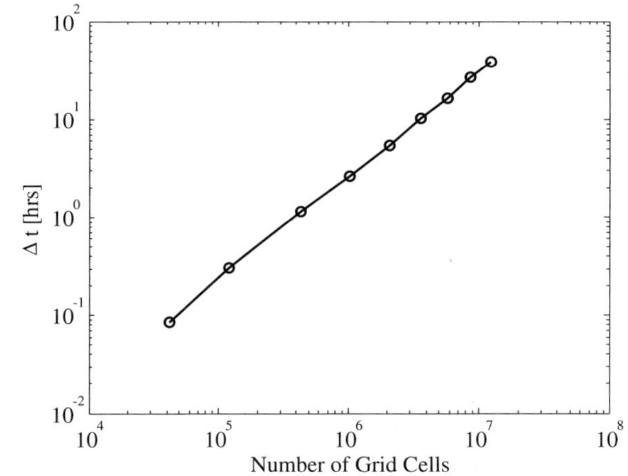

(b) Variation of simulation time

Fig. 8.20. Grid convergence study of the ONERA M6 wing at Mach number $M_\infty = 0.8395$ and angle of attack $\alpha = 3.06$ degrees.

8,563,000 grid cells are needed and the simulation time is around 27 hours.

The upper surface pressure contours are shown in Fig. 8.21(a), as well as the pressure distributions at the 20% and 60% span stations in Figs. 8.21(b) and 8.21(c), respectively. A lambda shock pattern is present on the upper surface.

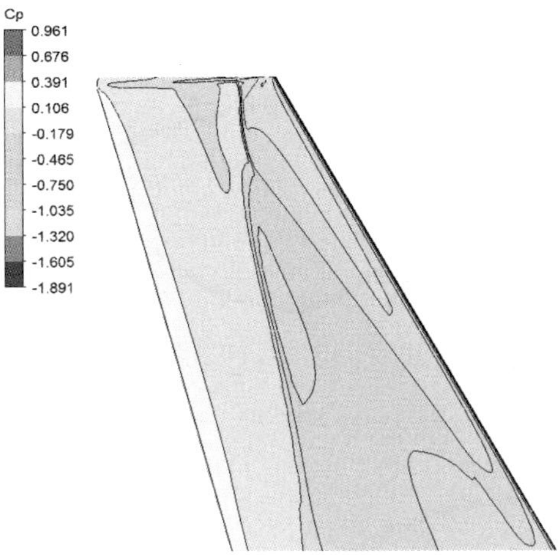

(a) Upper surface presssure contours

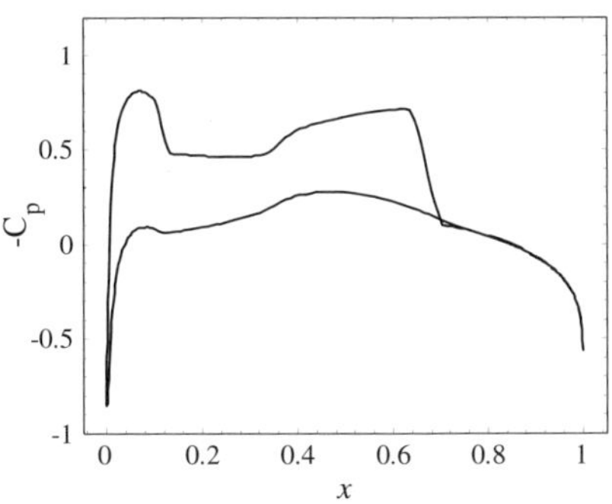

(b) Pressure distribution at $\eta = 0.2$

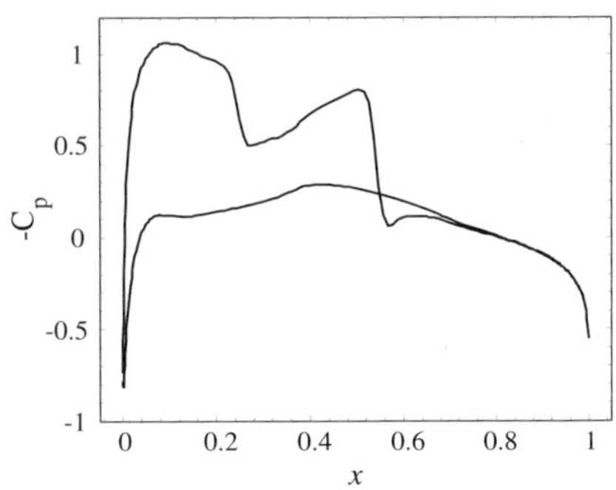

(c) Pressure distribution at $\eta = 0.6$

Fig. 8.21. Pressure distribution of the ONERA M6 at $M_\infty = 0.8395$ and $\alpha = 3.06$ degrees. The lift and drag coefficients are $C_L = 29.6$ l.c. and $C_D = 121.1$ d.c., respectively.

8.6 Viscous Wing CFD Model

A CFD model for the three-dimensional viscous flow analysis past a wing is described.

8.6.1 *Model setup*

The flow is governed by the steady compressible RANS equations with the Spalart–Allmaras turbulence model given in Section 8.4.1. FLUENT is used for the flow solution, as described in Section 8.2.2.

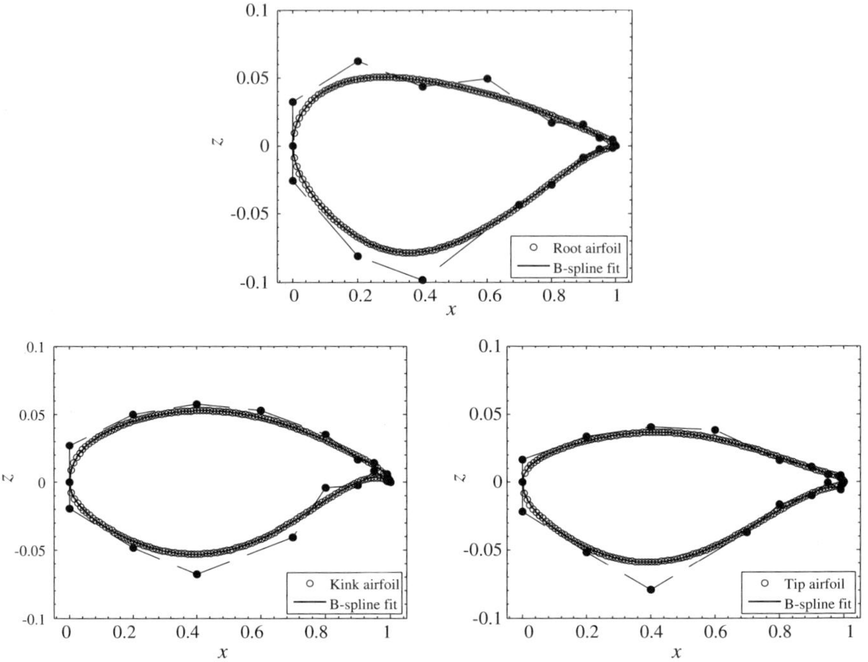

Fig. 8.22. Approximations to the NASA CRM airfoil profiles.

The computational grid is a three-dimensional extension of the hyperbolic C-grid by Kinsey and Barth (1984), as described in Section 8.5.1, and the distance from the wing surface to the first node is $5 \cdot 10^{-6} \times$ chord length.

8.6.2 *Grid convergence*

A grid convergence study of an approximation to the NASA common research model (CRM) wing is performed. The CRM is a typical transonic transport wing with an aspect ratio $AR = 9.0$, quarter chord sweep $\Lambda_{c/4} = 35$ degrees, wing span $b = 192.8$ ft and a taper ratio of $\lambda = 0.275$. The wing surface grid is shown in Fig. 8.5(c) and the airfoil shapes at the root, kink, and tip in Fig. 8.22.

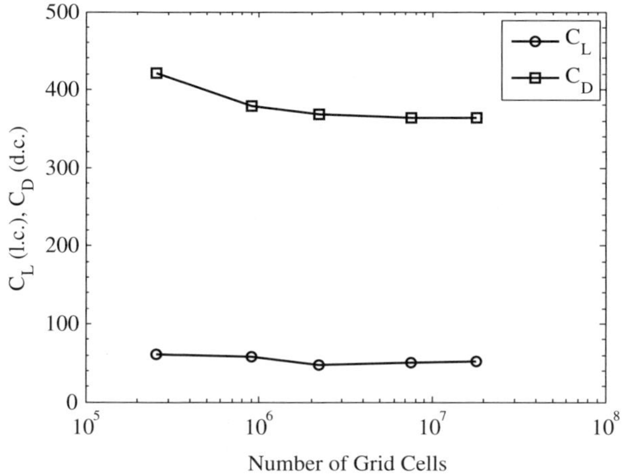

(a) Lift and drag coefficient convergence

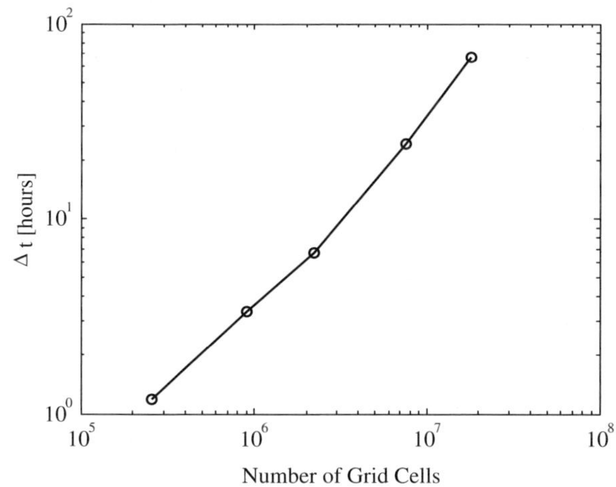

(b) Variation of simulation time

Fig. 8.23. Grid convergence study of an approximated NASA CRM wing at Mach number $M_\infty = 0.85$ and angle of attack $\alpha = 1.66$ degrees.

The results of the grid study, performed at $M_\infty = 0.85$ and $\alpha = 1.66$ degrees, are given in Fig. 8.23. The grid converges within one lift and drag count with roughly 7,554,000 grid cells, and the simulation time is around 27 hours.

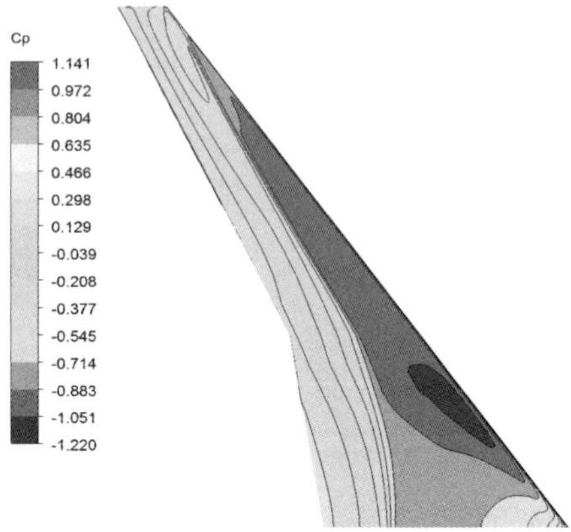

(a) Upper surface pressure contours

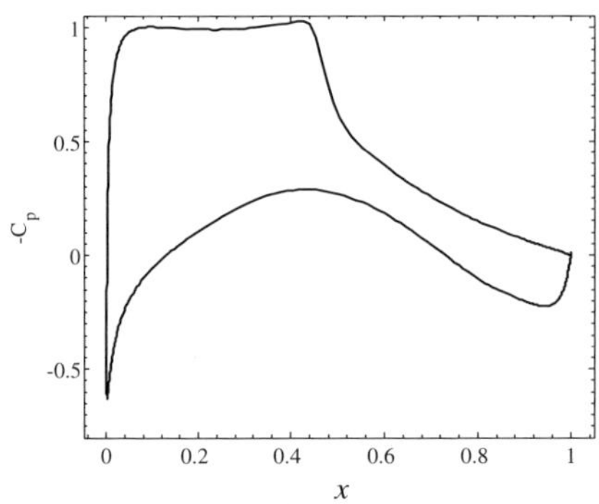

(b) Pressure distribution at $\eta = 0.4$

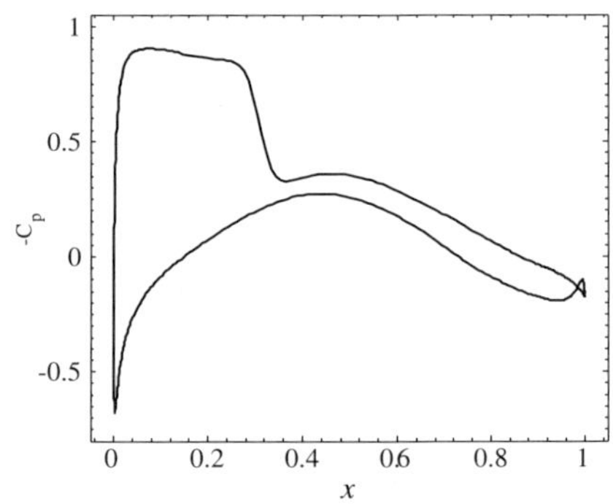

(c) Pressure distribution at $\eta = 0.8$

Fig. 8.24. Pressure distribution of an approximated NASA CRM wing at $M_\infty = 0.5$ and $\alpha =$ degrees. Here $C_L = 50.0$ l.c. and $C_D = 364.2$ d.c.

At the given flow condition the CRM wing has $C_L = 50$ l.c. and $C_D = 364.2$ d.c. The upper surface pressure contours are shown in Fig. 8.24(a), as well as the pressure distributions at the 40% and 80% span stations in Figs. 8.24(b) and 8.24(c), respectively.

Chapter 9

Low-Fidelity Aerodynamics Models

Low-fidelity aerodynamics models are fundamental components of the design optimization methodologies considered in this book. This chapter describes several ways of developing low-fidelity models as well as demonstrating specific examples for both airfoils and wings.

9.1 Low-Fidelity Modeling

9.1.1 *Principal properties*

We aim at constructing a physics-based surrogate model to replace a computationally expensive and accurate high-fidelity model in a design and optimization process. The optimization algorithm uses information from the surrogate to navigate the high-fidelity design space and search for the optimum. This means that the surrogate needs to be able to capture the changes in the measures of merit with respect to the design parameters in the same way as the high-fidelity model. For example, if a measure of merit obtained from the high-fidelity model increases due to a change in the design variables, the measure of merit obtained by the surrogate model also needs to increase for the same change in the design variables, but not necessarily by the same amount. In other words, the surrogate and the high-fidelity model have to be sufficiently well correlated.

Typically, the surrogate model is a combination of a low-fidelity model and a correction (as described in Section 4.5). In order for the above to be realized, the low-fidelity model needs to capture the governing physics of the high-fidelity one, because the correction is, in general, not intended to capture any nonlinear physics.

Therefore, the most important property of the low-fidelity model is to be capable of following the high-fidelity model within the design space. Moreover, to accelerate the design process the surrogate has to be fast.

9.1.2 *Methods*

A low-fidelity aerodynamics model can be developed based on one of, or a combination of, the following principles.

- *Simplified physics.* Replace the set of governing fluid flow equations with a set of simplified equations. For example, using the transonic small-disturbance equation in place of the Euler equations (see Fig. 8.8). These are often referred to as *variable-fidelity physics models.*
- *Coarse discretization.* Use the same fluid-flow model as in the high-fidelity model, but with a coarser computational grid discretization; often referred to as *variable-resolution models.*
- *Relaxed convergence criteria.* Reduce the number of maximum allowable iterations and/or reduce the convergence tolerance, sometimes referred to as *variable-accuracy models.*

In the following sections, we demonstrate how to construct both airfoil and wing aerodynamics models using variable-fidelity modeling.

9.2 Variable-Resolution Airfoil Model

We illustrate the development of a variable-resolution model by considering the inviscid airfoil computational fluid dynamics (CFD) model described in Section 8.2. The low-fidelity model is constructed in the same manner as the high-fidelity one, but with a coarser discretization grid and relaxed flow-solver convergence criteria.

Currently, there is no rigorous method of finding the appropriate discretization for the low-fidelity model. Therefore, the discretization parameters are found by examining the results of the grid convergence study (Fig. 8.10). We select grid c_3. This particular grid has roughly 32,000 cells and takes about two minutes to run. Although

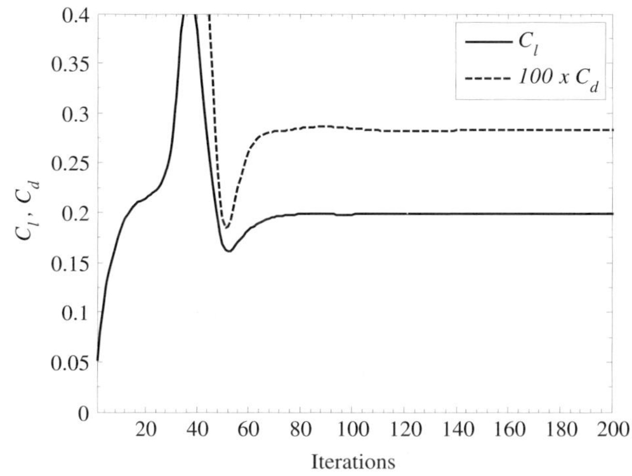

Fig. 9.1. Evolution of the lift and drag coefficients obtained by the low-fidelity model for NACA 0012 at $M_\infty = 0.75$ and $\alpha = 1$ degree.

grid c_4 is twice as fast as c_3, it is too inaccurate in terms of both lift and drag.

The flow solution history for the low-fidelity model is shown in Fig. 9.1 and indicates that the lift and drag coefficients are nearly converged after 80 to 100 iterations. The maximum number of iterations is set to 100 for the low-fidelity model. This reduces the overall simulation time to 45 seconds.

A comparison of the pressure distributions of the high- and low-fidelity models, shown in Fig. 9.2, indicates that the low-fidelity model, in spite of being based on much coarser mesh and reduced flow-solver iterations, captures the main features of the high-fidelity model pressure distribution quite well. The biggest discrepancy in the distributions is around the shock on the upper surface, leading to an over estimation of both lift and drag (Fig. 9.1).

The ratio of simulation times of the high- and low-fidelity model in this case study is 44. In many cases, the solver does not fully converge with respect to the residuals and goes on up to 1,000 iterations. The overall evaluation time of the high-fidelity model then goes up to 170 minutes. In those cases the ratio of simulation times of the high- and low-fidelity models is around 110.

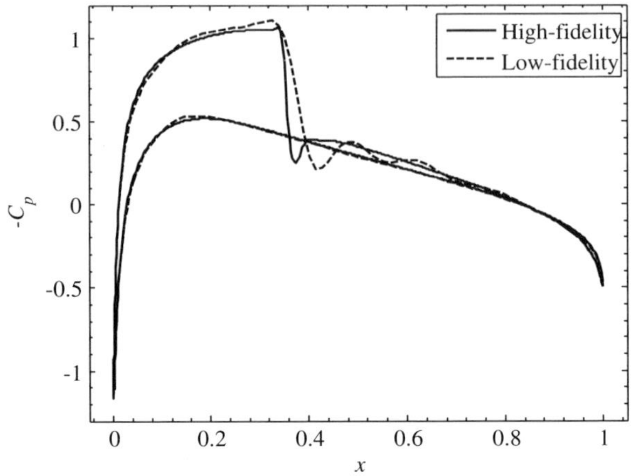

Fig. 9.2. Comparison of the pressure distributions obtained by the high- and low-fidelity models for NACA 0012 at $M_\infty = 0.75$ and $\alpha = 1$ degree.

The same approach can be applied to three-dimensional aerodynamics models, such as the one described in Sections 8.5 and 8.6. This is carried out in Chapter 11.

9.3 Variable-Fidelity Physics Airfoil Model

Let us consider the viscous airfoil CFD model described in Section 8.4 as the high-fidelity model. This model solves the compressible RANS equations along with the Spalart–Allmaras turbulence model. The model can predict the performance characteristics of transonic airfoils, including viscous effects such as boundary layer separation.

A low-fidelity model needs to pick up on any nonlinear physics associated with the high-fidelity model. In this case, a viscous-inviscid model can be used. In particular, we solve the compressible Euler equations coupled with a two-equation integral boundary-layer formulation using the displacement thickness concept. The computer code MSES (Drela and Giles, 1987) is used to solve the equations. The entire discrete equation set is solved as a fully coupled nonlinear system by a global Newton method. We set a fixed transition at the leading edge. The method is rapid and can handle strong viscous-inviscid interactions, such as shock-induced separation.

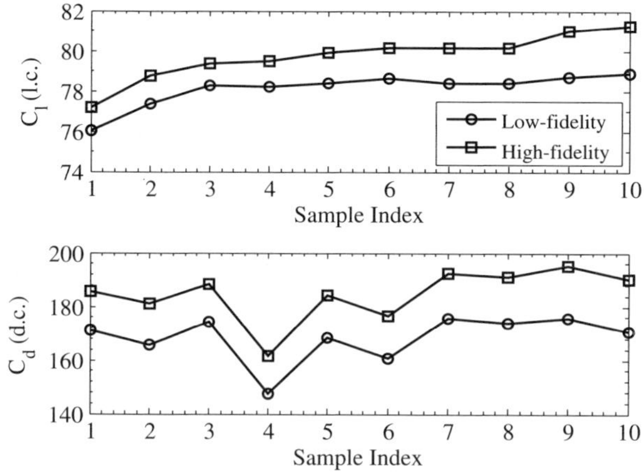

Fig. 9.3. Lift and drag coefficients for several perturbed designs of the RAE 2822 at Mach = 0.734 and Reynolds number of 6.5 million.

The MSES code solves the flow equations on a conservative streamline grid. We configure the grid to be coarse. A typical evaluation time of the MSES code for a supercritical airfoil section in transonic flow is less than 15 seconds. The high-fidelity model evaluation time is roughly 40 minutes. The ratio of high- to low-fidelity simulation time is about 160; in other words, the low-fidelity model is 160 times faster than the high-fidelity one.

To discover whether the low-fidelity model follows the high-fidelity one, a study of the RAE 2822 airfoil is undertaken at a Mach number of 0.734 and a Reynolds number of 6.5 million. The airfoil shape is perturbed slightly to generate nine new designs (locally around the original one). For each design the high- and low-fidelity models are evaluated at the prescribed condition. Figure 9.3 shows the variation in the lift and drag coefficients for all the designs, in which the lift has been sorted in ascending order. We can see that the low-fidelity model follows the high-fidelity one in terms of both lift and drag.

9.4 Variable-Fidelity Physics Wing Model

Now let us consider the viscous wing CFD model described in Section 8.6 as the high-fidelity model. This model solves the three-dimensional compressible RANS equations along with the Spalart–Allmaras turbulence model and is a good model for transonic flow past wings.

The low-fidelity model is defined as:

$$c(\boldsymbol{x}) = [C_{L.c}(\boldsymbol{x}) \quad C_{D.c}(\boldsymbol{x})], \tag{9.1}$$

where $C_{L.c}(\boldsymbol{x})$ and $C_{D.c}(\boldsymbol{x})$ are the wing lift and drag coefficients, respectively. We follow Grasmeyer (1999) and model the low-fidelity wing-drag coefficient in a component-wise fashion as:

$$C_{D.c}(\boldsymbol{x}) = C_{D0}(\boldsymbol{x}) + C_{Di}(\boldsymbol{x}) + C_{Dw}(\boldsymbol{x}), \tag{9.2}$$

where C_{D0} is the zero-lift drag coefficient, C_{Di} is the induced drag coefficient, and C_{Dw} is the compressibility drag coefficient. The zero-lift drag coefficient, estimated at zero lift, provides an estimate of the surface friction drag and the pressure drag due to the wing form. The induced drag coefficient yields an estimate of the drag due to lift. Drag due to pressure shocks is represented by the compressibility drag coefficient. Each component of the low-fidelity model needs to be modeled appropriately to capture the main behavior of the high-fidelity model with respect to any changes in the model parameters, such as the operating condition and the design variables. The low-fidelity wing lift coefficient is set equal to the high-fidelity wing lift coefficient, i.e. $C_{L.c} = C_{L.f}$.

Each component of Equation (9.2) is estimated in the following way. The zero-lift coefficient is modeled by the strip theory (see, for example, Mason (1990)), in which the planform is divided into several streamwise strips and the sum of the contribution of each strip yields the total wing drag coefficient as:

$$C_{D0}(\boldsymbol{x}) = \sum C_{d0}(\boldsymbol{x}) \frac{S_{strip}}{S_{ref}}, \tag{9.3}$$

where C_{d0} is the strip section zero-lift drag coefficient, S_{strip} is the area of the strip, and S_{ref} is the planform reference area. The section zero-lift drag coefficient is calculated using a two-dimensional CFD model. The induced drag coefficient is calculated using a vortex lattice method (VLM) (see, for example, Grasmeyer (1997)).

The input to the VLM analysis is the planform shape and the wing lift coefficient ($C_{L.c} = C_{L.f}$). The VLM analysis will yield the induced drag (C_{Di}), as well as the corresponding wing spanload distribution. Now the compressibility drag coefficient is calculated in a similar way as the zero-lift drag coefficient, but using the lift coefficient distribution from the VLM analysis as an input to the two-dimensional CFD analysis of each spanwise strip. The compressibility drag coefficient is, therefore, calculated as:

$$C_{Dw}(\boldsymbol{x}) = \sum C_{dw}(\boldsymbol{x}) \frac{S_{strip}}{S_{ref}}, \tag{9.4}$$

where C_{dw} is the strip section compressibility drag coefficient.

Simple sweep theory (see, for example, Mason (1990)) is used to account for wing-sweep effects. For each strip, the CFD analysis is performed at a Mach number $M_n = M_\infty cos\Lambda_{c/4}$ (where M_∞ is the free-stream Mach number and $\Lambda_{c/4}$ is the strip quarter-chord sweep angle), chord length $c_n = c\ cos\Lambda_{c/4}$ (c is the strip chord length), and thickness-to-chord ratio $(t/c)_n = (t/c)/cos\Lambda_{c/4}((t/c)$ is the strip thickness-to-chord ratio).

To demonstrate the properties of the low-fidelity model we consider the ONERA M6 wing, which has a swept trapezoidal shape with constant airfoil profile. The wing and airfoil shape is shown in Figs. 8.18 and 8.19. A description of the geometry is given in Section 8.5.2.

The wing is divided uniformly into six strips. The induced drag coefficient, C_{Di}, is calculated using a vortex lattice code (see, for example, Grasmeyer (1997)). The zero-lift and wave-drag coefficients, C_{D0} and C_{Dw}, respectively, are calculated using MSES (Drela and Giles, 1987).

The drag polar of the ONERA M6 at Mach = 0.8395 is shown in Fig. 9.4. One can observe that the drag polars of each model are similar.

The low-fidelity model predicts lower drag than the high-fidelity one for a given lift. Furthermore, the minimum drag value is obtained at different lift values for each model. The variation of the drag coefficient with Mach number at a constant lift is shown in Fig. 9.5. The

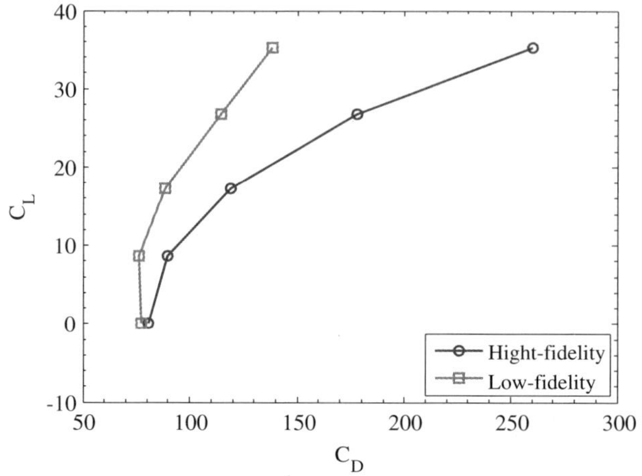

Fig. 9.4. Drag polar of the ONERA M6 wing at Mach = 0.8395.

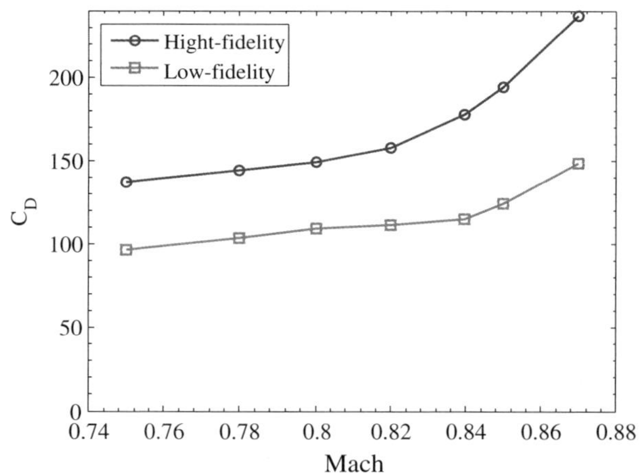

Fig. 9.5. Variation of the drag coefficient with Mach number at constant lift coefficient ($C_L \approx 26$ cts).

low-fidelity model follows the high-fidelity one, but is shifted in drag values.

Several wing shapes were generated by perturbing (moderately) the original ONERA M6 wing. The airfoil shapes are shown in Fig. 9.6. Flow analyses of the wings were performed at both low-

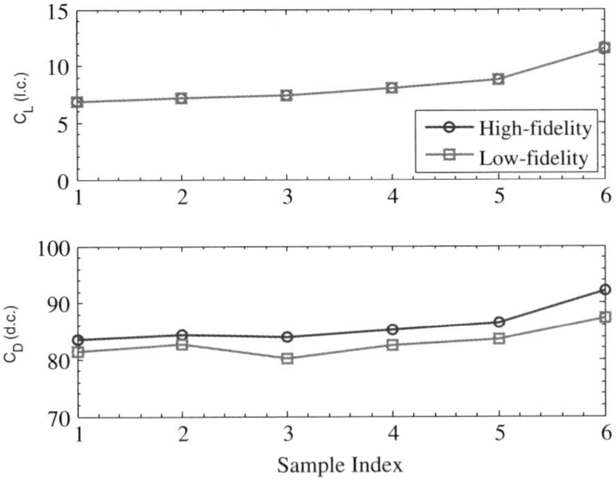

Fig. 9.6. Lift and drag coefficients for several perturbed designs of the ONERA M6 wing at Mach = 0.80 and angle of attack of 1 degree.

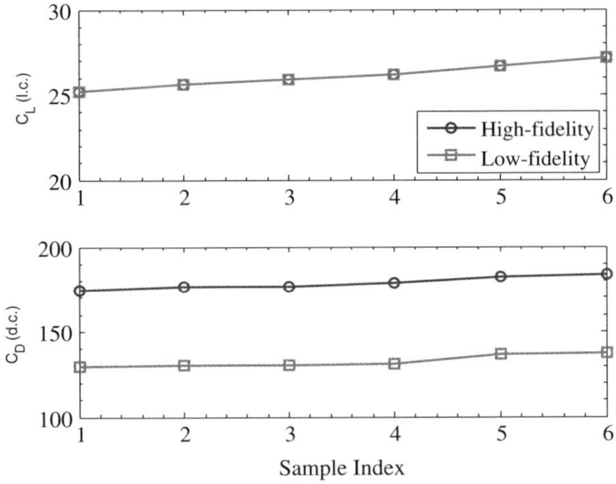

Fig. 9.7. Lift and drag coefficients for several perturbed designs of the ONERA M6 wing at Mach = 0.84 and angle of attack of 3 degrees.

and high-lift coefficient values. The results are shown in Fig. 9.7. The low-fidelity model drag values follow the high-fidelity model quite well, i.e. the same trend in variation of the drag is observed in both models.

Part IV

Applications

Chapter 10

Transonic Airfoil Shape Design

In this chapter, we apply the variable-fidelity aerodynamic shape optimization methods described in the previous chapters to a set of model airfoil design problems. The objective is to demonstrate the usage and the characteristics of the methods, as well as highlight their efficiency.

10.1 Lift Maximization with Variable-Resolution Models

The first problem considered involves airfoil lift maximization in inviscid flow for a given maximum drag and a minimum cross-sectional area. The surrogates are constructed using variable-resolution models and shape-preserving response prediction (SPRP), space mapping (SM), adaptive response correction (ARC), and multi-level optimization (MLO) (all described in Chapter 6). For comparison purposes, the problem is also solved using a direct approach (see pattern search in Chapter 3) and a functional surrogate (see kriging in Chapter 5).

10.1.1 *Problem formulation*

At a free-stream Mach number $M_\infty = 0.75$ and an angle of attack $\alpha = 1$ degree, solve:

$$\min_{\boldsymbol{x}} -C_l(\boldsymbol{x}), \tag{10.1}$$

subject to the following constraints:

$$g_1 = C_d(\boldsymbol{x}) - C_d^* \leq 0, \tag{10.2}$$

$$g_2 = A^* - A(\boldsymbol{x}) \leq 0, \tag{10.3}$$

where $\mathbf{x}$ are the design variables, C_d is the drag coefficient, C_l is the lift coefficient, C_d^* is the maximum drag coefficient, A is the airfoil cross-sectional area, and A^* is the minimum cross-sectional area. The maximum drag coefficient is set to $C_d^* = 0.0041$ and the minimum cross-sectional area is set to $A^* = 0.065$ (non-dimensionalized with the chord length squared).

The area constraint is handled directly because the high-fidelity model and the surrogate model yield exactly the same cross-sectional values. However, due to unavoidable misalignment between the high-fidelity model and its surrogate in terms of drag coefficient values, it is not convenient to handle the drag constraint directly. The reason for this is that the design that is feasible for the surrogate model may not be feasible for the high-fidelity model. In particular, the design obtained as a result of optimizing the surrogate model $s^{(i)}$, i.e. $\boldsymbol{x}^{(i+1)}$, will be feasible for $s^{(i)}$, but if $\boldsymbol{x}^{(i+1)}$ is not feasible for the high-fidelity model it will not be feasible for $s^{(i+1)}$, because we have $s^{(i+1)}(\boldsymbol{x}^{(i+1)}) = f(\boldsymbol{x}^{(i+1)})$ by the definition of the surrogate model. In order to alleviate this problem, the lift constraint is implemented through a penalty function.

More specifically, the objective function is defined as:

$$H(\boldsymbol{x}) = -C_l(\boldsymbol{x}) + \beta[\Delta C_{d.s}(\boldsymbol{x})]^2, \tag{10.4}$$

where $\Delta C_{d.s} = 0$ if $C_{d.s}(\boldsymbol{x}) \leq C_{d.s}^*$ and $\Delta C_{d.s} = C_{d.s}(\boldsymbol{x}) - C_{d.s}^*$ otherwise. Here, $C_{d.s}$ is the drag coefficient obtained by the surrogate model and $C_{d.s}^* = C_d^*$. In our numerical experiments we use $\beta = 1,000$.

10.1.2 *Design variables*

The airfoil shape is parameterized with the NACA (National Advisory Committee for Aeronautics) four-digit method (see Chapter 7). There are three design variables, namely the maximum ordinate of the camber line (m), the location of the maximum ordinate (p), and the maximum thickness (t). The design variable vector is written as:

$$\boldsymbol{x} = [m \quad p \quad t]^T. \tag{10.5}$$

Obviously, the NACA four-digit airfoil shapes are not suitable for transonic applications. However, the NACA four-digit parameterization is a convenient approach for numerical experiments of the optimization algorithms since there are only three well-defined parameters controlling the airfoil shape.

The initial design is set to NACA 3210. The design variable bounds are $0 \leq m \leq 0.1$, $0.2 \leq p \leq 0.8$, and $0.05 \leq t \leq 0.20$. Or, in vector form, the lower bounds are $\boldsymbol{l} = [0.0\ 0.20\ 0.05]^T$ and upper bounds $\boldsymbol{u} = [0.10\ 0.80\ 0.20]^T$.

10.1.3 *Optimization algorithms*

The problem is solved using the SPRP, SM, ARC, and MLO algorithms (described in Chapter 6). The pattern-search algorithm is used for the surrogate model optimization (described in Chapter 3). Surrogate model optimization is constrained using the trust region approach. The termination condition for the algorithms is the convergence in the argument, $\|\boldsymbol{x}^{(i)} - \boldsymbol{x}^{(i-1)}\| < 10^{-4}$, or the reduction of the trust region radius below a user-defined threshold, $\lambda < 10^{-4}$.

The results are compared with direct design optimization of the high-fidelity computational fluid dynamics (CFD) model using the pattern-search algorithm and kriging with updates. The kriging-based approach is implemented as described in Chapter 5. The initial kriging model is constructed using 20 fine-model samples distributed in the design space using Latin hypercube sampling (LHS). The fine model is evaluated at the optimum of the kriging model and these data are used to update the surrogate.

10.1.4 *Models*

The flow is taken to be inviscid and variable-resolution models are used in the optimization process. The high-fidelity model solves the compressible Euler equations on a body-fitted structured grid with roughly 400,000 mesh cells. The simulation time is around 67 minutes. See Section 8.2 for details.

The low-fidelity models are constructed in the same manner as the high-fidelity model, but with a coarser grid and relaxed convergence criteria. The low-fidelity model used in SPRP, SM, and ARC use a grid with roughly 30,000 cells. The low-fidelity model is around 80 times faster than the high-fidelity one. See Section 9.1 for details of the model setup.

The MLO algorithm utilizes two levels of low-fidelity models: a coarse mesh (c_1) and a medium mesh (c_2). The coarse mesh is the same mesh as used in the other algorithms. The medium mesh has around 100,000 mesh cells and is around 40 times faster than the high-fidelity model.

10.1.5 *Results*

Results of the optimization runs are presented in Table 10.1. All the optimization algorithms are able to meet the constraints. Overall, the optimized designs are comparable. Nevertheless, the shape obtained by the ARC algorithm is slightly better than the others. However, in general, the following three fundamental shape changes have been

Table 10.1 Lift maximization of NACA 3210 in inviscid flow using several optimization algorithms.

Variable	Initial	Direct	Kriging	SPRP	SM	ARC	MLO
m	0.0300	0.0080	0.0109	0.0091	0.0100	0.0053	0.0076
p	0.2000	0.6859	0.6633	0.7397	0.6929	0.8000	0.6902
t	0.1000	0.1044	0.0966	0.0981	0.0980	0.1080	0.1057
C_l	0.8035	0.4641	0.5381	0.5244	0.5281	0.4237	0.4544
C_d	0.0410	0.0041	0.0041	0.0041	0.0041	0.0041	0.0041
A	0.0675	0.0703	0.0650	0.0660	0.0659	0.0727	0.0712
N_{c1}^{*}	—	—	—	180	260	240	60
$N_{c2}^{\$}$	—	—	—	—	—	—	67
$N_{f}^{\%}$	—	120	29	7	5	8	2
Cost[&]	—	120	29	<11	<9	11	<6

[*]Number of evaluations of the low-fidelity model with the coarse grid.
[$]Number of evaluations of the low-fidelity model with the medium grid.
[%]Number of evaluations of the high-fidelity model.
[&]Number of equivalent evaluations of the high-fidelity model based on the simulation time ratio of the high- to low-fidelity models.

made compared to the initial design:

1. the maximum ordinate of the mean camber line (m) is reduced;
2. the location of the maximum ordinate of the mean camber line (p) is moved aft, thus increasing the trailing-edge (TE) camber; and
3. the thickness (t) is reduced.

Shape changes (1) and (3) reduce the shock strength and, thus, reduce the wave-drag coefficient. The associated change in the pressure distribution reduces the lift coefficient.

However, shape change (2) improves (or recovers a part of) the lift by opening up the pressure distribution behind the shock. These effects are clearly illustrated in the pressure distribution plot and airfoil shape plots in Fig. 10.1, and the Mach contour plots in Fig. 10.2.

A typical optimization history is shown in Fig. 10.3 and the evolution of the lift and drag coefficient is shown in Fig. 10.4. It follows that the algorithms exhibit, in general, good convergence patterns, and the mechanisms introduced in the algorithm (in particular the trust region approach and the penalty function) enforce that the drag limitation be satisfied while increasing the lift coefficient as much as possible.

The optimization cost of the direct method is 120 high-fidelity function evaluations. Although solving the problem directly is not a very good approach in this case, especially without adjoint sensitivity information, it is interesting to compare the results against the surrogate-based approaches. Kriging requires 29 high-fidelity function evaluations (20 of which are for the initial surrogate setup). The variable-fidelity approaches need six to 11 equivalent high-fidelity evaluations. In this particular case, the MLO algorithm needed the fewest evaluations. This reduced computational cost is obtained by exploiting a sequence of CFD models of increasing fidelity. Each of the lower fidelity models is optimized only once. Also, the high-fidelity model is not evaluated until the very last stage of the design process; that is, the design refinement.

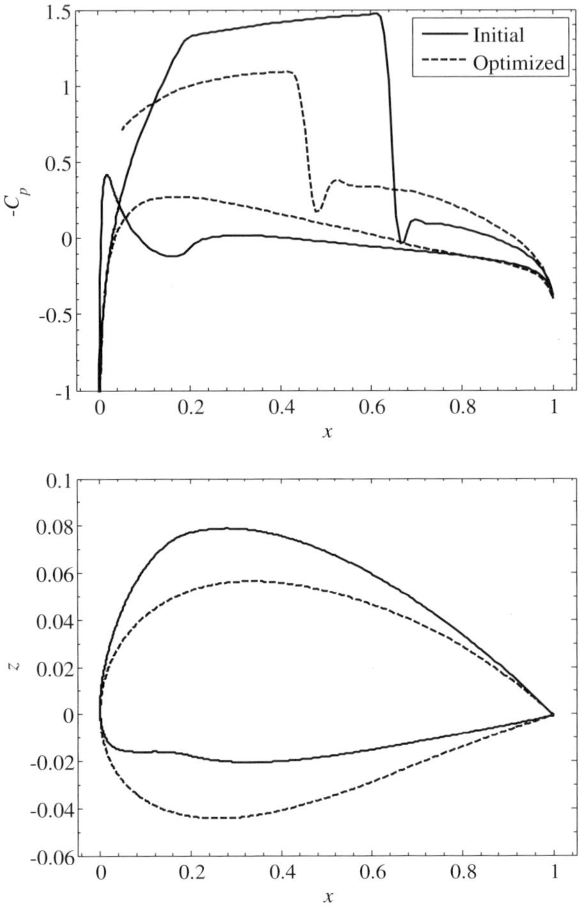

Fig. 10.1. Initial and optimized airfoil shapes and pressure distributions.

10.1.6 *Algorithm robustness*

An important performance characteristic of any optimization routine is its robustness; in particular, sensitivity of the final results to the particular choice of the starting point. Table 10.2 shows the results of six optimization runs, with SPRP starting from randomly selected starting points.

It turns out that the SPRP algorithm gives consistent results both in terms of the final values of the figures of interest (lift and drag

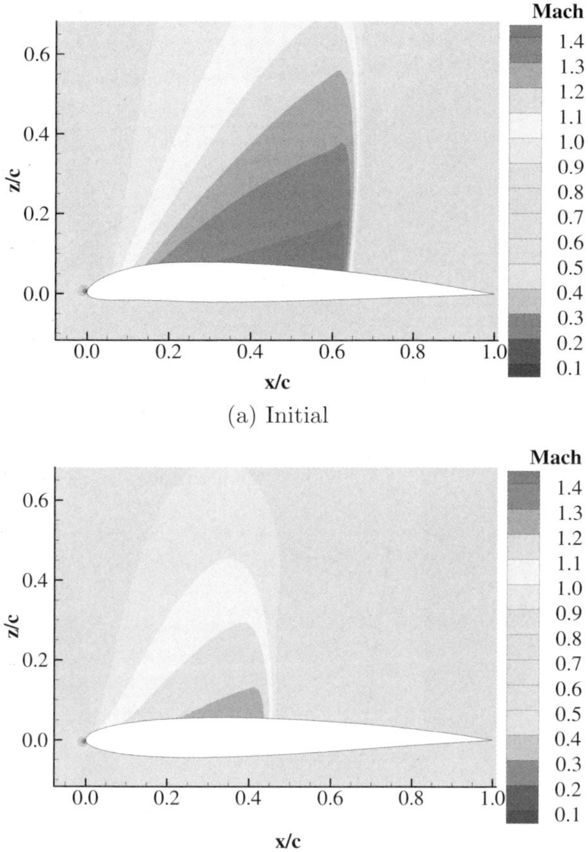

(a) Initial

(b) Optimized

Fig. 10.2. Contours of Mach number for the initial and optimized designs.

coefficients and cross-sectional area) and in terms of the values of the design parameters. This is indicated in Figs. 10.5–10.7, in which the spread of the initial and SPRP-optimized designs in the design space is shown, as well as the pressure distributions for the initial and optimized airfoil shapes.

10.1.7 *Algorithm scalability*

Another relevant issue is algorithm scalability, i.e. dependence of the optimization cost on the design space dimensionality.

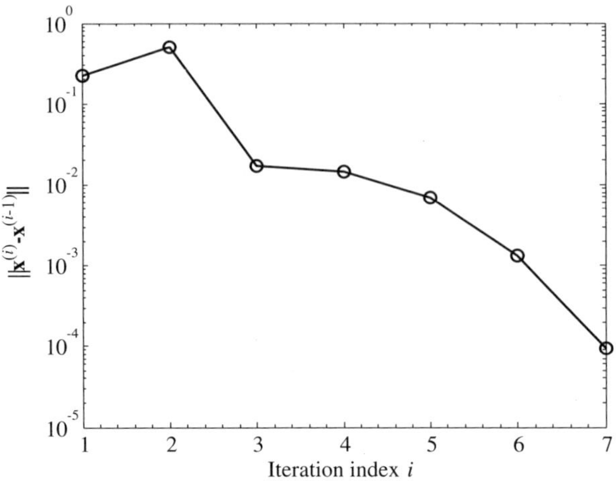

(a) Convergence plot

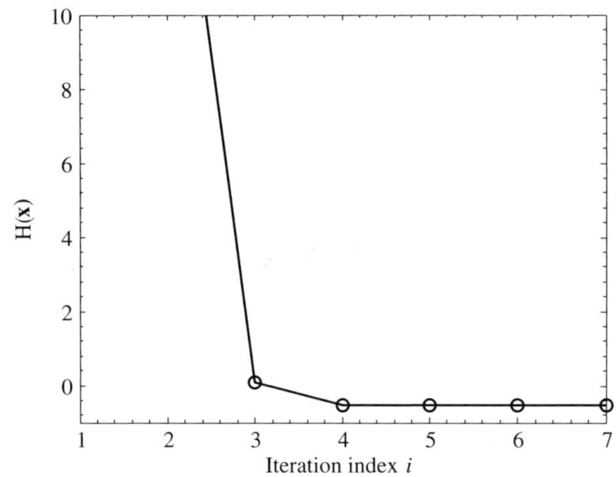

(b) Evolution of the objective function

Fig. 10.3. Optimization history.

To investigate this relationship for the SPRP technique, a lift maximization case is solved with 3, 7, and 11 design variables. The case is defined the same manner as in Section 10.1.1, except that $C_d^* = 0.0060$ and $A^* = 0.075$. The Bezier parameterization approach, described in Chapter 7, is used to define the airfoil shapes with 7

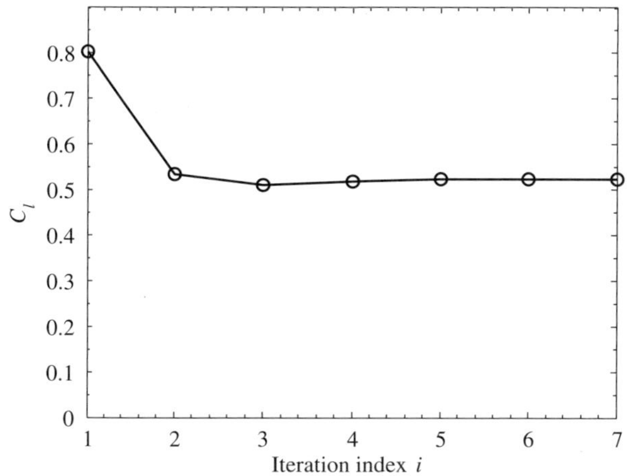

(a) Lift coefficient

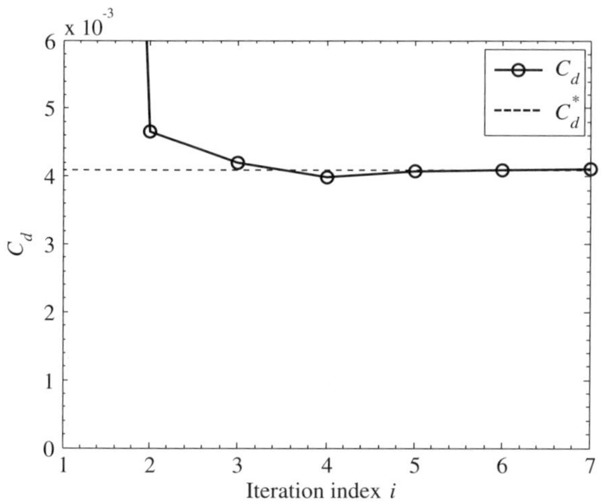

(b) Drag coefficient

Fig. 10.4. Evolution of lift and drag coefficients.

and 11 design variables, and the NACA four-digit method is used to define the three-design variable case.

The results are shown in Fig. 10.8. It can be observed that the number of algorithm iterations, i.e. the number of high-fidelity model evaluations, grows slowly (at most linearly) with the dimensionality

Table 10.2 Results for the design optimization runs, with SPRP starting from six randomly chosen initial designs.

Initial designs								
M	p	t	C_l	C_d	A			
0.0150	0.7150	0.1007	0.7333	0.0102	0.0678			
0.0140	0.6908	0.0989	0.6696	0.0078	0.0666			
0.0064	0.4444	0.0964	0.3512	0.0017	0.0649			
0.0075	0.4530	0.1062	0.3862	0.0037	0.0715			
0.0086	0.7591	0.0775	0.4960	0.0014	0.0522			
0.0209	0.7479	0.0922	1.0410	0.0188	0.0621			
Optimized designs								
M	p	T	C_l	C_d	A	N_c	N_f	Cost
0.0091	0.7607	0.0966	0.5382	0.0041	0.0650	180	6	<10
0.0091	0.7524	0.0970	0.5352	0.0041	0.0653	150	5	8
0.0109	0.6396	0.0975	0.5276	0.0041	0.0656	180	6	<10
0.0106	0.6831	0.0965	0.5395	0.0041	0.0650	240	8	<13
0.0087	0.7658	0.0974	0.5303	0.0041	0.0655	120	4	<7
0.0090	0.7598	0.0970	0.5347	0.0041	0.0653	180	6	<10

of the design space. The worse-than-linear increase of the overall cost is mostly due to the surrogate model optimization.

10.2 Drag Minimization with Variable-Resolution Models

In this problem, the airfoil drag will be minimized in inviscid flow for a given minimum lift and cross-sectional area. The solution approach is the same as in the previous problem, i.e. the same CFD models and the same optimization algorithms.

10.2.1 *Problem formulation*

At $M_\infty = 0.70$ and $\alpha = 1$ deg, solve:

$$\min_{\boldsymbol{x}} C_d(\boldsymbol{x}), \tag{10.6}$$

subject to the following constraints:

$$g_1 = C_l^* - C_l(\boldsymbol{x}) \leq 0, \tag{10.7}$$

$$g_2 = A^* - A(\boldsymbol{x}) \leq 0, \tag{10.8}$$

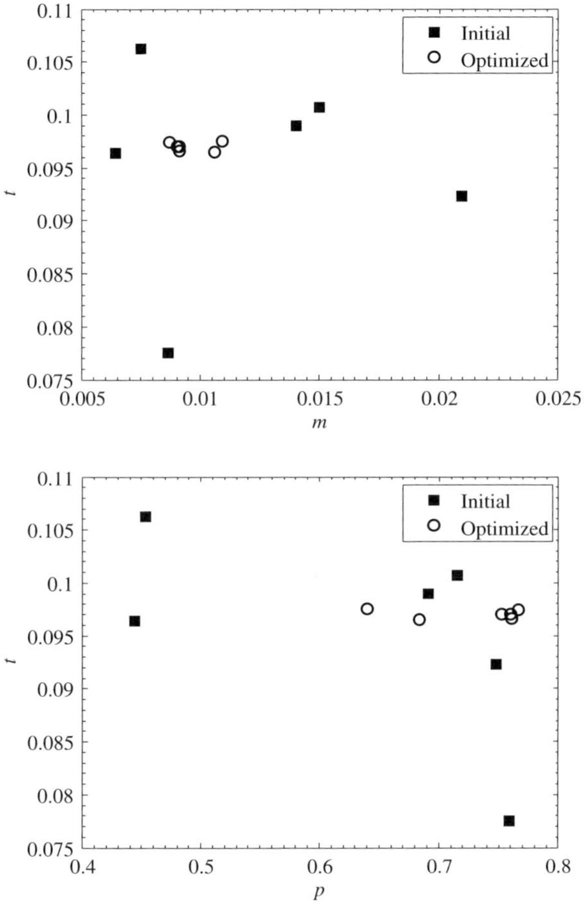

Fig. 10.5. The spread of the initial and optimized designs in the NACA four-digit airfoil design space.

where C_l^* is the minimum lift coefficient, set to $C_l^* = 0.60$, and the minimum cross-sectional area is set to $A^* = 0.075$.

For the same reasons as discussed in the last section, the objective function is defined as:

$$H(\boldsymbol{x}) = C_d(\boldsymbol{x}) + \beta[\Delta C_{l.s}(\boldsymbol{x})]^2, \tag{10.9}$$

where $\Delta C_{l.s} = 0$ if $C_{l.s}(\boldsymbol{x}) \geq C_{l.s}^*$ and $\Delta C_{l.s} = C_{l.s}^* - C_{l.s}(\boldsymbol{x})$ otherwise. Here, $C_{l.s}$ is the lift coefficient obtained by the surrogate model and $C_{l.s}^* = C_l^*$.

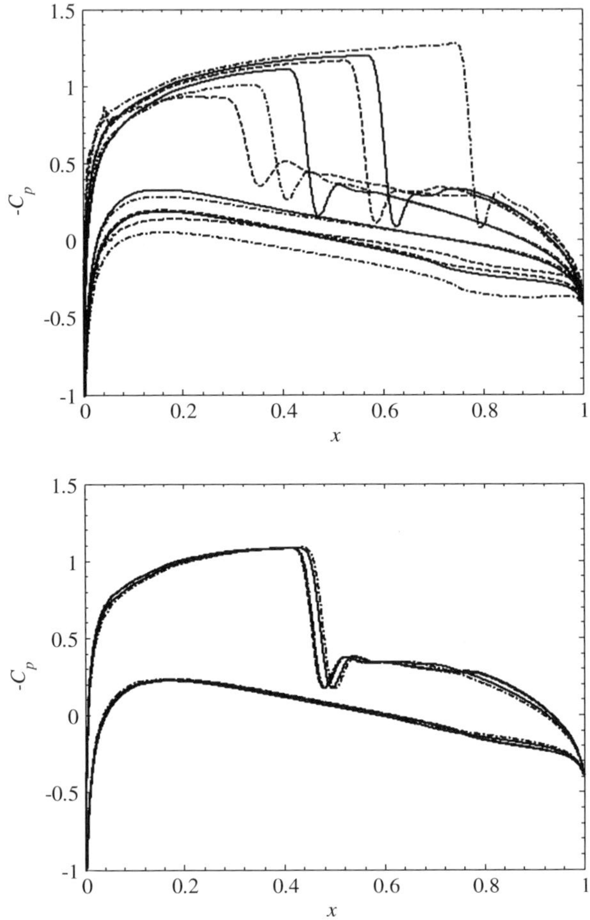

Fig. 10.6. Pressure distributions of the initial (above) and optimized (below) designs.

10.2.2 *Design variables*

Again, the airfoil shape is parameterized with the NACA four-digit method, with the same variable bounds. The initial design is set to NACA 2412.

10.2.3 *Results*

The results are given in Table 10.3. The best airfoil designs are obtained by direct, SM, and ARC, with the drag coefficient between

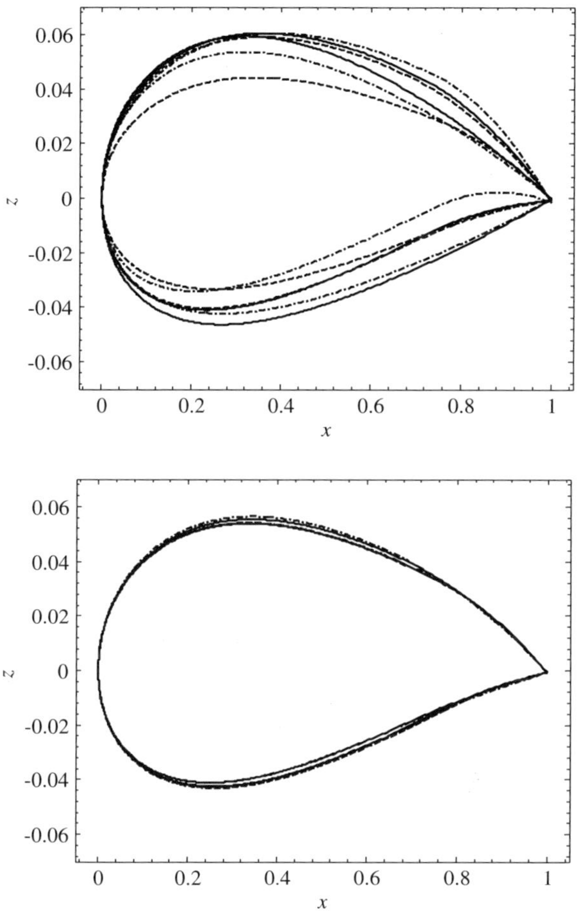

Fig. 10.7. Airfoil shapes of the initial (above) and optimized (below) designs.

17 and 19 counts and the lift and cross-sectional area close to the constraint boundaries. Figure 10.8 shows the shapes and the pressure distributions. The designs obtained by SPRP and MLO are slightly worse, which can be explained in part by the fact that they are further from the constraint boundaries. The kriging method yields a design that violates both constrains.

In terms of the optimization cost, the MLO and SM need the fewest equivalent high-fidelity evaluations (less than five). The ARC needs less than seven and the SPRP less than 12. The direct and

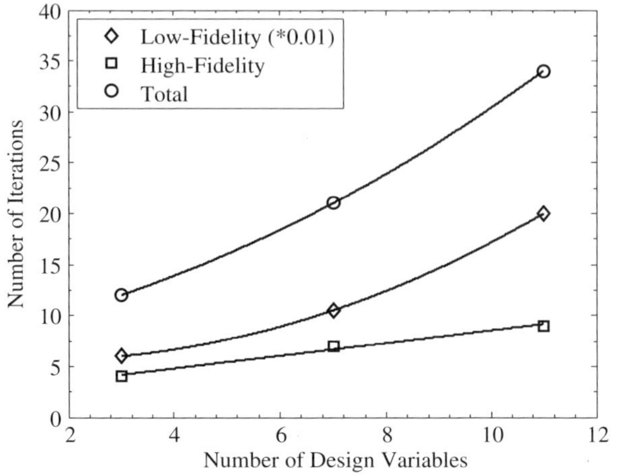

Fig. 10.8. Variation of the computational cost with the size of the design space.

Table 10.3 Drag minimization of NACA 2412 in inviscid flow using several optimization algorithms.

Variable	Initial	Direct	Kriging	SPRP	SM	ARC	MLO
m	0.0200	0.0180	0.0164	0.0203	0.0180	0.0180	0.0180
P	0.4000	0.5207	0.5998	0.4107	0.5290	0.5275	0.6100
T	0.1200	0.1141	0.1091	0.1117	0.1113	0.1115	0.1164
C_l	0.5963	0.6001	0.5946	0.6006	0.6002	0.6002	0.6508
C_d	0.0047	0.0019	0.0015	0.0032	0.0017	0.0017	0.0024
A	0.0808	0.0768	0.0735	0.0753	0.0749	0.0751	0.0784
N_{c1}	—	—	—	210	160	150	60
N_{c2}	—	—	—	—	—	—	47
N_f	—	110	30	7	3	5	2
Cost	—	110	30	<12	5	<7	<5

kriging methods require similar numbers of evaluations as in the previous problem, i.e. over 100 for direct and 30 for kriging.

Based on the design problems in this section and the previous one, it appears that the variable-fidelity approaches offer a significant speed-up when compared to a conventional surrogate-based approach, such as kriging with updates. Moreover, the efficiency of the variable-fidelity approaches varies as well. For example, the MLO requires the fewest evaluations for both problems, whereas SPRP needs the most. However, it turns out that SPRP is more suitable for a different problem formulation, namely inverse design, the subject of the next section.

10.3 Inverse Design with SPRP

We want to minimize the difference between a target pressure distribution, $C_p^*(x)$, where x is the x-axis coordinate, and the airfoil pressure distribution, $C_p(x, \boldsymbol{x})$, where $\boldsymbol{x}$ is the design variable vector.

10.3.1 *Problem formulation*

The flow Mach number is $M_\infty = 0.75$ and the angle of attack is fixed to $\alpha = 1$ degree. The function to be minimized can be written as:

$$J(C_p(x, \boldsymbol{x})) = \|C_p(x, \boldsymbol{x}) - C_p^*(x)\|. \qquad (10.10)$$

A cross-sectional area constraint, such as Equation (10.8), is included to avoid, to some extent, infeasible shapes. The minimum cross-sectional area is $A^* = 0.065$.

The SPRP algorithm is applied to this problem for airfoil design in inviscid flow. We use the same high- and low-fidelity models as in the two previous problems. The airfoil shape is parameterized with the NACA four-digit method, since we are dealing with those types of airfoils. The design variable bounds are $\boldsymbol{l} = [0.00.200.05]^T$ and $\boldsymbol{u} = [0.100.800.20]^T$. The initial design is set to the pressure distribution of NACA 0012 and the target design is the pressure distribution of NACA 3210.

The objective function is defined as:

$$H(\boldsymbol{x}) = J^2, \qquad (10.11)$$

and the area constraint is handled directly.

10.3.2 *Results*

The initial, target, and optimized airfoil shapes and pressure distributions are shown in Fig. 10.9. The SPRP algorithm is able to match airfoil shape and pressure distribution well on the upper surface.

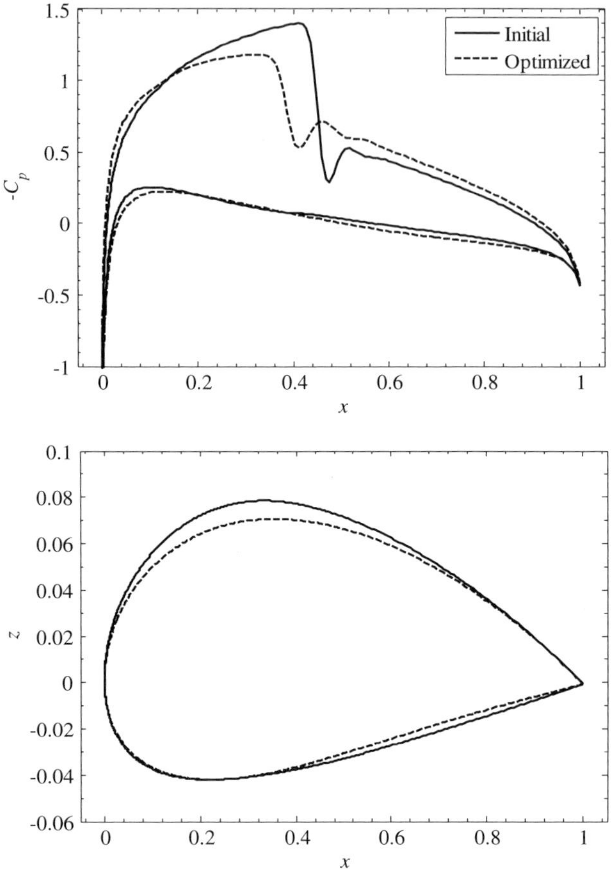

Fig. 10.9. Initial and optimized airfoil shapes and pressure distributions.

There is a slight difference on the lower surface. The overall optimization cost is less than 16 equivalent high-fidelity model evaluations. The pattern-search algorithm achieves a similar solution as the SPRP algorithm, but requires 202 high-fidelity model evaluations.

10.4 Drag Minimization of Modified NACA 0012 with Variable-Fidelity Physics Models

The objective is to minimize the drag of the modified NACA 0012 airfoil section in inviscid flow at a zero-degree angle of attack subject

Table 10.4 Inverse airfoil design in inviscid flow using the SPRP and the pattern-search algorithms.

Variable	Initial	Target	Pattern-search	SPRP
m	0.0000	0.0300	0.0287	0.0293
p	0.0000	0.2000	0.2067	0.2057
t	0.1200	0.1000	0.1083	0.1022
C_l	0.2082	0.8035	0.7786	0.7905
C_d	0.0024	0.0410	0.0424	0.0405
A	0.0808	0.0675	0.0731	0.0689
J	—	—	0.0492	0.0182
N_c	—	—	—	457
N_f	—	—	202	6
Cost	—	—	202	<16

to a minimum-thickness constraint. The problem is solved using variable-fidelity physics models and SM.

10.4.1 *Problem formulation*

The free-stream Mach number is set to $M_\infty = 0.85$ and the angle of attack is fixed at $\alpha = 0$ degrees. The optimization problem is stated as:

$$\min_{\boldsymbol{x}} C_d(\boldsymbol{x}), \tag{10.12}$$

subject to:

$$g_1 = t^*(x) - t(x, \boldsymbol{x}) \leq 0, \tag{10.13}$$

where $t(x,\mathbf{x})$ is the airfoil thickness distribution and $t^*(x)$ is the thickness distribution of the baseline airfoil. The baseline airfoil is a modified version of the NACA 0012, defined as:

$$t^*(x) = \pm 0.6 \begin{pmatrix} 0.2969\sqrt{x} - 0.1260x - 0.3516x^2 \\ +0.2843x^3 - 0.1036x^4 \end{pmatrix}. \tag{10.14}$$

The objective function is implemented through a penalty function approach as:

$$H(\boldsymbol{x}) = C_{d.s}(\boldsymbol{x}) + \beta[\Delta t(x, \boldsymbol{x})]^2, \tag{10.15}$$

where $\Delta t(x, \boldsymbol{x}) = 0$ if $t(x, \mathbf{x}) \geq t^*(x)$ for $x \in [0, 1]$, and $\Delta t(x, \boldsymbol{x}) = (t(x, \mathbf{x}) - t^*(x))/t^*(x)$ otherwise. We use $\beta = 1,000$.

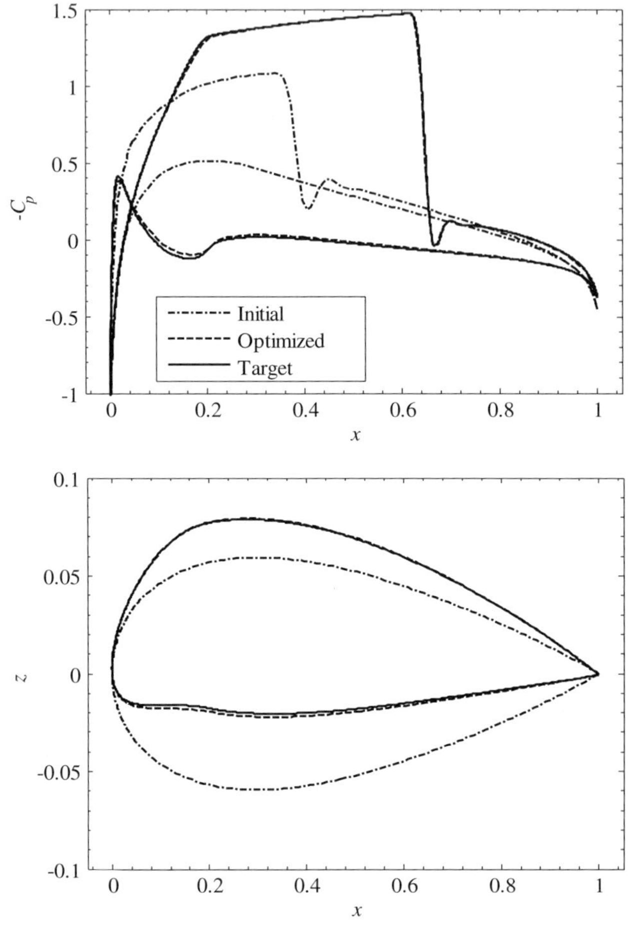

Fig. 10.10. Initial, target, and optimized airfoil shapes and pressure distributions.

10.4.2 *Design variables*

The airfoil shape is described by a B-spline surface (see Chapter 7), as shown in Fig. 10.10. There are two fixed control points, one at the leading-edge (LE) and the other at the TE. Six free control points govern the airfoil shape. The chordwise location of the control points are $x = [0.0, 0.05 + \Delta x, 0.2 + \Delta x, 0.4 + \Delta x,\ 0.6 + \Delta x,\ 0.97 + \Delta x]$, where $\Delta x \in [-0.02, 0.02]$. The control points are allowed to move along the z-direction.

There are seven design parameters and the design variable vector can be written as $\boldsymbol{x} = [z_1 z_2 z_3 z_4 z_5 z_6\ \Delta x]^T$, where z_i is the coordinate of control point i along the z-direction. The lower bound is $\boldsymbol{l} = [0\ 0\ 0\ 0\ 0\ 0\ -0.02]^T$ and the upper bound is $\boldsymbol{u} = [0.1\ 0.1\ 0.1\ 0.1\ 0.1\ 0.1\ 0.02]^T$.

10.4.3 *Optimization algorithm*

The problem is solved using the SM algorithm (described in Chapter 6). The pattern-search algorithm is used for the surrogate model optimization (described in Chapter 3). Surrogate model optimization is constrained using the trust-region approach. The termination condition for the algorithms is the convergence in argument, $\|\boldsymbol{x}^{(i)} - \boldsymbol{x}^{(i-1)}\| < 10^{-4}$, or reduction of the trust region radius below a user-defined threshold; here, $\lambda < 10^{-4}$.

10.4.4 *Models*

The high-fidelity model solves the steady compressible Euler equations on a structured mesh. The mesh is constructed using the hyperbolic C-mesh generator by Kinsey and Barth (1984). The mesh has around 160,000 cells. FLUENT (2012) is used as the flow solver. A typical evaluation time of the high-fidelity model is around 15 minutes. Model details are given in Section 8.3.

The computer system MSES (Drela and Giles, 1987) is the low-fidelity model and is configured to solve a hybrid formulation of the standard momentum-conserving Euler equations and standard entropy-conserving full-potential equations. The mesh is set coarse and a typical evaluation time is around 5 seconds. The simulation time ratio of high- to low-fidelity models is approximately 180. Model details can be found in Chapter 9.

10.4.5 *Results*

The optimization results are shown in Table 10.5 and Fig. 10.11. The drag coefficient is quickly reduced in the first iteration from 471.4 counts to 96.1 counts. This required one high-fidelity model evaluation and 125 low-fidelity model evaluations. One more iteration

Table 10.5 Drag minimization of modified NACA 0012 in inviscid flow with SM and variable-fidelity physics models.

Variable	$\mathbf{x}_0$	$\mathbf{x}_1$	$\mathbf{x}_2$
C_d (d.c.)	471.4	96.1	84.2
N_c	—	125	153
N_f	—	1	1

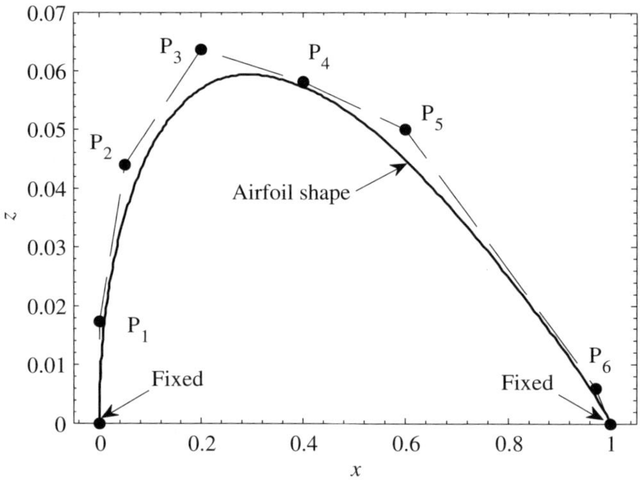

Fig. 10.11. A B-spline setup with six free control points for a symmetric airfoil.

reduces the drag coefficient to 84.2 counts. This iteration required one high-fidelity model evaluation and 153 low-fidelity model evaluations. The algorithm stops because the trust region radius is reduced below the specified threshold value of 10^{-4}. The overall optimization cost corresponds to less than four high-fidelity model evaluations.

We can see from Figs. 10.12 and 10.13 that the initial airfoil has a very strong shock at the 75% chord and the optimized airfoil has a very weak shock close to the TE. This is the source of the reduced drag. The optimized airfoil has a fuller shape than the initial airfoil. This means that the airfoil is flatter and thereby prevents the flow from accelerating into the shock at the aft.

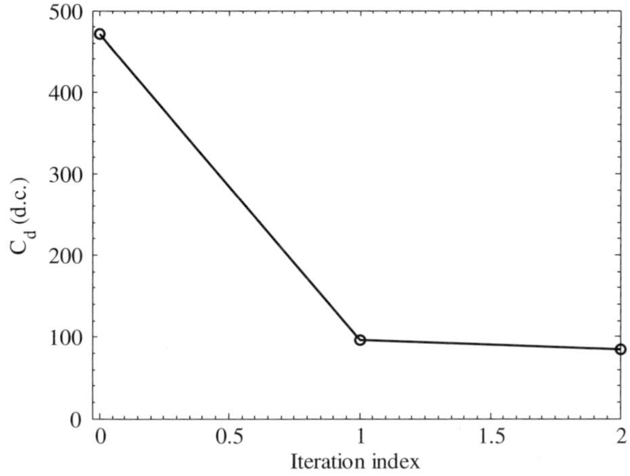

Fig. 10.12. Evolution of the drag coefficient in the optimization.

10.5 Drag Minimization of RAE 2822 in Viscous Flow with Variable-Fidelity Physics Models

The objective is to minimize the drag of the RAE 2822 at a constant lift coefficient and Reynolds number, subject to constraints on the cross-sectional area and the pitching moment.

10.5.1 *Problem formulation*

At a free-stream Mach number $M_\infty = 0.734$ and a Reynolds number $Re = 6.5 \times 10^6$, solve:

$$\min_{\boldsymbol{x}} C_d(\boldsymbol{x}), \tag{10.16}$$

subject to:

$$h_1 = C_l(\boldsymbol{x}) = 0.824, \tag{10.17}$$

$$g_1 = C_m^* - C_m(\boldsymbol{x}) \leq 0, \tag{10.18}$$

$$g_2 = A^* - A(\boldsymbol{x}) \leq 0, \tag{10.19}$$

where $C_m(\boldsymbol{x})$ is the pitch moment, $C_m^* = -0.092$ is the minimum pitching moment, and A^* is the cross-sectional area of the baseline airfoil, i.e. the RAE 2822.

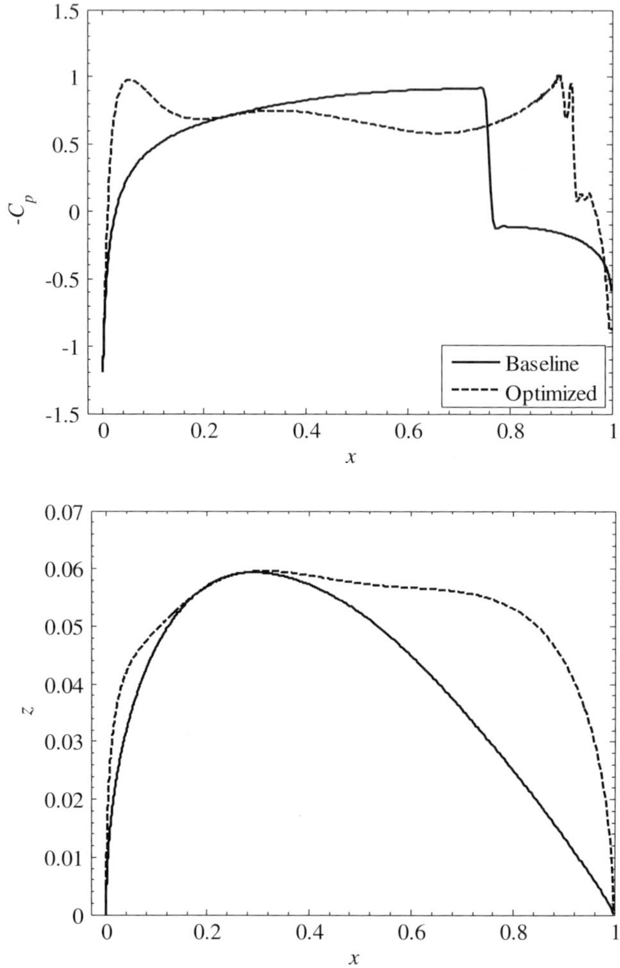

Fig. 10.13. Initial and optimized airfoil shapes and pressure distributions.

10.5.2 *Design variables*

The PARSEC parameterization (described in Chapter 7) is used in this case study. The airfoil has a sharp TE ($t_{TE} = 0$) located at the x-axis ($t_{off} = 0$), leaving 10 parameters to control the airfoil shape. The design variable vector is $\boldsymbol{x} = [R_U \; X_U \; Z_U \; Z_{xxU} \; R_L \; X_L \; Z_L \; Z_{xxL} \; \theta_{TE} \; \beta_{TE}]^T$.

10.5.3 *Optimization algorithm*

The optimization algorithm follows the SM approach (as described in Chapter 6), with the following modification. Let us denote the high-fidelity model as $f(\boldsymbol{x}) = [C_{l.f}(\boldsymbol{x})\ C_{d.f}(\boldsymbol{x})C_{m.f}(\boldsymbol{x})]^T$, where $C_{l.f}$, $C_{d.f}$, and $C_{m.f}$ are high-fidelity lift and drag coefficients, and the pitching moment (a similar convention is used for the low-fidelity model c). The surrogate model at iteration i is defined as:

$$s^{(i)}(\boldsymbol{x}) = [C_{l.s}^{(i)}(\boldsymbol{x})C_{d.s}^{(i)}(\boldsymbol{x})C_{m.s}^{(i)}(\boldsymbol{x})]^T, \tag{10.20}$$

where

$$C_{l.s}^{(i)}(\boldsymbol{x}) = C_{l.c}(\boldsymbol{x}) + [C_{l.f}(\boldsymbol{x}^{(i)}) - C_{l.c}(\boldsymbol{x}^{(i)})], \tag{10.21}$$

$$C_{d.s}^{(i)}(\boldsymbol{x}) = C_{d.c}(\boldsymbol{x}) + [C_{d.f}(\boldsymbol{x}^{(i)}) - C_{d.c}(\boldsymbol{x}^{(i)})], \tag{10.22}$$

$$C_{m.s}^{(i)}(\boldsymbol{x}) = C_{m.c}(\boldsymbol{x}) + [C_{m.f}(\boldsymbol{x}^{(i)}) - C_{m.c}(\boldsymbol{x}^{(i)})]. \tag{10.23}$$

Thus, the surrogate model is obtained by correcting the low-fidelity model response components using output SM based on the last available evaluation of both high- and low-fidelity models. The algorithm is embedded in the trust-region framework for improved convergence.

10.5.4 *Models*

The high-fidelity model solves the steady compressible Reynolds-averaged Navier–Stokes (RANS) equations with the Spalart–Allmaras turbulence model on a structured mesh constructed using the hyperbolic C-mesh generator by Kinsey and Barth (1984). The mesh has around 400,000 cells. FLUENT (2012) is used as the flow solver. A typical evaluation time is around 40 minutes. Model details are provided in Section 8.4.

MSES (Drela and Giles, 1987) is utilized as the low-fidelity model in the viscous-inviscid analysis mode with a fixed transition. A typical evaluation time is roughly 15 seconds. The ratio of high- to low-fidelity simulation time is about 160. Details on the model are included in Chapter 9.

Table 10.6 Drag minimization of RAE 2822 in viscous flow with SM and variable-fidelity physics models.

Variable	$\mathbf{x}_0$	$\mathbf{x}_1$	$\mathbf{x}_2$	$\mathbf{x}_3$
C_d (d.c)	165.2	125.8	128.1	127.0
C_l (l.c.)	77.68	80.41	82.67	82.46
C_m	−0.08703	−0.08118	−0.08195	−0.08222
A	0.07788	0.08084	0.08075	0.08073
N_c	—	302	224	150
N_f	—	1	1	1

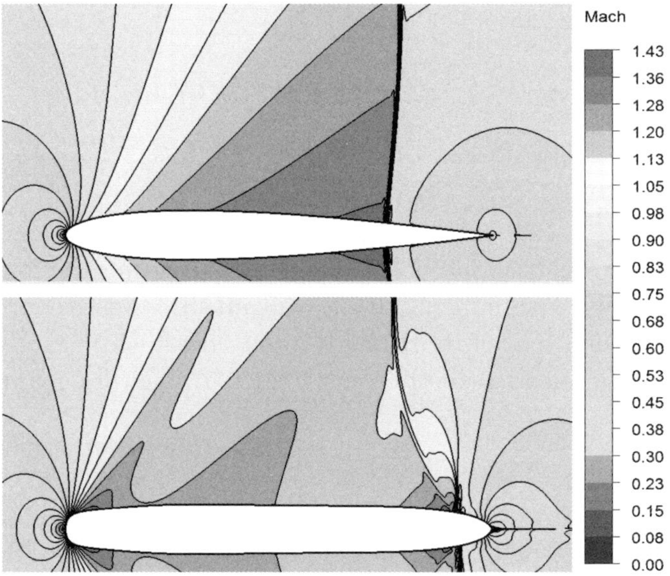

Fig. 10.14. Contours of Mach number of the initial (top) and optimized (bottom) designs.

10.5.5 *Results*

The results of the optimization run are given in Table 10.6. A plot indicating the evolution of the drag and lift coefficients is shown in Fig. 10.14. The algorithm needs three iterations to converge to an optimized shape, shown in Fig. 10.15. Initially, the high-fidelity solution is infeasible, as the lift is below the required constraint. The low-fidelity model solution is initially feasible. After three iterations the lift of the high-fidelity model has increased to the constraint

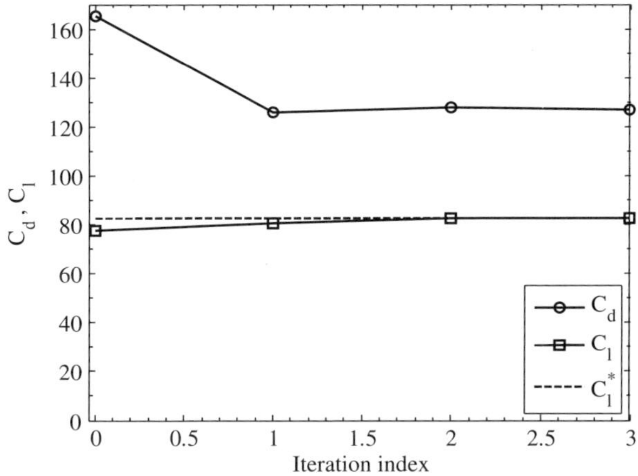

Fig. 10.15. Evolution of the drag and lift coefficients in the optimization.

value of 82.4 counts. At the same time the drag has been reduced from 165.2 counts to 127.0 counts, or by 38.2 counts.

The drag reduction comes from a much weaker shock on the upper surface and the improved lift comes from a high-pressure peak at the LE as well as an opening of the pressure distribution behind the shock due to added aft camber (see Figs. 10.15 and 10.16).

The algorithm needed three high-fidelity model evaluations and 676 low-fidelity model evaluations. The equivalent number of function calls is less than eight.

10.6 Multi-Objective Airfoil Design with SM

By nature, aerodynamic shape optimization is a multi-objective problem. In many cases, a priori preference articulation is possible, so that only one specific objective (e.g. drag coefficient) is to be explicitly optimized, whereas the others (e.g. lift coefficient) can be handled through design constraints. This allows the solving of the problem as a single-objective one. Sometimes, however, this is neither possible nor convenient, e.g. when gaining knowledge about possible trade-offs between competing objectives is important. In those situations genuine multi-objective optimization is necessary.

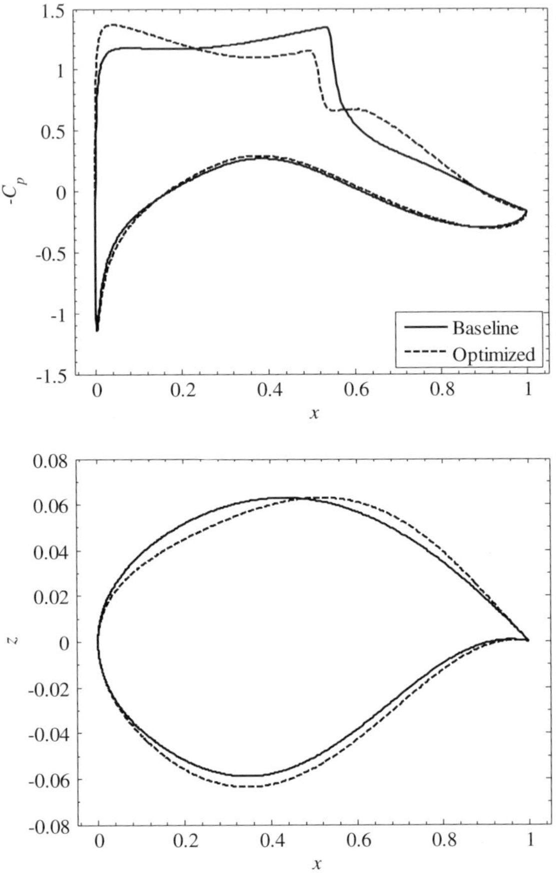

Fig. 10.16. Initial and optimized airfoil shapes and pressure distributions.

Typically, a solution to multi-objective design is represented as a so-called Pareto front, which represents the set of the best possible designs that are non-commensurable in the conventional (single-objective) sense. The most popular multi-objective optimization methods include multi-objective evolutionary algorithms (MOEA) because these methods are capable of yielding the entire representation of the Pareto front in a single algorithm run. Unfortunately, the computational complexity of MOEA is high due to the fact that a (usually large) set of designs is being processed in each iteration of the algorithm. This may be prohibitive, particularly if high-fidelity CFD simulations are used to evaluate the airfoil/wing under design.

In this section, a fast surrogate model, constructed by approximating the low-fidelity CFD simulation data of the airfoil, is exploited to reduce the computational cost of the design process. The surrogate is corrected using sparsely sampled high-fidelity CFD data and parameterized output SM and, subsequently, optimized using MOEA to obtain the Pareto front. The quality of the Pareto front is improved with an iterative procedure by focusing the SM correction of the low-fidelity model in the Pareto-optimal region of the design space and re-optimizing the surrogate model.

10.6.1 *Problem formulation*

Let $f(\boldsymbol{x}) = [C_{l.f}(\boldsymbol{x})C_{d.f}(\boldsymbol{x})]^T$ be the high-fidelity CFD model, i.e. the lift and drag coefficients $C_{l.f}$ and $C_{d.f}$, respectively, are obtained by a high-fidelity CFD simulation. $\boldsymbol{x}$ is the vector of designable parameters. Let $F_k(\boldsymbol{x}), k = 1, \ldots, N_{obj}$ be a kth design objective of interest. A typical performance objective would be to minimize the drag coefficient, in which case $F_k(\boldsymbol{x}) = C_{d.f}$. Another objective would be to maximize lift, in which case $F_k(\boldsymbol{x}) = 1/C_{l.f}$ or $F_k(\boldsymbol{x}) = -C_{l.f}$.

If $N_{obj} > 1$ then any two designs, $\boldsymbol{x}^{(1)}$ and $\boldsymbol{x}^{(2)}$, for which $F_k(\boldsymbol{x}^{(1)}) < F_k(\boldsymbol{x}^{(2)})$ and $F_l(\boldsymbol{x}^{(2)}) < F_l(\boldsymbol{x}^{(1)})$ for at least one pair $k \neq l$, are not commensurable, i.e. none is better than the other in the multi-objective sense. We define Pareto dominance relation $\angle$ (see, for example, Fonseca (1995)), saying that for the two designs $\boldsymbol{x}$ and $\boldsymbol{y}$, we have $\boldsymbol{x} \angle \boldsymbol{y}$ ($\boldsymbol{x}$ dominates $\boldsymbol{y}$) if $F_k(\boldsymbol{x}) < F_k(\boldsymbol{y})$ for all $k = 1, \ldots, N_{obj}$. The goal of the multi-objective optimization is to find a representation of a so-called Pareto front (of Pareto-optimal set), X_P, of the design space X, such that for any $\boldsymbol{x} \in X_P$, there is no $\boldsymbol{y} \in X$ for which $\boldsymbol{y} \angle \boldsymbol{x}$ (Fonseca, 1995).

In practice, the Pareto front gives us information about possible trade-offs between the competing objectives, e.g. what is the minimum drag for given value of the lift coefficient. Having reasonable representation of the Parcto front is therefore indispensable in making various design decisions.

It should be emphasized that an important part of the multi-objective design is a decision making process, i.e. the selection of a specific design from a number of alternatives contained in the

Pareto front, based on given preferences concerning, among others, the importance of particular objectives.

10.6.2 *Design variables*

The airfoil shape is parameterized with the NACA four-digit method. The lower bounds are $\mathbf{l} = [0.0\ 0.20\ 0.08]^T$ and upper bounds $\mathbf{u} = [0.03\ 0.80\ 0.14]^T$.

10.6.3 *Surrogate model construction with SM*

The Pareto front generation is realized using evolutionary algorithms. Evolutionary algorithms iteratively process the entire (usually large) set of potential solutions to the problem at hand, so that they require numerous evaluations of the objective function. For that reason, using the expensive high-fidelity model f directly in the multi-objective optimization would be prohibitive.

To alleviate this problem, we utilize a surrogate model, constructed using SM with the following modifications. Let $c(\boldsymbol{x}) = [C_{l.c}(\boldsymbol{x})\ C_{d.c}(\boldsymbol{x})]^T$ denote the low-fidelity airfoil model, where $C_{l.c}$ and $C_{d.c}$ are the lift and drag coefficients obtained by the low-fidelity CFD model. The surrogate model is constructed in two steps.

Step 1. SM correction of the low-fidelity model. In this step, the initial surrogate model, $s_0(\boldsymbol{x})$, is obtained by applying a parameterized output SM. The mapping uses the correction terms that are directly applied to the response components, $C_{l.c}(\boldsymbol{x})$ and $C_{d.c}(\boldsymbol{x})$, of the low-fidelity model. The surrogate model is defined as follows:

$$\begin{aligned} s_0(\boldsymbol{x}) &= \boldsymbol{A}(\boldsymbol{x}) \circ c(\boldsymbol{x}) + \boldsymbol{D} \\ &= [a_l(\boldsymbol{x})C_{l.c}(\boldsymbol{x}) + d_l(\boldsymbol{x}) a_d(\boldsymbol{x})C_{d.c}(\boldsymbol{x}) + d_d(\boldsymbol{x})]^T, \end{aligned} \tag{10.24}$$

where $\circ$ denotes component-wise multiplication. Both the multiplicative and additive correction terms are design-variable-dependent and take the form of:

$$\begin{aligned} \boldsymbol{A}(\boldsymbol{x}) = [&a_{l.0} + [a_{l.1} a_{l.2} \ldots a_{l.n}] \cdot (\boldsymbol{x} - \boldsymbol{x}^0) a_{d.0} \\ &+ [a_{d.1} a_{d.2} \ldots a_{d.n}] \cdot (\boldsymbol{x} - \boldsymbol{x}^0)]^T, \end{aligned} \tag{10.25}$$

$$\boldsymbol{D}(\boldsymbol{x}) = [d_{l.0} + [d_{l.1}d_{l.2}\dots d_{l.n}]\cdot(\boldsymbol{x}-\boldsymbol{x}^0)d_{d.0} \\ + [d_{d.1}d_{d.2}\dots d_{d.n}]\cdot(\boldsymbol{x}-\boldsymbol{x}^0)]^T, \tag{10.26}$$

where $\boldsymbol{x}^0$ is the center of the design space. Response correction parameters $\boldsymbol{A}$ and $\boldsymbol{D}$ are obtained as:

$$[\boldsymbol{A},\boldsymbol{D}] = \arg\min_{[\bar{\boldsymbol{A}},\bar{\boldsymbol{D}}]} \sum_{k=1}^{N} \|f(\boldsymbol{x}^k) - (\bar{\boldsymbol{A}}(\boldsymbol{x}^k)\circ c(\boldsymbol{x}^k) + \boldsymbol{D}(\boldsymbol{x}^k))\|^2, \tag{10.27}$$

i.e. the response scaling is supposed to (globally) improve the matching for all training points, $\boldsymbol{x}^k$, $k = 1,\dots,N$.

The training set consists of the following.

1. Factorial design with $2n+1$ training points (n being the number of design variables) allocated at the center of the design space $\boldsymbol{x}^0 = (\boldsymbol{l}+\boldsymbol{u})/2$ ($\boldsymbol{l}$ and $\boldsymbol{u}$ being the lower and upper bound for the design variables, respectively) and the centers of its faces, i.e. points with all coordinates but one equal to those of $\boldsymbol{x}^0$, and the remaining one equal to the corresponding component of $\boldsymbol{l}$ or $\boldsymbol{u}$; this sampling scheme is also referred to as the star distribution.
2. Design space corners.
3. Additional 10 points allocated using the LHS. Because we use the NACA four-digit parameterization, the entire training set consists of 25 points.

The correction parameters $\boldsymbol{A}$ and $\boldsymbol{D}$ can be calculated analytically as:

$$\begin{bmatrix} a_{l.0} \\ a_{l.1} \\ \vdots \\ a_{l.n} \\ d_{l.0} \\ \vdots \\ d_{l.n} \end{bmatrix} = (\boldsymbol{C}_l^T\boldsymbol{C}_l)^{-1}\boldsymbol{C}_l^T\boldsymbol{F}_l \quad \begin{bmatrix} a_{d.0} \\ a_{d.1} \\ \vdots \\ a_{d.n} \\ d_{d.0} \\ \vdots \\ d_{d.n} \end{bmatrix} = (\boldsymbol{C}_d^T\boldsymbol{C}_d)^{-1}\boldsymbol{C}_d^T\boldsymbol{F}_d, \tag{10.28}$$

where

$$\boldsymbol{C}_l = \left[\begin{matrix} C_{l.c}(\boldsymbol{x}^1) & C_{l.c}(\boldsymbol{x}^1)\cdot(x_1^1 - x_1^0) & \cdots & C_{l.c}(\boldsymbol{x}^1)\cdot(x_n^1 - x_n^0) \\ C_{l.c}(\boldsymbol{x}^2) & C_{l.c}(\boldsymbol{x}^2)\cdot(x_1^2 - x_1^0) & \cdots & C_{l.c}(\boldsymbol{x}^2)\cdot(x_n^2 - x_n^0) \\ \vdots & \vdots & \ddots & \vdots \\ C_{l.c}(\boldsymbol{x}^N) & C_{l.c}(\boldsymbol{x}^N)\cdot(x_1^N - x_1^0) & \cdots & C_{l.c}(\boldsymbol{x}^N)\cdot(x_n^N - x_n^0) \end{matrix}\right.$$

$$\left.\begin{matrix} 1 & (x_1^1 - x_1^0) & \cdots & (x_n^1 - x_n^0) \\ 1 & (x_1^1 - x_1^0) & \cdots & (x_n^1 - x_n^0) \\ \vdots & \vdots & \vdots & \vdots \\ 1 & (x_1^1 - x_1^0) & \cdots & (x_n^1 - x_n^0) \end{matrix}\right],$$

$$\boldsymbol{F}_l = [C_{l.f}(\boldsymbol{x}^1)C_{l.f}(\boldsymbol{x}^2)\ldots C_{l.f}(\boldsymbol{x}^N)]^T,$$

$$\boldsymbol{C}_d = \left[\begin{matrix} C_{d.c}(\boldsymbol{x}^1) & C_{d.c}(\boldsymbol{x}^1)\cdot(x_1^1 - x_1^0) & \cdots & C_{d.c}(\boldsymbol{x}^1)\cdot(x_n^1 - x_n^0) \\ C_{d.c}(\boldsymbol{x}^2) & C_{d.c}(\boldsymbol{x}^2)\cdot(x_1^2 - x_1^0) & \cdots & C_{d.c}(\boldsymbol{x}^2)\cdot(x_n^2 - x_n^0) \\ \vdots & \vdots & \ddots & \vdots \\ C_{d.c}(\boldsymbol{x}^N) & C_{d.c}(\boldsymbol{x}^N)\cdot(x_1^N - x_1^0) & \cdots & C_{d.c}(\boldsymbol{x}^N)\cdot(x_n^N - x_n^0) \end{matrix}\right.$$

$$\left.\begin{matrix} 1 & (x_1^1 - x_1^0) & \cdots & (x_n^1 - x_n^0) \\ 1 & (x_1^1 - x_1^0) & \cdots & (x_n^1 - x_n^0) \\ \vdots & \vdots & \vdots & \vdots \\ 1 & (x_1^1 - x_1^0) & \cdots & (x_n^1 - x_n^0) \end{matrix}\right],$$

$$\boldsymbol{F}_d = [C_{d.f}(\boldsymbol{x}^1)C_{d.f}(\boldsymbol{x}^2)\ldots C_{d.f}(\boldsymbol{x}^N)]^T,$$

which is a least-square optimal solution to the linear regression problems $[a_{l.0}\ a_{l.1}\ \ldots\ a_{l.n}\ d_{l.0}\ d_{l.1}\ \ldots\ d_{l.n}]^T\ \boldsymbol{C}_l = \boldsymbol{F}_l$ and $[a_{d.0}\ a_{d.1}\ \ldots\ a_{d.n}\ d_{d.0}\ d_{d.1}\ \ldots\ d_{d.n}]^T\ \boldsymbol{C}_d = \boldsymbol{F}_d$, equivalent to Equation (10.27). Note that the matrices $\boldsymbol{C}_l^T\boldsymbol{C}_l$ and $\boldsymbol{C}_d^T\boldsymbol{C}_d$ are non-singular for $N > n+1$, which is the case for our choice of the training set.

Step 2. Construction of the response surface approximation (RSA) surrogate. Having the low-fidelity SM-corrected, we sample the initial surrogate and create the RSA model of it. Here we use kriging interpolation. The design of experiments used at this step is a uniform grid with $6 \times 6 \times 6 = 216$ samples. The kriging surrogate model s is very fast, smooth, and easy to optimize.

Step 1 of the surrogate modeling process allows us to reduce misalignment between the low- and high-fidelity models. The surrogate model created in Step 2 is a response surface approximation one, making it fast and yet accurate because a relatively dense sampling of the design space is utilized. Nevertheless, the computational cost of constructing the surrogate is low because it is based on low-fidelity model data: the cost of 216 low-fidelity data samples is equivalent to less than three high-fidelity ones.

10.6.4 *Optimization algorithm*

The Pareto front is obtained by optimizing the surrogate model. The surrogate is optimized in the multi-objective sense using a MOEA. Here we use a standard MOEA with fitness sharing, Pareto-dominance tournament selection, and mating restrictions. The algorithm is modified in order to handle nonlinear constraints. The modification include:

1. an initialization procedure that only generated feasible individuals; and
2. crossover and mutation procedures that maintain the feasibility of individuals.

The Pareto set is obtained by the following iterative procedure.

1. Correct the low-fidelity model c using parameterized output space mapping as in Equation (10.24).
2. Sample the design space and acquire the s_0 data.
3. Construct the kriging interpolation surrogate model, s.
4. Obtain the Pareto front by optimizing s using MOEA.
5. Evaluate high-fidelity models at selected geometries from the Pareto front.
6. If the termination condition is not satisfied, add the new high-fidelity data set to the existing one, and go to Step 1.
7. END.

It should be emphasized that the high-fidelity model f is only evaluated in Step 1 of the above procedure. The overall computational cost of obtaining the initial Pareto front representation corresponds

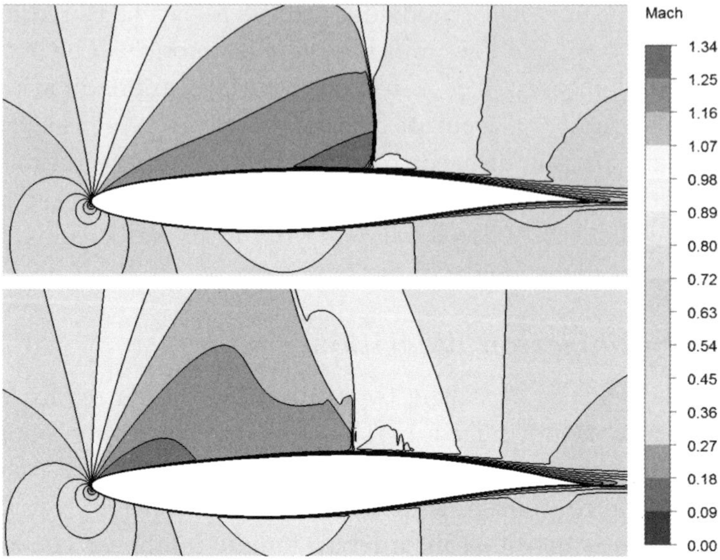

Fig. 10.17. Contours of Mach number of the initial (top) and optimized (bottom) designs.

to about 30 evaluations of f (25 high-fidelity model samples are necessary to perform the SM correction, and around five equivalent evaluations are necessary to construct the response surface surrogate).

A few comments should be made regarding Steps 5 and 6 of the above algorithm. Since the goal is to improve the quality of the surrogate model in the vicinity of the Pareto front, additional high-fidelity data points are allocated along the Pareto set obtained in Step 4. As a result, after a few iterations the focus of the SM correction will be on the portion of the design space that contains the Pareto set. Consequently, the surrogate model accuracy will increase in that region, at the expense of the rest of the design space. The latter is not a problem because the designs of interest are Pareto-optimal ones only.

10.6.5 *Models*

The flow is taken to be inviscid and the variable-resolution models described in Section 10.1.4 are used.

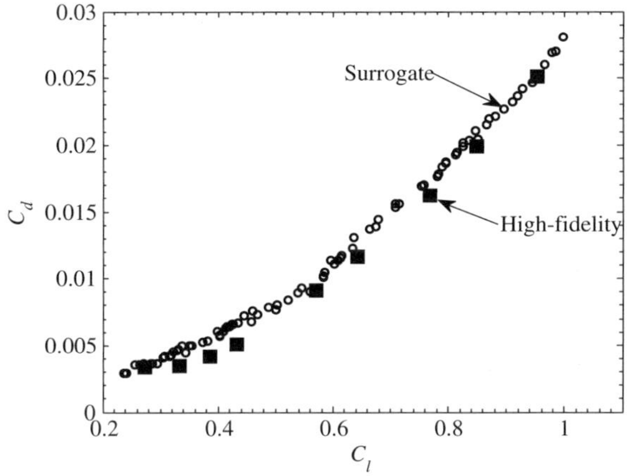

(a) First iteration (initial Pareto set)

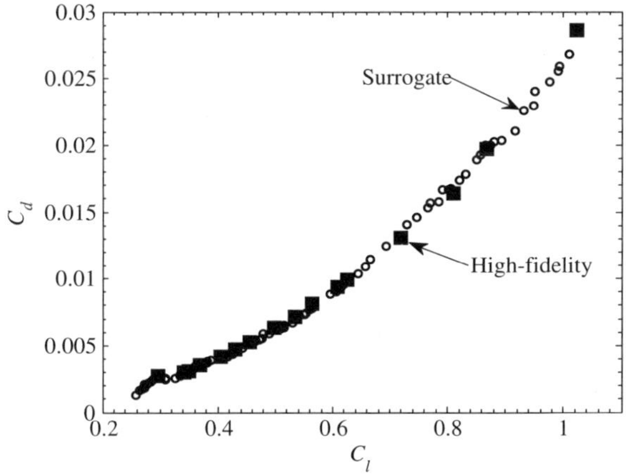

(b) Last iteration (final Pareto set)

Fig. 10.18. Pareto-front representation obtained by using surrogate-based optimization with multi-objective evolutionary algorithms used as a main-optimization engine and SM-utilized for low-fidelity model correction. The surrogate model is constructed using sampled data from the corrected low-fidelity model.

10.6.6 *Results*

We have two objectives, $F_1(\boldsymbol{x}) = 1/C_{l.f}$ and $F_2(\boldsymbol{x}) = C_{d.f}$, i.e. we aim to maximize the lift coefficient and minimize the drag coefficient at the same time. The Pareto front representation obtained using the multi-objective evolutionary algorithm is shown in Fig. 10.17. The population size used by the algorithm was $N = 50$.

The Pareto set obtained in the first iteration is shown in Fig. 10.17(a). It is generated by optimizing the initial kriging model, i.e. the one constructed from the SM-corrected low-fidelity model, with the training samples allocated as described below Equation (10.27). The new set of 10 high-fidelity samples allocated along that initial Pareto front representation (only nine of them are shown in Fig. 10.17(a)) is used for verification purposes, but also to enhance the surrogate model by rerunning the SM correction Equations (10.25)–(10.28). It can be observed that there is some discrepancy between the Pareto-optimized surrogate model and the sampled high-fidelity model, which means that the optimization process has to be continued. The final Pareto set, shown in Fig. 10.17(b), is obtained after four iterations of the algorithm. Its verification, carried out using an additional set of high-fidelity model samples, indicates that the surrogate model is a sufficiently good representation of the high-fidelity model (both data sets are practically aligned). The total optimization cost is about 60 high-fidelity model evaluations: 30 to obtain the initial set as mentioned above, and 10 per each additional iteration of the algorithm.

Chapter 11

Transonic Wing Shape Design

This chapter continues to demonstrate the variable-fidelity aerodynamic shape optimization method described in this book. In particular, the applications in this chapter involve transonic wings.

11.1 Drag Minimization of an Unswept Rectangular Wing with Variable-Fidelity Physics Models

This problem involves an unswept rectangular wing with a constant airfoil profile along the wingspan. The airfoil shape is optimized using variable-fidelity physics models. In particular, the high-fidelity model is a three-dimensional computational fluid dynamics (CFD) model and the low-fidelity model is based on a two-dimensional CFD model. The design is carried out using a surrogate-based optimization (SBO) algorithm with local response surface approximations (RSA).

11.1.1 *Problem formulation*

At $M_\infty = 0.80$, solve:

$$\min_{\boldsymbol{x}} C_D(\boldsymbol{x}), \tag{11.1}$$

subject to:

$$h_1 = C_L^* - C_L(\boldsymbol{x}) = 0, \tag{11.2}$$

where $C_L^* = 0.30$ is the target lift coefficient. The objective function is defined as:

$$H(\boldsymbol{x}) = C_{D.f}(\boldsymbol{x}) + \beta[(C_{L.f}(\boldsymbol{x}) - C_L^*)/C_L^*]^2, \tag{11.3}$$

where $C_{D.f}$ and $C_{L.f}$ are the drag and lift coefficients, respectively, obtained by the high-fidelity model. The objective function for the surrogate model is defined analogously. The penalty coefficient, β, is set to 100 in our numerical experiments. Note that handling the equality constraint, $C_{L.f}(\boldsymbol{x}) = C_L^*$, through a penalty function is convenient because we are using the surrogate model. More specifically, the design that is feasible for the surrogate model may not be feasible for the high-fidelity one, which would be troublesome when making the transition from one optimization algorithm iteration to another.

11.1.2 *Design variables*

The chord is fixed to $c_r = c_t = 1$ and the span is set $b = 10$. The wing section profile is the same throughout the span and is described by the three design variables of the NACA (National Advisory Committee for Aeronautics) four-digit parameterization (see Section 7.7). The design variable vector is $\boldsymbol{x} = [m\ p\ t/c\alpha]^T$, where m is the maximum ordinate of the mean camberline as a fraction of chord, p is the chordwise position of the maximum ordinate, t/c is the thickness-to-chord ratio, and α is the angle of attack. The bounds on the design variables are $0 \leq m \leq 0.1$, $0.2 \leq p \leq 0.8$, $0.05 \leq t/c \leq 0.2$, and $0 \leq \alpha \leq 3$ deg.

11.1.3 *High-fidelity model*

The flow is assumed to be steady, viscous, and adiabatic with no body forces. We solve the Reynolds-averaged Navier–Stokes equations with the one-equation Spalart–Allmaras turbulence model. The farfield is configured in a box-topology, in which the airfoil/wing geometry is placed in the center of the box. The leading edge (LE) is placed at the origin, with the farfield extending 100 chord lengths away from the origin in every direction. To reflect the compressible nature of this problem, two types of boundaries are used. The pressure-farfield is applied to the boundary on all surfaces, except where the wing penetrates the symmetry boundary. A sketch of the setup in three-dimensions is shown in Fig. 8.7(a).

An unstructured tri/tetra shell grid is created on all surfaces. The shell grid from the wing is then extruded into the volume, where the volume is flooded with tri/tetra elements. The grid is made dense close to the wing where it then gradually grows in size as it moves away from the wing surfaces. To capture the viscous boundary layer, an inflation layer or a prism layer is created on the wing surfaces. In the streamwise direction, the number of elements on the wing is set to 100 on both upper and lower surfaces. The bi-geometric bunching law with a growth ratio of 1.2 is employed in the streamwise direction over the wing to obtain a denser element distribution at the LE and trailing edge (TE). This is done in order to capture the high-pressure gradient at the LE and the separation at the TE. The minimum element size of the wing in the streamwise direction is set to $0.1\%c$, where c is the wing chord length, and it is located at the LE and TE. The elements in the spanwise direction are distributed uniformly and the number of elements is set to 100 over the semi-span. A prism layer is used to capture the viscous boundary layer. This layer consists of a number of structured elements that grow in size normal to the wing surface into the domain volume. The inflation layer has an initial height of $5 \times 10^{-6}c$, where it has grown 20 layers into the volume using an exponential growth law with a ratio of 1.2. The initial layer height is chosen so that $y^+ < 1$ at all nodes on the wing. An example grid is shown in Figs. 8.7(b) and 8.7(c).

The flow solver is an implicit density-based one using the Roe-FDS flux type. The spatial discretization schemes are set to second order for all variables, and the gradient information is found using the Green–Gauss node-based method. The residuals, which are the sum of the L^2 norm of all governing equations in each cell, are monitored and checked for convergence. The convergence criterion for the high-fidelity model is such that a solution is considered to be converged if the residuals have dropped by six orders of magnitude or the total number of iterations has reached 1,000. The lift and drag coefficients are obtained by integrating the pressure and skin friction coefficients over the surface of the wing.

The grid parameters were obtained by a grid independence study. The grid has approximately 4 million cells and the wall-clock simulation time is approximately 5 hours.

11.1.4 *Low-fidelity model*

The low-fidelity model is constructed along the lines of the setup described in Section 9.4. In this particular case, the wing is unswept, rectangular, and with constant airfoil profiles. So, the low-fidelity model is entirely based on two-dimensional CFD analyses. We set $C_{L.c}(\boldsymbol{x}) = C_l(\boldsymbol{x})$, where C_l is the section lift coefficient obtained by the two-dimensional CFD model. By neglecting the three-dimensional induced drag effects, we set $C_{D.c}(\boldsymbol{x}) = C_d(\boldsymbol{x})$, where C_d is the section drag coefficient. Since the wing is unswept, we have $M_n = M_\infty$, $c_n = c$, and $(t/c)_n = (t/c)$.

The two-dimensional CFD model is configured in a similar way to the three-dimensional one, i.e. the same grid-generation methodology and flow-solver approach. Figure 8.6 shows the details of the two-dimensional solution domain and an example grid. The two-dimensional CFD model uses a fine grid (obtained by a grid-independence study) with approximately 400,000 cells and a wall-clock simulation time of around 20 minutes. The simulation time ratio of the high- to low-fidelity models is about 15.

11.1.5 *Optimization by local response surface approximations*

The design optimization method exploits a generic SBO scheme that generates a sequence, $\boldsymbol{x}^{(i)}$, $i = 0, 1, \ldots$, of approximate solutions to Equation (11.1) (here, $\boldsymbol{x}^{(0)}$ is the initial design) as:

$$\boldsymbol{x}^{(i+1)} = \arg\min_{\boldsymbol{x}} H(s^{(i)}(\boldsymbol{x})), \tag{11.4}$$

where $s^{(i)}(\boldsymbol{x}) = [C_{L.s}^{(i)}(\boldsymbol{x})\ C_{D.s}^{(i)}(\boldsymbol{x})]^T$ is a surrogate model at iteration i. The surrogate model is constructed using a low-fidelity model, $c(\boldsymbol{x}) = [C_{L.c}(\boldsymbol{x}) C_{D.c}(\boldsymbol{x})]^T$, which is a fast aerodynamic analysis of the clean wing shape (described in the previous section). The surrogate model construction is described in the next section.

In many situations, the surrogate model is optimized directly because its evaluation cost is negligible compared to the high-fidelity one. In the current case, the cost of the surrogate model evaluation cannot be neglected because it may involve one or more low-fidelity computational models. The construction and handling of the surrogate model aims to reduce the overall design cost, having in mind non-negligible contributions of the low-fidelity model evaluations. For that purpose, Equation (11.4) is replaced by a constrained version:

$$\boldsymbol{x}^{(i+1)} = \arg \min_{\boldsymbol{x}^{(i)} - \boldsymbol{d}^{(i)} \leq \boldsymbol{x} \leq \boldsymbol{x}^{(i)} + \boldsymbol{d}^{(i)}} H(s^{(i)}(\boldsymbol{x})), \tag{11.5}$$

where $\boldsymbol{x}^{(i)} = [x_1^{(i)}\ \mathrm{x}_2^{(i)} \ldots \mathrm{x}_n^{(i)}]^T$ is a design obtained as a result of iteration $i - 1$ of Equation (11.5), whereas $\boldsymbol{d}^{(i)} = [d_1^{(i)}\ d_2^{(i)} \ldots d_n^{(i)}]^T$ is the size parameter that is used to define the vicinity of the vector $\boldsymbol{x}^{(i)}$ in which the optimization of the surrogate model $s^{(i)}$ is carried out. The size parameter is updated in each iteration, as explained later.

The optimization algorithm workflow is the following (here, $\boldsymbol{x}^{(0)}$ is the initial design and $\boldsymbol{d}^{(0)}$ is the initial size parameter, normally a fraction of the design space size).

1. Set $i \quad = 0$; evaluate high-fidelity model at $\boldsymbol{x}^{(0)}$.
2. Sample the interval $[\boldsymbol{x}^{(i)} - \boldsymbol{d}^{(i)}, \boldsymbol{x}^{(i)} + \boldsymbol{d}^{(i)}]$ to generate the training set $X_T^{(i)}$.
3. Evaluate the low-fidelity model c at $X_T^{(i)}$.
4. Identify the RSA model $q^{(i)}$.
5. Identify the surrogate model $s^{(i)}$.
6. Find a candidate design $\boldsymbol{x}^{tmp}$ as $\boldsymbol{x}^{tmp} = \arg \min_{\boldsymbol{x}^{(i)} - \boldsymbol{d}^{(i)} \leq \boldsymbol{x} \leq \boldsymbol{x}^{(i)} + \boldsymbol{d}^{(i)}} H(s^{(j)}(\boldsymbol{x}))$.
7. Evaluate high-fidelity model at $\mathbf{x}^{tmp}$ and calculate the gain ratio,
$$r = \frac{H(f(\boldsymbol{x}^{tmp})) - H(f(\boldsymbol{x}^{(i)}))}{H(s^{(j)}(\boldsymbol{x}^{tmp})) - H(s^{(j)}(\boldsymbol{x}^{(i)}))}.$$
8. If $r > r_{incr}$, set $\boldsymbol{d}^{(j+1)} = \boldsymbol{d}^{(j)} \cdot m_{incr}$.
9. If $r < r_{decr}$, set $\boldsymbol{d}^{(j+1)} = \boldsymbol{d}^{(j)} / m_{decr}$.
10. If $r > 0$, set $\boldsymbol{x}^{(j+1)} = \boldsymbol{x}^{tmp}$; otherwise $\boldsymbol{x}^{(j+1)} = \boldsymbol{x}^{(j)}$.
11. If the termination condition is satisfied, END; or else set $j = j+1$ and go to Step 2.

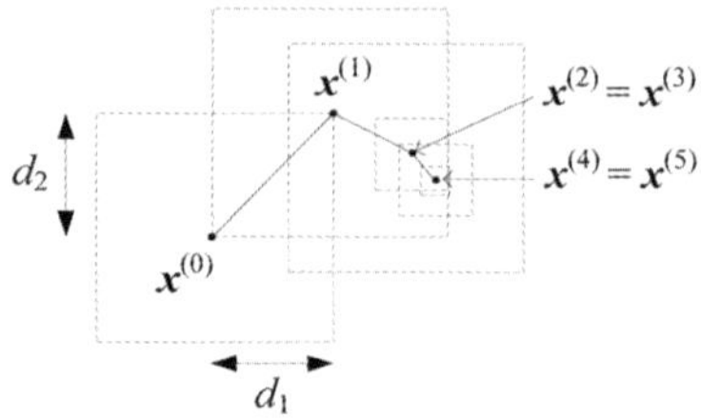

Fig. 11.1. A conceptual illustration of the optimization procedure ($n = 2$).

In the above procedure, the updating parameters for the trust region size, i.e. r_{incr}, m_{incr}, r_{decr}, and m_{decr}, are set by the user. Here, we set $r_{incr} = 0.75$, $m_{incr} = 1.5$, $r_{decr} = 0.25$, and $m_{decr} = 2$. The algorithm is terminated if either $\|\boldsymbol{x}^{(i+1)} - \boldsymbol{x}^{(i)}\| < 0.001$ or $\|\boldsymbol{d}^{(i)}\| < 0.001$. Due to numerical noise associated with the CFD simulations, a pattern-search algorithm (see Section 3.3.1) is used to optimize the surrogate model. The initial step size is set as 10% of the search domain. Figure 11.1 illustrates the operation of the optimization procedure for $n = 2$.

11.1.6 *Surrogate model*

The surrogate model is constructed as follows. In the first step, we build a local RSA model of the low-fidelity model $\boldsymbol{c}$ in the interval $[\boldsymbol{x}^{(i)} - \boldsymbol{d}^{(i)}, \boldsymbol{x}^{(i)} + \boldsymbol{d}^{(i)}]$. We denote by $X_T^{(i)} = \{\boldsymbol{x}_t^{(i.1)}, \ldots, \boldsymbol{x}_t^{(i.N)}\}$ the training set obtained by sampling the aforementioned interval. The RSA model is obtained by approximating the data pairs $\{\boldsymbol{x}_t^{(i.k)}, c(\boldsymbol{x}_t^{(i.k)})\}$, $k = 1, \ldots, N$. Here, we use factorial design with $N = 2n+1$ samples allocated in the center of the interval, i.e. $\boldsymbol{x}_t^{(i.1)} = \boldsymbol{x}^{(i)}$, and in the centers of all interval faces, i.e. $\boldsymbol{x}_t^{(i.k)} = [x_1^{(i)} \ldots x_k^{(i)} + (-1)^k \cdot d\lceil (k-1)/2 \rceil^{(i)} \ldots x_n^{(i)}]^T$, $k = 2, \ldots, 2n+1$. The RSA surrogate is based on the second order polynomial model, $q^{(j)}$, without mixed terms:

$$q^{(i)}(\boldsymbol{x}) = \lambda_0^{(i)} + \sum_{k=1}^{n} \lambda_k^{(i)} x_k + \sum_{k=1}^{n} \lambda_{n+k}^{(i)} x_k^2. \tag{11.6}$$

The unknown coefficients $\lambda = [\lambda_0\ \lambda_1 \ldots \lambda_{2n}]$ are found by solving the linear regression problem $q(\boldsymbol{x}_t^{(i.k)}) = c(\boldsymbol{x}_t^{(i.k)})\}$, $k = 1, \ldots, N$. A unique solution to this problem exists and can be found analytically. It should be noted that the RSA model represents both lift and drag coefficients, so its components are $q^{(i)}(\boldsymbol{x}) = [C_{L.q}^{(i)}(\boldsymbol{x}) C_{D.q}^{(i)}(\boldsymbol{x})]^T$.

In the second step, the RSA model is corrected using output space mapping (SM) as follows:

$$\begin{aligned} s^{(i)}(\boldsymbol{x}) &= \boldsymbol{A}^{(i)} \circ q^{(i)}(\boldsymbol{x}) + \boldsymbol{D}^{(i)} + \boldsymbol{r}^{(i)} \\ &= [a_L^{(i)} C_{L.q}^{(i)}(\boldsymbol{x}) + d_L^{(i)} + r_L^{(i)}\ a_D^{(i)} C_{D.q}^{(i)}(\boldsymbol{x}) + d_D^{(i)} + r_D^{(i)}]^T, \end{aligned} \tag{11.7}$$

where $\circ$ denotes component-wise multiplication. The response correction parameters $\boldsymbol{A}^{(i)}$ and $\boldsymbol{D}^{(i)}$ are obtained by solving:

$$[\boldsymbol{A}^{(i)}, \boldsymbol{D}^{(i)}] = \arg \min_{[\boldsymbol{A},\boldsymbol{D}]} \sum_{k=0}^{i} \| f(\boldsymbol{x}^{(k)}) - \boldsymbol{A} \circ q^{(i)}(\boldsymbol{x}^{(k)}) + \boldsymbol{D} \|^2, \tag{11.8}$$

i.e. the response scaling is supposed to (globally) improve the matching for all previous iteration points. The additive response correction term $\boldsymbol{r}^{(i)}$ is defined as:

$$\boldsymbol{r}^{(i)} = f(\boldsymbol{x}^{(i)}) - [\boldsymbol{A}^{(i)} \circ q^{(i)}(\boldsymbol{x}^{(i)}) + \boldsymbol{D}^{(i)}], \tag{11.9}$$

i.e. it ensures perfect matching between the surrogate and the high-fidelity model at the current design, $\boldsymbol{x}^{(i)}$, $s^{(i)}(\boldsymbol{x}^{(i)}) = f(\boldsymbol{x}^{(i)})$ (so-called zero-order consistency).

$\boldsymbol{A}^{(i)}$, $\boldsymbol{D}^{(i)}$, and $\boldsymbol{r}^{(i)}$ can be calculated analytically so that there is no need to carry out a nonlinear-minimization-like parameter extraction (PE) process (Equation (11.8)). The term $\boldsymbol{r}^{(i)}$ can be calculated using Equation (11.9). Terms $\boldsymbol{A}^{(i)}$ and $\boldsymbol{D}^{(i)}$ can be obtained as:

$$\begin{bmatrix} a_L^{(i)} \\ d_L^{(i)} \end{bmatrix} = (\boldsymbol{C}_L^T \boldsymbol{C}_L)^{-1} \boldsymbol{C}_L^T \boldsymbol{F}_L, \quad \begin{bmatrix} a_D^{(i)} \\ d_D^{(i)} \end{bmatrix} = (\boldsymbol{C}_D^T \boldsymbol{C}_D)^{-1} \boldsymbol{C}_D^T \boldsymbol{F}_D, \tag{11.10}$$

where:

$$\boldsymbol{C}_L = \begin{bmatrix} C_{L.q}^{(i)}(\boldsymbol{x}^{(0)}) & C_{L.q}^{(i)}(\boldsymbol{x}^{(1)}) & \cdots & C_{L.q}^{(i)}(\boldsymbol{x}^{(i)}) \\ 1 & 1 & \cdots & 1 \end{bmatrix}^T,$$

$$\boldsymbol{F}_L = [C_{L.f}(\boldsymbol{x}^{(0)}) \quad C_{L.f}(\boldsymbol{x}^{(1)}) \quad \ldots \quad C_{L.f}(\boldsymbol{x}^{(i)})]^T,$$

$$\boldsymbol{C}_D = \begin{bmatrix} C_{D.q}^{(i)}(\boldsymbol{x}^{(0)}) & C_{D.q}^{(i)}(\boldsymbol{x}^{(1)}) & \cdots & C_{D.q}^{(i)}(\boldsymbol{x}^{(i)}) \\ 1 & 1 & \cdots & 1 \end{bmatrix}^T,$$

$$\boldsymbol{F}_D = [C_{D.f}(\boldsymbol{x}^{(0)}) \quad C_{D.f}(\boldsymbol{x}^{(1)}) \quad \ldots \quad C_{D.f}(\boldsymbol{x}^{(i)})]^T,$$

which is a least-square optimal solution to the linear regression problem $\boldsymbol{C}_L a_L^{(i)} + d_L^{(i)} = \boldsymbol{F}_L$ and $\boldsymbol{C}_D a_D^{(i)} + d_D^{(i)} = \boldsymbol{F}_D$, equivalent to Equation (11.8). Note that the matrices $C_L^T C_L$ and $C_D^T C_D$ are nonsingular for $i > 1$. For $i = 1$ only the multiplicative correction with $\boldsymbol{A}^{(i)}$ components are used, which can be calculated in a similar way

11.1.7 *Results*

Table 11.1 shows the results obtained after five iterations of the algorithm (Equation (11.5)). The optimizer reduces the maximum camber from 0.02 to 0.01, the location of maximum camber from 0.4 to 0.44, and the thickness-to-chord ratio from 12% to 11.5%. The resulting section profile therefore has less overall camber but it is located more aft. The profile is also thinner. The drag coefficient is reduced

Table 11.1 Numerical results of optimization at $M_\infty = 0.80$.

Variable	Initial	Final
m	0.0200	0.0100
p	0.4000	0.4432
t/c	0.1200	0.1148
α	1.0000	0.9837
C_L	0.3772	0.3046
C_D	0.0422	0.0286
N_c	—	45
N_f	—	6
Cost	—	~9

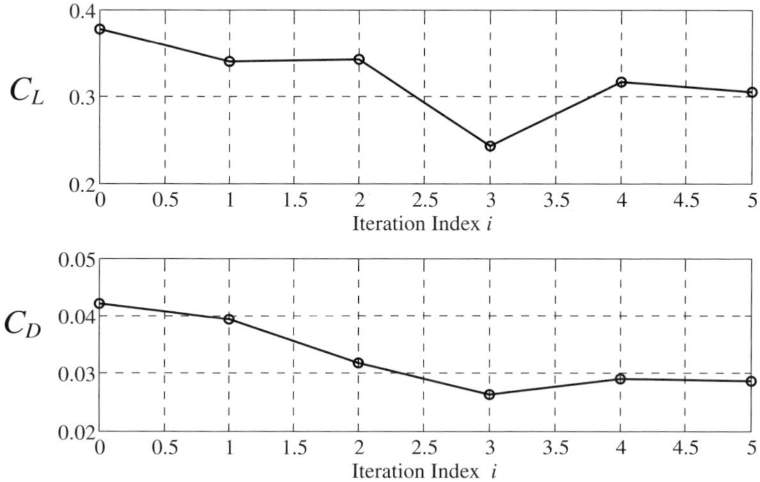

Fig. 11.2. Evolution of the lift and drag coefficient for the high-fidelity model vs. optimization algorithm iteration index.

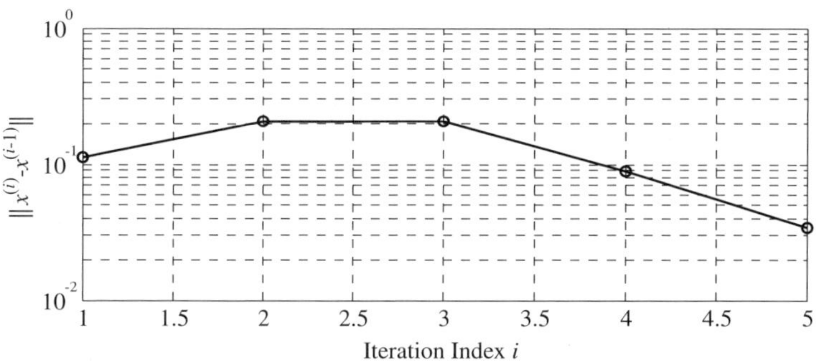

Fig. 11.3. Convergence plot, i.e. the distance between two subsequent iteration designs vs. optimization algorithm iteration index.

from the initial value of 0.0422 to 0.0286, which is a substantial improvement. At the same time, the lift coefficient is reduced from 0.3772 to 0.3046, which is close to the target value. Further iterations are expected to bring the lift coefficient closer to C_L^*, although only five iterations were completed at the time of submitting this abstract.

Figure 11.2 shows the evolution of the lift and drag coefficients, and Figure 11.3 shows the convergence plot of the optimization process. It can be observed that both quantities become converging towards their final values. The cost of each iteration of the algorithm is $N_c = 9$ evaluations of the low-fidelity model and $N_f = 1$ evaluation of the high-fidelity model. Thus, a total of five iterations yields $N_c = 45$ evaluations of the low-fidelity model and $N_f = 5+1 = 6$ evaluations of the high-fidelity model (accounting for the initial design). The total cost then corresponds to only about nine evaluations of f.

11.2 Drag Minimization of the ONERA M6 Wing with Variable-Resolution Models

11.2.1 *Problem formulation*

At $M_\infty = 0.8395$, minimize the drag coefficient for a given lift coefficient, i.e. solve:

$$\min_{\boldsymbol{x}} C_D(\boldsymbol{x}), \tag{11.11}$$

subject to:

$$h_1 = C_L^* - C_L(\boldsymbol{x}) = 0, \tag{11.12}$$

$$g_1 = A^* - A(\boldsymbol{x}) \leq 0, \tag{11.13}$$

where $C_L^* = 0.30$ is the target lift coefficient and A^* is the baseline cross-sectional area. The objective function is defined in the same way as in Equation (11.3), where the lift constraint (Equation (11.12)) is handled through a penalty function. However, the area constraint (Equation (11.13)) is handled directly, since the geometry of the surrogate is exactly the same as that of the high-fidelity model.

11.2.2 *Design variables*

The wing planform shape is fixed. The wing has a semi-span b/2 = 1193.6 mm, root chord $c_r = 805.9$ mm, LE sweep $\Lambda_{LE} = 30$ degrees, taper ratio $\lambda = 0.56$, and aspect ratio $AR = 3.8$. The wing surface is defined by a loft between the airfoil profiles at the root and the tip. In this case, we set the root and the tip profiles to the same shape.

The airfoil shape is described by a B-spline surface (see Chapter 7), as shown in Fig. 11.4. There are two fixed control points, one at the

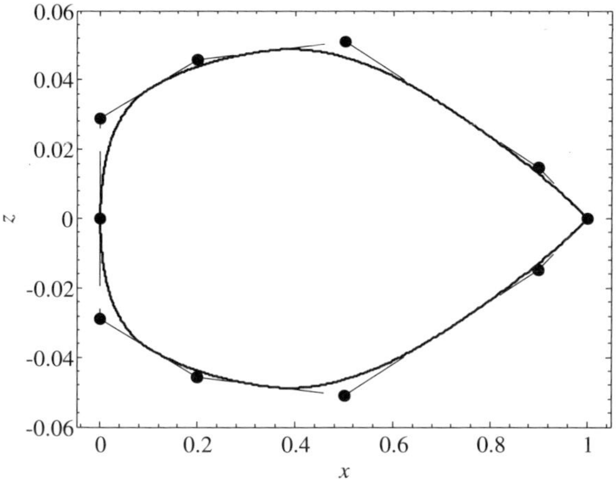

Fig. 11.4. The ONERA M6 airfoil profile is parameterized with B-spline curves.

LE and the other at the TE, the other eight control points govern the airfoil shape. Thus, there are eight design variables altogether. The chordwise location of the control points, for both upper and lower surfaces, are $\boldsymbol{x} = [0.0\ 0.2\ 0.5\ 0.9]$. The control points are allowed to move along the z-direction. The upper surface points can move between 0 and 0.1, and the lower surface points between -0.1 and 0.

11.2.3 *Optimization algorithm*

The problem is solved using the SM algorithm (described in Chapter 6). The pattern-search algorithm is used for the surrogate model optimization (described in Chapter 3). Surrogate model optimization is constrained using the trust-region approach. The termination condition for the algorithms is the convergence in argument, $\|\boldsymbol{x}^{(i)} - \boldsymbol{x}^{(i-1)}\| < 10^{-4}$, or the reduction of the trust-region radius below a user-defined threshold, here $\lambda < 10^{-4}$. The maximum number of surrogate model evaluations per design iteration is set to 200.

11.2.4 *Models*

The flow is taken to be inviscid and we use the high-fidelity wing CFD model described in Section 8.5. Based on the grid study (shown in

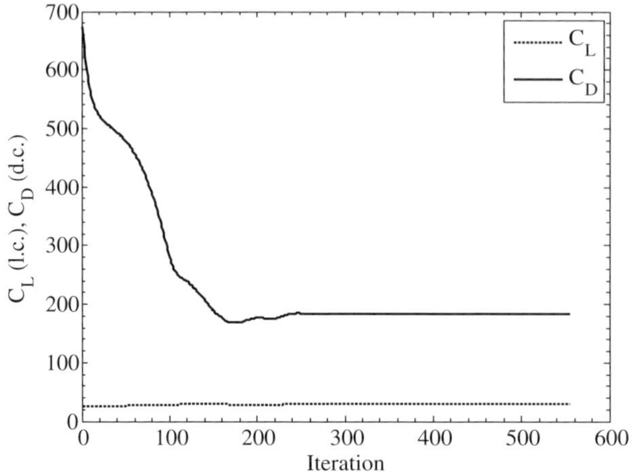

Fig. 11.5. Evolution of the lift and drag coefficients obtained by the low-fidelity model for ONERA M6 at $M_\infty = 0.8395$ and $\alpha = 3.06$ degrees.

Fig. 8.20), we use the grid with 8,563,000 cells and a simulation time of 27 hours.

The variable-resolution modeling approach (see Section 9.2) is used in this problem, i.e. the low-fidelity model is based on the same model as the high-fidelity model but with a coarser grid discretization and relaxed convergence criteria. We use the grid with 122,000 grid cells, based on the grid study in Fig. 8.20. The simulation time is about 20 minutes. However, based on the flow-solver history, shown in Fig. 11.5, the maximum number of iterations is set to 300. This reduces the overall simulation time to about 12 minutes. The simulation time ratio of the high to low models is 108.

11.2.5 *Results*

The optimization history is shown in Fig. 11.6. We see that the optimization converged on the argument (top-left graph) in two iterations. The lift constraint is active, i.e. it is within 1.5% of the target lift coefficient (bottom-left graph). The drag coefficient is reduced from 126.4 counts to 115.9 counts, a reduction of 10.5 counts. The cross-sectional area constraints are active, but not violated.

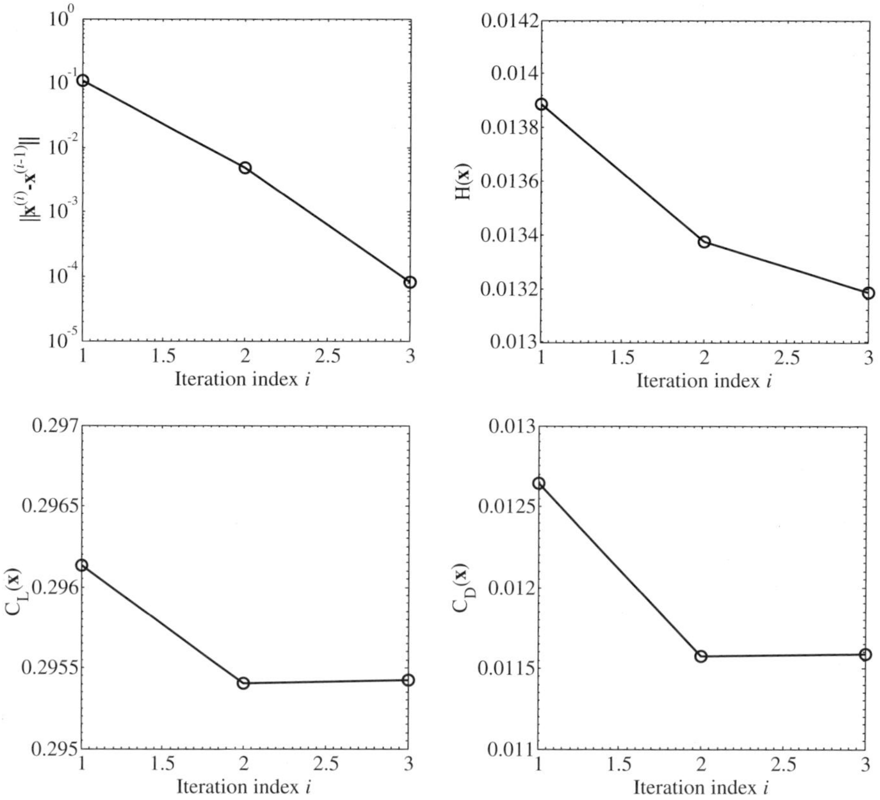

Fig. 11.6. Optimization history.

The optimization cost comprises two high-fidelity model evaluations (excluding evaluation of the initial design) and 400 low-fidelity model evaluations. The overall optimization cost is equivalent to less than six high-fidelity model evaluations.

The initial and optimized airfoil shapes are shown in Fig. 11.7. The optimizer reduces the thickness at the LE of the upper surface. The source of the drag reduction is the reduction in the shock strength of the shock closer to the LE, as can be seen from the pressure contours in Fig. 11.8 and the pressure distributions in Fig. 11.9.

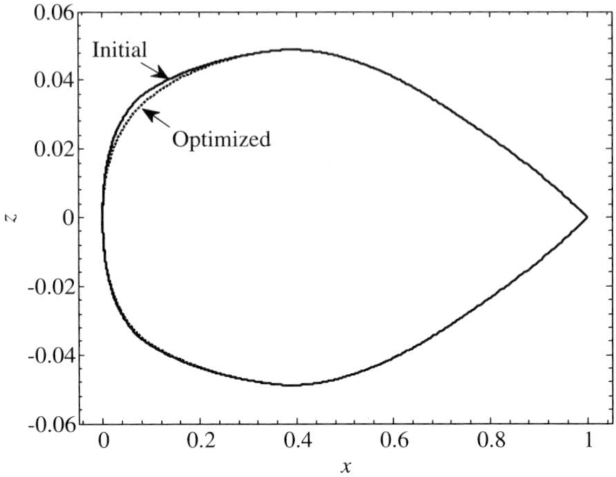

Fig. 11.7. Initial (solid) and optimized (dashed) airfoil shapes.

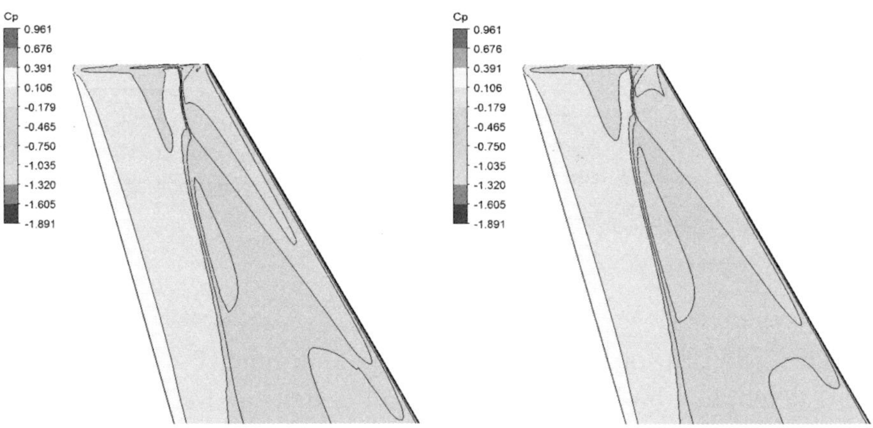

Fig. 11.8. Contours of pressure coefficient of the initial (left) and optimized (right) wings.

11.3 Drag Minimization of the NASA CRM Wing with Variable-Resolution Models

11.3.1 *Problem formulation*

At $M_\infty = 0.85$, solve:

$$\min_{\boldsymbol{x}} C_D(\boldsymbol{x}), \tag{11.14}$$

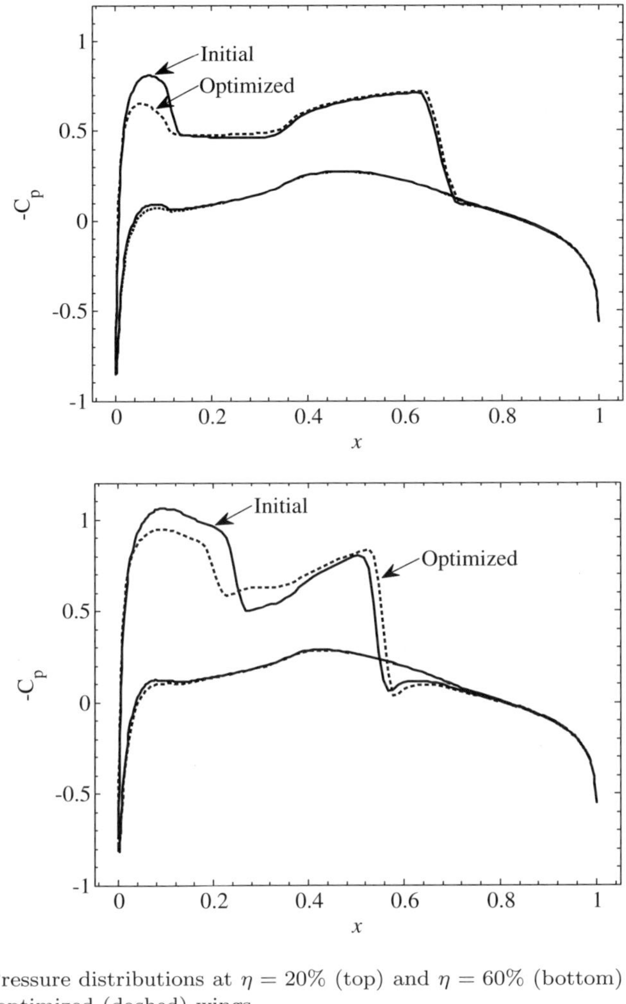

Fig. 11.9. Pressure distributions at $\eta = 20\%$ (top) and $\eta = 60\%$ (bottom) of the initial (solid) and optimized (dashed) wings.

subject to:

$$h_1 = C_L^* - C_L(\boldsymbol{x}) = 0, \tag{11.15}$$

$$g_1 = A^* - A(\boldsymbol{x}) \leq 0, \tag{11.16}$$

where $C_L^* = 0.50$ is the target lift coefficient and A^* is the baseline cross-sectional area. The constraints are handled in the same manner as discussed in Section 11.2.1.

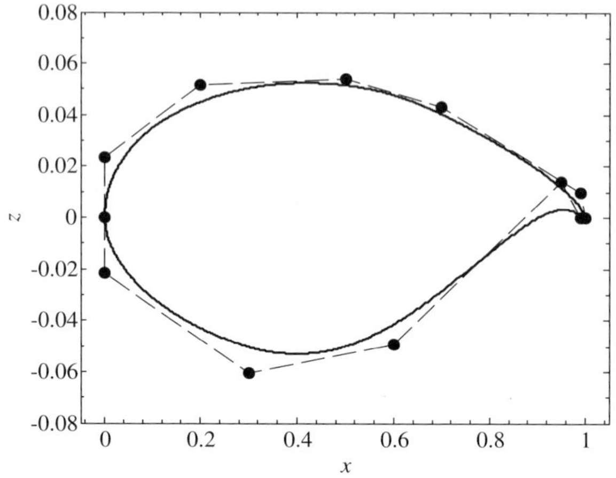

Fig. 11.10. The upper surface of the airfoil at the kink of the NASA CRM wing is parameterized with B-spline curves.

11.3.2 *Design variables*

The planform shape of the CRM wing, described in Section 8.6.2, is kept fixed. The airfoil profiles at the root and the tip are also kept fixed. The airfoil shape at the kink is to be optimized. In fact, only the upper surface will be subject to geometry adjustment. We describe the upper surface by a B-spline surface (see Chapter 7), as shown in Fig. 11.10. There are five free control points on the upper surface, at $x = [0.0\ 0.2\ 0.5\ 0.7\ 0.99]$, and they can move in the vertical direction between 0 and 0.1.

11.3.3 *Optimization algorithm*

The problem is solved using the SM algorithm in exactly the same manner as described in Section 11.2.3.

11.3.4 *Models*

The flow is taken to be viscous and we use the high-fidelity wing CFD model described in Section 8.6. Based on the grid study (shown in Fig. 8.23), we use the finest grid, which has 18 million cells and a simulation time of around 67.5 hours.

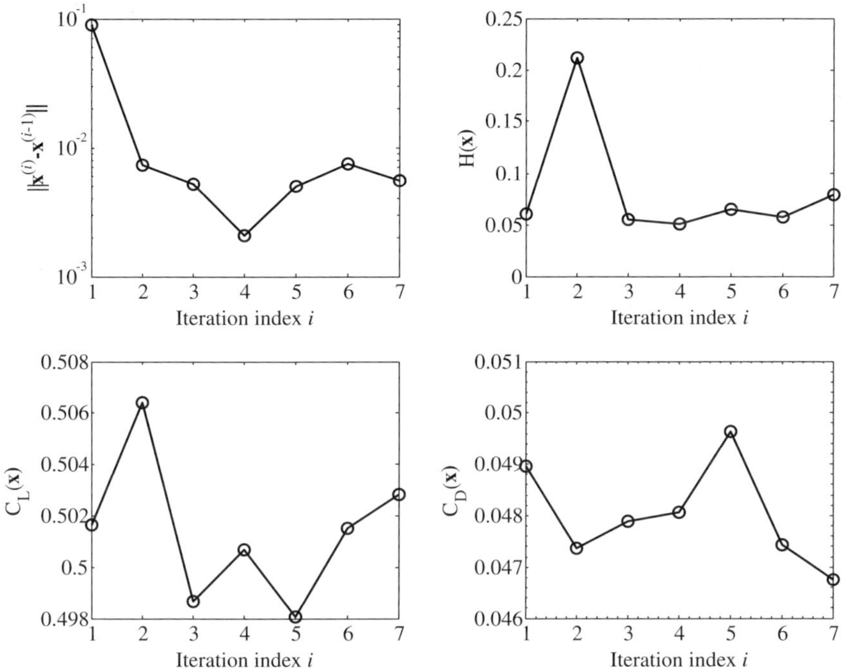

Fig. 11.11. Optimization history.

Again, we use the variable-resolution modeling approach to solve the problem. In particular, we use the grid with 259,000 grid cells, based on the grid study in Fig. 8.23. The simulation time is about 70 minutes. Based on the flow-solver history, the maximum number of iterations is set to 300. This reduces the overall simulation time to about 40 minutes. The simulation time ratio of the high to low models is about 102.

11.3.5 *Results*

The optimization algorithm stops after seven iterations when the trust-region radius is reduced below the threshold (see Fig. 11.11). The lift constraint is within 1% of the target lift coefficient (bottom-left graph). The drag coefficient is reduced from 489.6 counts to 467.8 counts (bottom-right graph), a reduction of 21.8 counts. The cross-sectional area constraints are active, but not violated.

The overall optimization cost is seven high-fidelity model evaluations (excluding evaluation of the initial design) and 6×150 low-fidelity model evaluations. The overall optimization cost is therefore equivalent to about 15 high-fidelity model evaluations.

The upper surface of the airfoil at the kink is shown in Fig. 11.12. The airfoil LE thickness is reduced, whereas the maximum thickness (around the 30% chord) and the TE thickness are increased slightly. As a result, the upper-surface shock strength is reduced considerably, especially in the outboard part of the wing (see Figs. 11.13 and 11.14).

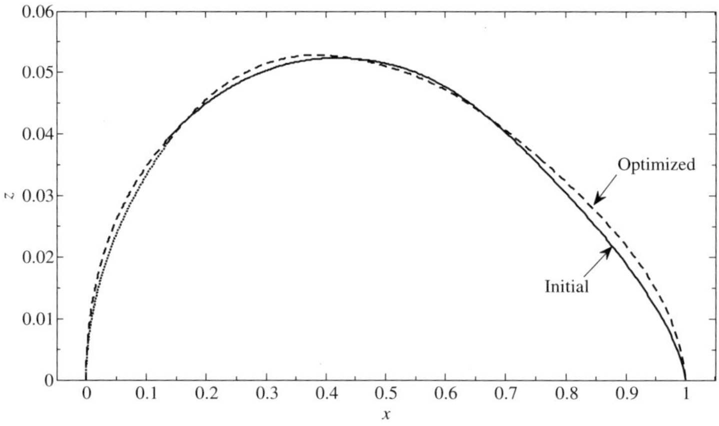

Fig. 11.12. The upper-surface shapes of the initial (solid) and optimized (dashed) airfoils.

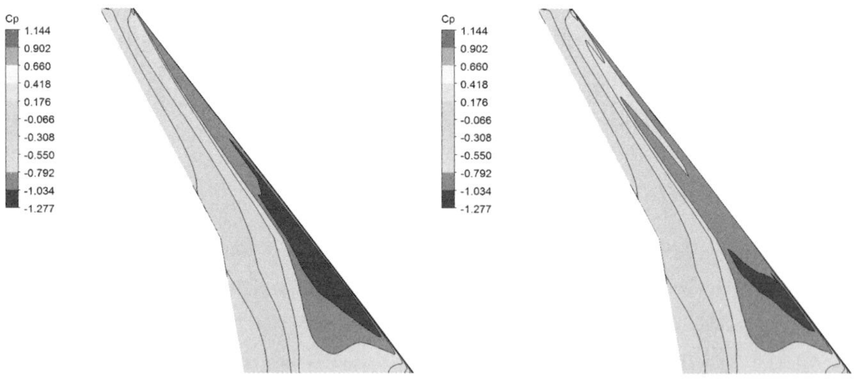

Fig. 11.13. Contours of pressure coefficient of the initial (left) and optimized (right) wings.

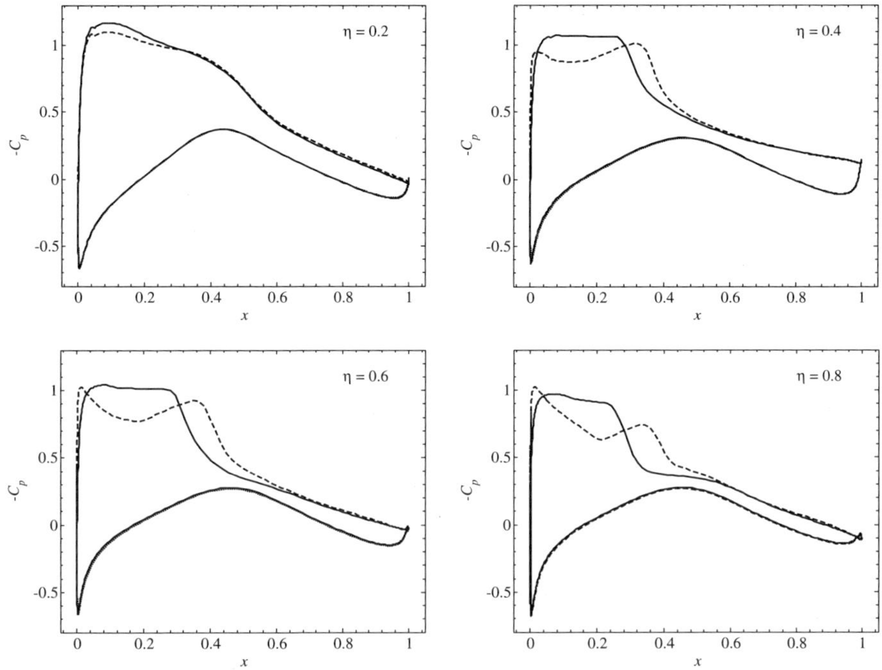

Fig. 11.14. Pressure distributions at several spanwise locations for the initial (solid) and optimized (dashed) wings.

Chapter 12

Subsonic Shape Design

Aerodynamic design can benefit from using surrogate-based optimization (SBO) whenever expensive computational tools are used in the process. The examples illustrating the advantages of physics-based SBO methods for aerodynamic shape optimization of transonic airfoils and wings were demonstrated in Chapters 10 and 11. In this chapter, we discuss the applications of SBO for subsonic shape design. More specifically, we consider the design of a low-speed wind tunnel contraction, design optimization of a trawl-door, and the hydrodynamic shape optimization of axisymmetric bodies using multiplicative response correction. In some of the considered cases, the surrogate model is constructed using an underlying low-fidelity model, which is suitably corrected to align with the high-fidelity model to be optimized. In one case (trawl-door design) a sequential approximate optimization (cf. Section 5.2) is utilized.

12.1 Design of Low-Speed Wind Tunnel Contraction

A low-speed wind tunnel, intended for fluid dynamics research, has been developed and constructed at Reykjavik University, Iceland. The utilization of the facility is twofold, involving the research on micro air vehicles (MAV), both with fixed and flapping wings, as well as fundamental research on turbulence. High flow quality is essential for both these research projects. The facility is expected to provide capability for the characterization of the aerodynamic performance of MAV at low Reynolds numbers ($Re < 100{,}000$). The required flow-speed range is 0 to 30 m/s. The work conducted in the tunnel will

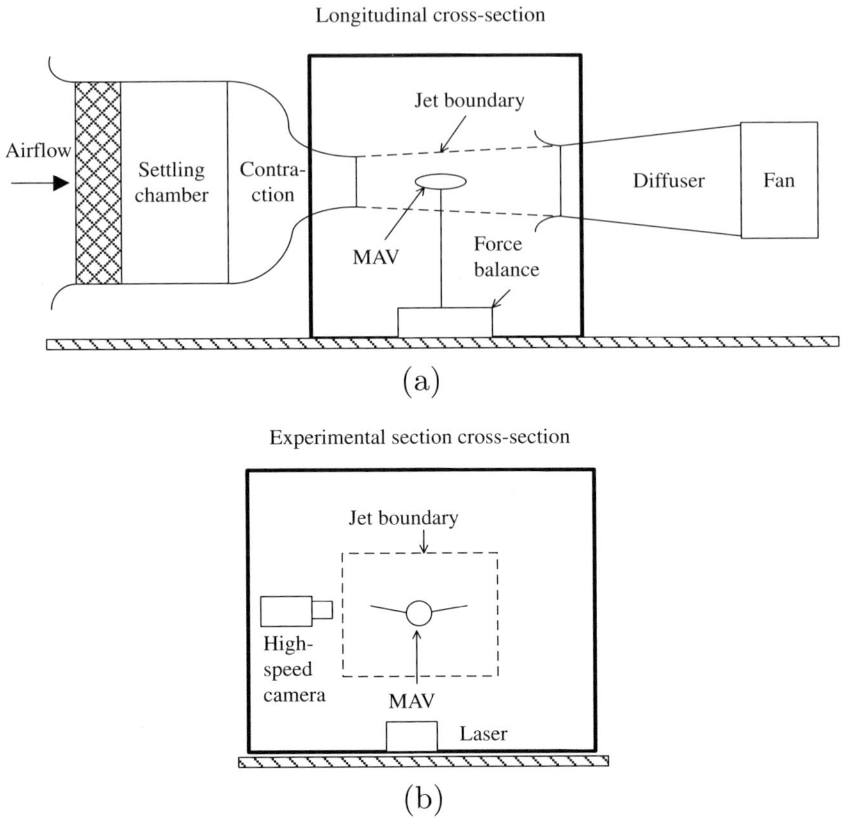

Fig. 12.1. A schematic of the wind-tunnel. (a) Longitudinal cross-section of the entire tunnel; (b) a cross-sectional view of the test section.

involve the analysis of various aspects of MAV, such as the wing shape, flapping parameters, and control effectiveness.

The wind tunnel will be of open suction type, as shown in Fig. 12.1, and it is composed of a square inlet with honeycomb and turbulence screens, settling chamber, contraction, experimental section housing, diffuser, and axial fan. The experimental section is located in a large enclosure which houses the experimental equipment (the MAV) and the instrumentation (such as a force balance and a particle image velocimetry (PIV) system).

The contraction is a critical component of the wind tunnel for the provision of high-quality test flow in the working section (Barlow *et al.*, 1999). The contraction shape controls the quality (such as flow angularity and uniformity, as well as turbulence intensity) of the flow at the outlet. As the effects of the shape on the flow are quite complicated, it is necessary to use computational fluid dynamics (CFD) to make the design decisions, such as the length, contraction ratio, and the wall shape. The process can be automated using a numerical optimization procedure; see, for example, Watmuff (1986), Su (1991), Mathew *et al.*, (2005), and Doolan and Morgans (2007).

Due to the computational expense of three-dimensional high-fidelity CFD simulations, and the fact that numerical optimization requires a large number of function calls, all of the aforementioned work uses traditional fluid-flow solvers (coupled potential and viscous fluid-flow solvers), which are computationally cheap, and traditional optimization techniques. The final design is then typically verified by using a high-fidelity flow solver.

Here we demonstrate the design of the contraction shape with the focus on improving the flow characteristics in the experimental section of the tunnel. To be able to capture the nonlinear flow physics and geometry of the problem accurately throughout the optimization process, the use of high-fidelity CFD simulation is necessary. The coupled potential and viscous flow solvers, as used in previous work (e.g. Doolan and Morgans (2007)), will not capture these effects with sufficient accuracy. As a matter of fact, high-fidelity simulations are essential. As shown in Section 12.1.4, the contraction design obtained based on low-fidelity CFD analyses does not ensure the high-quality flow.

Because of the high computational cost of high-fidelity CFD simulations, SBO plays an important role in reducing the CPU (central processing unit) time of the design process. The SBO approach utilized here exploits the high-fidelity model based on the steady Reynolds-averaged Navier–Stokes (RANS) equations and the low-fidelity model based on the same equations, but with coarse discretization and relaxed convergence criteria. We use a simple response correction technique to create a surrogate that is subsequently used as a prediction tool to yield an approximate high-fidelity model optimum. Experimental validation of the design was carried

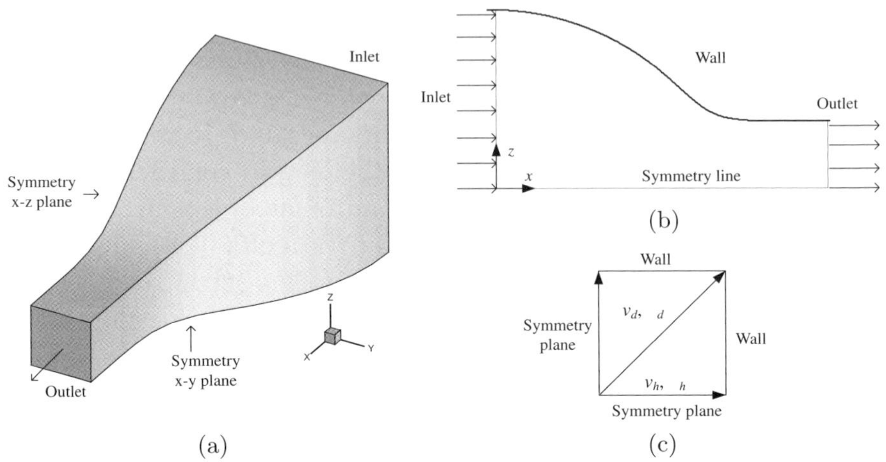

Fig. 12.2. (a) One fourth of a three-dimensional contraction shape; (b) view of the upper half of the x–z symmetry plane; (c) control directions in the outlet plane.

out on the fabricated tunnel. In particular, physical measurements of the flow quality in the experimental section have been performed and compared with our computational model.

12.1.1 *Design problem formulation*

A sketch of a typical square three-dimensional contraction shape is shown in Fig. 12.2. At the outlet, the flow is required to be as uniform as possible, and the boundary layer needs to be thin to maximize the usable cross-sectional area. The overall goal is thus to maximize the flow uniformity at the outlet in terms of speed and angularity. More specifically, we want to minimize the variation of the speed, v, and angularity, α, at the horizontal and diagonal control directions in the outlet plane (Fig. 12.2(c)). The vertical control direction is not needed due to symmetry.

The following variables control the contraction shape: the inlet size; the contraction area ratio; the length; and the wall shape. To simplify our problem, we fix the inlet size, the contraction area ratio, and length. This is necessary in this particular work, since the size of the laboratory that houses the wind tunnel is a limiting factor. Given

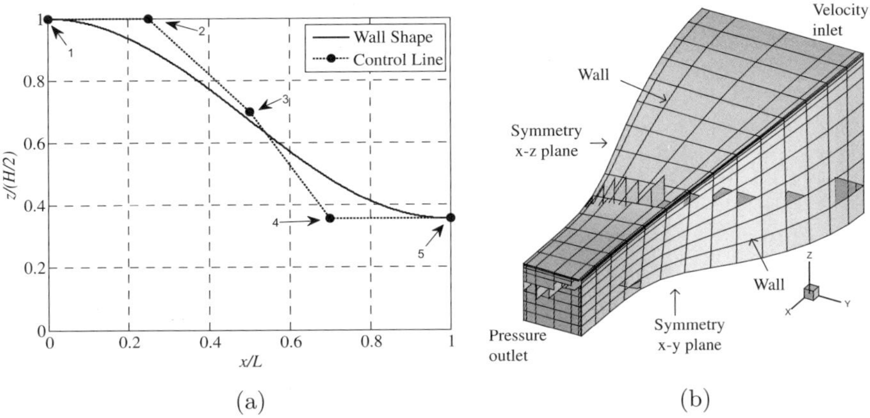

Fig. 12.3. (a) Bezier curve of the contraction wall shape (solid line) in the symmetry plane, and the control line (dashed line) with five control points (dots); (b) example computational mesh of the three-dimensional contraction.

these parameters, the design variables are the parameters controlling the wall shape.

The contraction wall shape is defined by Bezier curves on the symmetry planes. The Bezier curves, of order n, are defined as (Lepine *et al.*, 2001):

$$B(t) = \sum_{k=1}^{m} \sum_{i=0}^{n} \frac{n!}{i!(n-i)!}(1 - t(k))^{n-i} t(k)^i P(i), \tag{12.1}$$

where $P(i)$, $i = 0, 1, \ldots, n$ are the control points and t is a $1 \times m$ array with entries from 0 to 1. Here, we use five control points, as shown in Fig. 12.3(a). The first control point is fixed to the inlet height, H. Control points 2 and 4 are fixed in the horizontal direction. The third control point is free. The fifth control point is fixed to the outlet height. In total, this gives four design variables. The design variable vector is written as $\boldsymbol{x} = [x_2\ x_3\ z_3\ x_4]^T$, where x and z are the coordinates of the corresponding control points.

12.1.2 *Computational fluid dynamic model*

The flow is assumed to be steady, incompressible, and viscous. The steady RANS equations are taken to be the governing flow equations for the high-fidelity CFD model with the k-ω SST turbulence model by Menter (1994). Due to symmetry, the computational domain is only one-quarter of the contraction. The computational mesh is structured with brick elements and is generated using the computer code ICEM CFD (2013). The grid is made denser in the wall region. An example mesh is shown in Fig. 12.3(b), along with the boundary conditions. A grid-convergence study was performed to determine the appropriate mesh density. Initially, a relatively fine grid was developed and then the grid density was changed (the number of grid points halved or doubled) in several steps to observe the changes in the velocity field at the outlet.

The fine mesh, used in the high-fidelity CFD model, has 50 elements in the x-direction and 80 elements in the y- and z-directions. The total number of mesh cells is 320,000. The non-dimensional normal distance, y^+, from the wall to the first grid point is roughly 1.

The numerical simulations are performed using the commercial CFD solver FLUENT (2012). The flow solver is of implicit density-based formulation and the fluxes are calculated by an upwind-biased second-order spatially accurate Roe flux scheme. Asymptotic convergence to a steady-state solution is obtained for each case. The iterative convergence of each solution is examined by monitoring the overall residual, which is the sum (over all the cells in the computational domain) of the L^2 norm of all the governing equations solved in each cell. The convergence criteria used in this work for the high-fidelity CFD model is the one that occurs first of either a reduction in the residuals by six orders of magnitude or a maximum number of iterations of 1,000.

The low-fidelity model is based on the same CFD model as the high-fidelity one. However, as the low-fidelity model will be used in place of the high-fidelity model in the optimization process, it needs to be faster in order to reduce the computational cost of the design process. The simulation time can be substantially reduced by making the grid coarser. Here we used the medium mesh obtained from the grid convergence study. This mesh has 20 elements in the x-direction and 40 elements in the y- and z-directions, with a total number of mesh cells of 32,000. The distance to the first nodal point was kept fixed. The simulation time is reduced further by reducing the number of iterations. We, therefore, relax the convergence criteria for the low-fidelity model by setting the maximum number of iterations to 15. The ratio of simulation time of the high-fidelity model to the low-fidelity model is around 250.

12.1.3 *Design optimization methodology*

In this section, we describe our approach to optimizing the contraction shape. The design objective is to obtain as uniform a velocity field, v, as possible, where v is a function of four shape parameters, $\boldsymbol{x} = [x_2\ x_3\ z_3\ x_4]^T$. In the following, we describe the formulation of the objective function, the SBO approach that is adopted here to design the contraction shape, as well as the construction of the surrogate model, which is based on the low-fidelity CFD simulations.

We are interested in minimizing the variation of the velocity magnitude, i.e. the speed, as well as the velocity angle for the two control directions, horizontal and diagonal, denoted as $v_{h.f}(x,t)$, $\alpha_{h.f}(x,t), v_{d.f}(x,t)$, $\alpha_{d.f}(x,t)$, with $0 \leq t \leq 1$ (0 corresponds to the contraction center, 1 corresponds to the wall). The subscript "f" indicates that the models are associated with high-fidelity CFD simulations. For the horizontal direction we are interested in minimizing $v_{h.f}(x,t)$ and $\alpha_{h.f}(x,t)$ for $0 \leq t \leq 0.9$, for the diagonal direction we want to minimize $v_{d.f}(x,t)$ and $\alpha_{d.f}(x,t)$ for $0 \leq t \leq 0.85$. Because our two objectives (velocity magnitude and angle) are noncommensurable, only one of them (angle) is treated directly; the other one (magnitude) is handled through penalty function. The objective function is therefore defined as:

$$\begin{aligned} f(\boldsymbol{x}) = {} & \max\{\Delta\alpha_{h.f}(\boldsymbol{x}), \Delta\alpha_{d.f}(\boldsymbol{x})\} \\ & + \beta[(\max\{\Delta v_{h.f}(\boldsymbol{x})/v_{h.f.\max}, \Delta v_{d.f}(\boldsymbol{x})/v_{d.f.\max}\} \\ & - \Delta v_{\max})/\Delta v_{\max}]^2, \end{aligned} \tag{12.2}$$

where $\Delta v_{h.f}(\boldsymbol{x}) = \max\{0 \leq t \leq 0.9\colon |v_{h.f}(\boldsymbol{x},t)|\} - \min\{0 \leq t \leq 0.9\colon |v_{h.f}(\boldsymbol{x},t)|\}$, $\Delta v_{d.f}(\boldsymbol{x}) = \max\{0 \leq t \leq 0.85\colon |v_{d.f}(\boldsymbol{x},t)|\} - min\{0 \leq t \leq 0.85\colon |v_{d.f}(\boldsymbol{x},t)|\}$, $v_{h.f.\max} = \max\{0 \leq t \leq 0.9\colon |v_{h.f}(\boldsymbol{x},t)|\}$, and $v_{d.f.\max} = \max\{0 \leq t \leq 0.85\colon |v_{d.f}(\boldsymbol{x},t)|\}$. The definitions of

$\Delta\alpha_{h.f}(\boldsymbol{x})$ and $\Delta\alpha_{d.f}(\boldsymbol{x})$ are similar. In our numerical experiments we use $\beta = 1{,}000$. $\Delta v_{\max}$ is the maximum allowable variation of the velocity magnitude. Here we set $\Delta v_{\max} = 0.01$.

In order to reduce the design cost, we exploit SBO principles (Bandler *et al.*, 2004a; Queipo *et al.*, 2005). More specifically, the direct minimization of the high-fidelity objective function (Equation (12.2)) is replaced by an iterative procedure (Koziel *et al.*, 2011b) of the form $\boldsymbol{x}^{(i+1)} = \text{argmin}\{\boldsymbol{x}\colon s^{(i)}(\boldsymbol{x})\}$ (cf. Equation (4.2)), which generates a sequence of approximate solutions, $\boldsymbol{x}^{(i)}$, $i = 0, 1, \ldots$, to the original problem, $\text{argmin}\{\boldsymbol{x}\colon H_f(\boldsymbol{x})\}$. Each $\boldsymbol{x}^{(i+1)}$ is the optimal design of the surrogate model $s^{(i)}$, $i = 0, 1, \ldots$, which is assumed to be a computationally cheap and sufficiently reliable representation of f, particularly in the neighborhood of $\boldsymbol{x}^{(i)}$.

The surrogate model $s^{(i)}$ is defined as:

$$s^{(i)}(\boldsymbol{x}) = \max\{\Delta\alpha_{h.s}^{(i)}(\boldsymbol{x}), \Delta\alpha_{d.s}^{(i)}(\boldsymbol{x})\} + \beta[(\max\{\Delta v_{h.s}^{(i)}(\boldsymbol{x}), \Delta v_{d.s}^{(i)}(\boldsymbol{x})\} - dv_{\max})/dv_{\max}]^2, \quad (12.3)$$

where the definitions of $\Delta v_{h.s}^{(i)}(\boldsymbol{x})$, $\Delta v_{d.s}^{(i)}(\boldsymbol{x})$, $\Delta\alpha_{h.f}^{(i)}(\boldsymbol{x})$, and $\Delta\alpha_{d.f}^{(i)}(\boldsymbol{x})$ are based on the surrogate functions $v_{h.s}^{(i)}(\boldsymbol{x}, t)$, $v_{d.s}^{(i)}(\boldsymbol{x}, t)$, $\alpha_{h.s}^{(i)}(\boldsymbol{x}, t)$, and $\alpha_{d.s}^{(i)}(\boldsymbol{x}, t)$, and analogous to that for the high-fidelity model. Here the surrogate functions are obtained from low-fidelity CFD simulations. At iteration i, the surrogate, $v_{h.s}^{(i)}(\boldsymbol{x}, t)$, of the velocity magnitude, $v_{h.f}(\boldsymbol{x}, t)$, is constructed using the multiplicative response correction of the form:

$$v_{h.s}^{(i)}(\boldsymbol{x}, t) = v_{h.c}(\boldsymbol{x}, t)\frac{v_{h.f}(\boldsymbol{x}^{(i)}, t)}{v_{h.c}(\boldsymbol{x}^{(i)}, t)}, \quad (12.4)$$

where $v_{h.c}(\boldsymbol{x},t)$ is the low-fidelity velocity magnitude function. Similar definitions hold for the remaining surrogate functions, $v_{d.s}^{(i)}(\boldsymbol{x}, t)$, $\alpha_{h.s}^{(i)}(\boldsymbol{x}, t)$, and $\alpha_{d.s}^{(i)}(\boldsymbol{x}, t)$. Note that Equation (12.4) ensures zero-order consistency (Alexandrov and Lewis, 2001) between the surrogate and the high-fidelity model, i.e. $v_{h.s}^{(i)}(\boldsymbol{x}^{(i)}, t) = v_{h.f}(\boldsymbol{x}^{(i)}, t)$. While — in principle — this is not sufficient to ensure the convergence of Equation (5.2) to an optimum of the high-fidelity model, as demonstrated in Section 12.1.4, the algorithm performs very well. The primary reason is that the low-fidelity model is physically based, so the changes of $v_{h.f}(\boldsymbol{x}, t)$ and $v_{h.c}(\boldsymbol{x}, t)$ are correlated under the change of the shape parameters, $\boldsymbol{x}$. Figure 12.4 illustrates the surrogate model construction.

12.1.4 *Numerical results*

The wind tunnel contraction shape has been optimized using the methodology described in Section 12.1.3. The contraction length was set to 2 m, the inlet width was set to 2 m, whereas the contraction area ratio is 8:1. For comparison and verification, we carried out four optimization runs, using: (1) direct optimization of the low-fidelity model; (2) direct optimization of the high-fidelity model; (3) SBO using the methodology of Section 12.1.3; and (4) SBO using the methodology of Section 12.1.3 embedded in the trust-region framework (Conn *et al.*, 2000). In all cases, the optimization engine was a pattern-search algorithm (Koziel, 2010a). The numerical results are presented in Table 12.1.

The results show that the shape obtained by optimizing the low-fidelity model does not give much of an improvement in terms of reducing the velocity angle variation. As shown in Fig. 12.5, the corresponding contraction shape is actually quite similar to the initial one. This indicates that simply replacing the high-fidelity model with the low-fidelity one in the design process is not a good option. Direct optimization of the high-fidelity model substantially reduces $\Delta\alpha$ (from 4.6 to 1.6 degrees); however, the design cost is quite substantial: 134 high-fidelity model evaluations. The SBO enhanced by the trust-region framework yields the design with a slightly worse reduction of $\Delta\alpha$ (2.0 degree); however, the design cost is much lower: only 13 equivalent high-fidelity model evaluations.

The SBO working without trust region yields a much better design with respect to the velocity angle variation: $\Delta\alpha$ of only 0.7 degrees and at a very low cost of less than six equivalent high-fidelity model evaluations. However, the Δv constraint is slightly violated in this case, which is a result of not using the trust-region approach

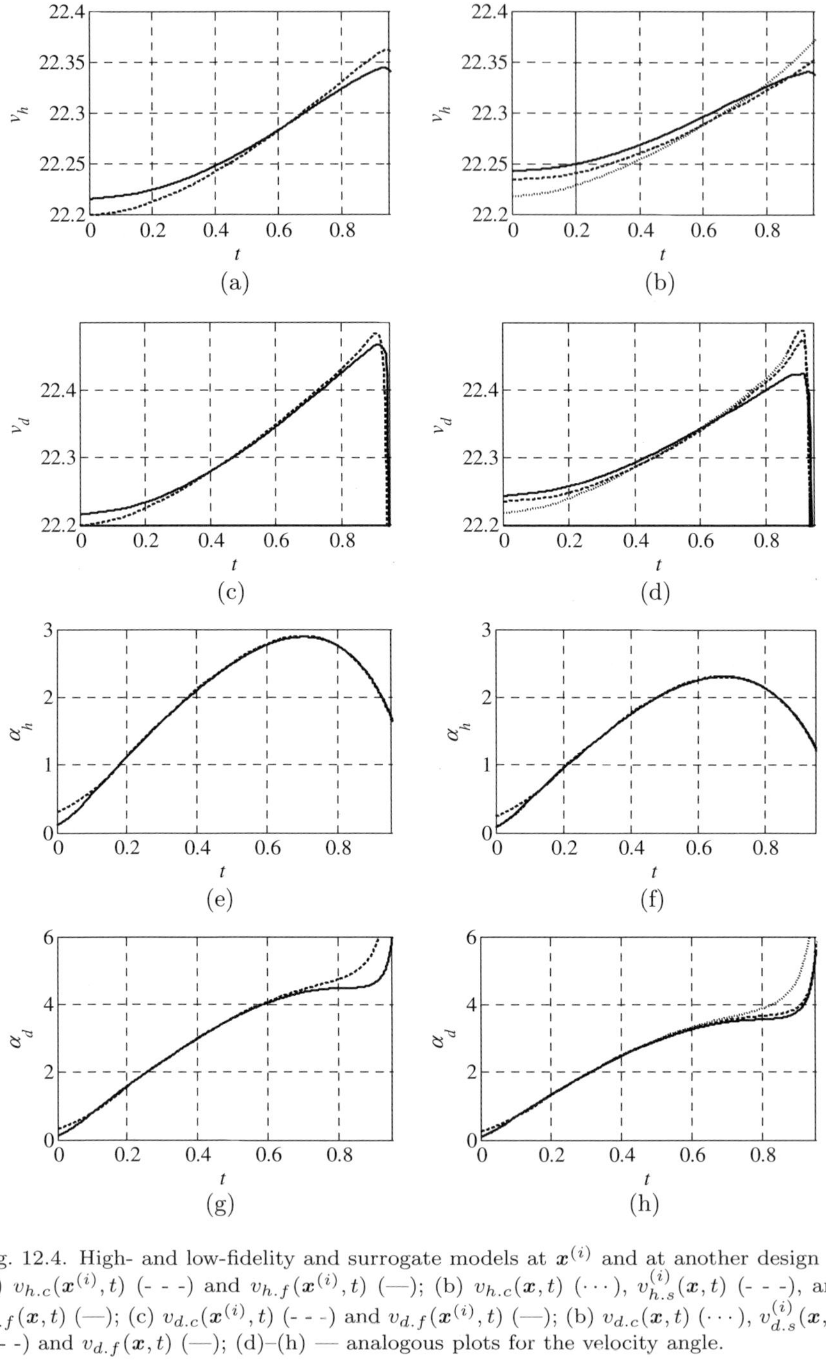

Fig. 12.4. High- and low-fidelity and surrogate models at $\boldsymbol{x}^{(i)}$ and at another design $\boldsymbol{x}$. (a) $v_{h.c}(\boldsymbol{x}^{(i)}, t)$ (- - -) and $v_{h.f}(\boldsymbol{x}^{(i)}, t)$ (—); (b) $v_{h.c}(\boldsymbol{x}, t)$ (· · ·), $v_{h.s}^{(i)}(\boldsymbol{x}, t)$ (- - -), and $v_{h.f}(\boldsymbol{x}, t)$ (—); (c) $v_{d.c}(\boldsymbol{x}^{(i)}, t)$ (- - -) and $v_{d.f}(\boldsymbol{x}^{(i)}, t)$ (—); (b) $v_{d.c}(\boldsymbol{x}, t)$ (· · ·), $v_{d.s}^{(i)}(\boldsymbol{x}, t)$ (- - -) and $v_{d.f}(\boldsymbol{x}, t)$ (—); (d)–(h) — analogous plots for the velocity angle.

Table 12.1 Wind tunnel contraction shape design: numerical results.

Variable	Initial	Case 1*	Case 2$	Case 3%	Case 4&
x_1	0.25	0.3333	0.5000	0.3925	0.4977
x_2	0.50	0.6333	0.5038	0.6236	0.6840
x_3	0.70	0.6667	0.5556	0.5545	0.2691
x_4	0.70	0.6327	0.5025	0.5262	0.5305
$\Delta\alpha$ [deg]	4.62	3.8	1.6	2.0	0.7
Δv [%]	1.1	1.0	1.0	1.0	1.2
N_f	N/A	0	134	9	4
N_c	N/A	120	0	900	400
Total cost#	N/A	<0.5	134	<13	<6

*Design obtained by direct low-fidelity model optimization using the pattern-search algorithm.
$Design obtained by direct high-fidelity model optimization using the pattern-search algorithm.
%Design obtained by SBO with trust region; surrogate model optimization performed using the pattern-search algorithm.
&Same as Case 3, but without the trust region framework.
#The total optimization cost is expressed in terms of the equivalent number of high-fidelity model evaluations. The ratio of the high-fidelity model evaluation time to the corrected low-fidelity model evaluation time is around 250.

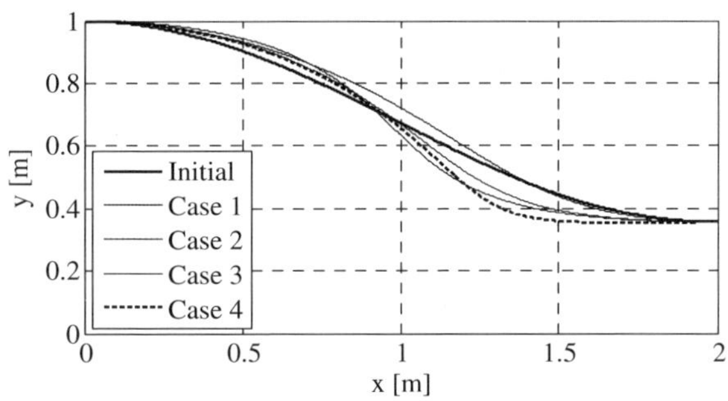

Fig. 12.5. Optimized nozzle shape obtained using four different design approaches: initial shape (—); the shape resulting from direct optimization of the low-fidelity model (– – –); the shape resulting from the direct optimization of the high-fidelity model (– · – ·); the shape obtained by SBO within trust-region framework (· · · ·); the shape obtained by SBO (no trust region) (- - - -).

(trust-region enforces satisfaction of Δv constraint in Case 3). The contraction shapes obtained in Cases 2 through 4 are rather similar to each other. Note that the lowest value of $\Delta\alpha$ corresponds to the flattest design of the nozzle outlet (cf. Fig. 12.6).

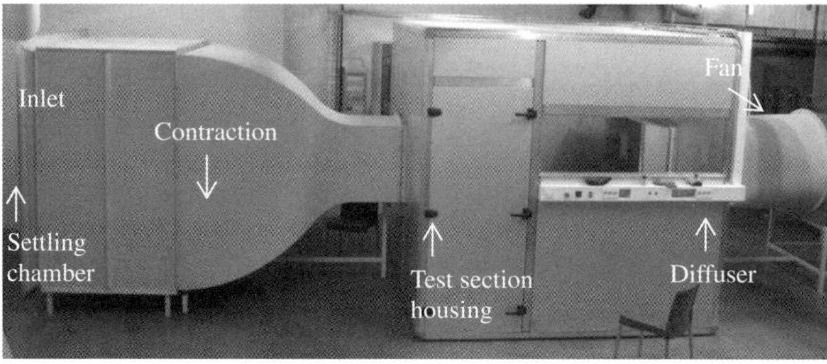

(a)

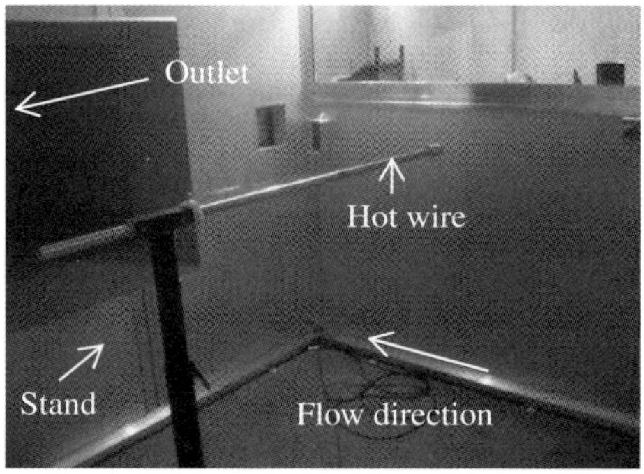

(b)

Fig. 12.6. (a) A side view of the low-speed wind tunnel showing (from the left): the inlet (includes the bell mouth and turbulence screens), settling chamber, contraction, experimental section, diffuser, and axial fan. The overall length is 8 m. (b) A view inside the test-section housing showing the single-component constant-temperature hot-wire probe on a stand.

12.1.5 *Experimental validation*

A single-component constant-temperature hot-wire anemometer system was used to investigate the flow uniformity, as well as the flow speed fluctuating levels, in the test section. The sensing element of the hot-wire probe was made from tungsten and had a diameter of 5 microns and length of nominally 1.2 mm. The sensing element was connected to the MiniCTA bridge. The output of the bridge was then digitized with a National Instruments data acquisition card with a

16-bit resolution. The sampling time was 6.5 seconds at 10 kHz. For the low order statistics presented here, we estimate the uncertainty in the hot-wire measurements to be less than 1%.

Calibration of the hot-wire probe was accomplished by placing a pitot-static probe in close proximity to the hot-wire and collecting data for a series of wind-tunnel settings that extended over the full range of the VFD (variable frequency drive), i.e. 0–50 Hz. For each of the free-stream velocity settings, mean values of both the hot-wire voltage and speed (calculated from the pitot-static probe) were determined, and a curve fit was used to relate the two. The error of the calibration curve fit of the mean velocities was less than 1% for all velocities.

In the test section, the hot-wire probe was placed on a stand facing the flow coming from the contraction exit, as shown in Fig. 12.6(b). By traversing the probe, measurements were taken in one cross-sectional plane, located 60 cm from the contraction exit, which is one-third of the test section length. The measurements were taken at nine equally spaced locations, separated by 10 cm in each direction, yielding a total of 81 measurements. The cross-sectional area in which the measurements were taken was 80 cm by 80 cm and extended well outside the contraction exit area (71 cm by 71 cm).

Contour maps of the mean velocity field and the turbulence intensity, calculated from the hot-wire probe measurements, are shown in Fig. 12.7 for three different free-stream velocity settings. The VFD was set to 10, 20, and 30 Hz, which yielded mean free-stream speeds of 5.4, 11.1, and 16.5 m/s (according to the calibration curve), respectively.

The mean velocity field contours indicate that the flow is uniform within the region of interest, the 85% area of the contraction exit, and the flow speed within the cross-section matches well with the free-stream velocity. However, there is a slight variation at the lowest free-stream velocity, where the speed is reduced to 5.3 m/s in the upper-left-hand corner (Fig. 12.7(a)). This variation of the flow speed is not within the measurement uncertainty. The authors believe that the reason for this variation is due to unsymmetrical inflow at the inlet caused by objects located in the experimental room being

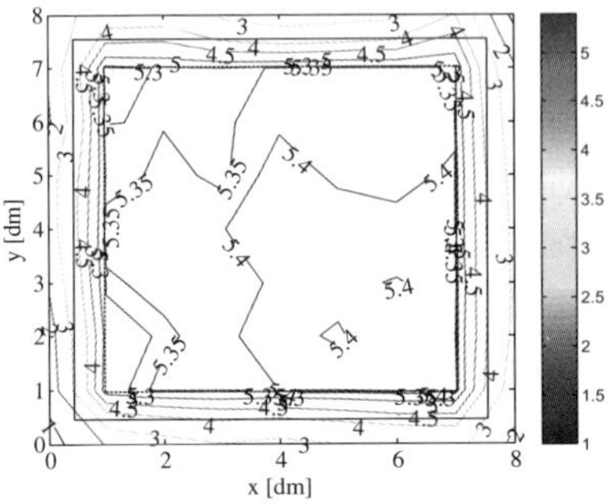

(a) Mean flow speed contours at $V_\infty = 5.4$ m/s

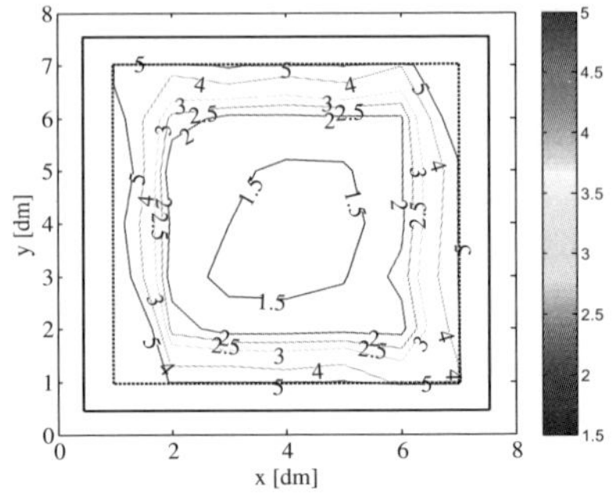

(b) Turbulence intensity contours at $V_\infty = 5.4$ m/s

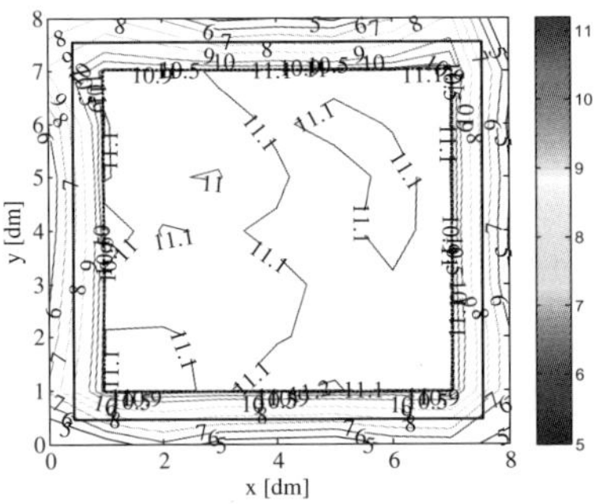

(c) Mean flow speed contours at $V_\infty = 11.1$ m/s

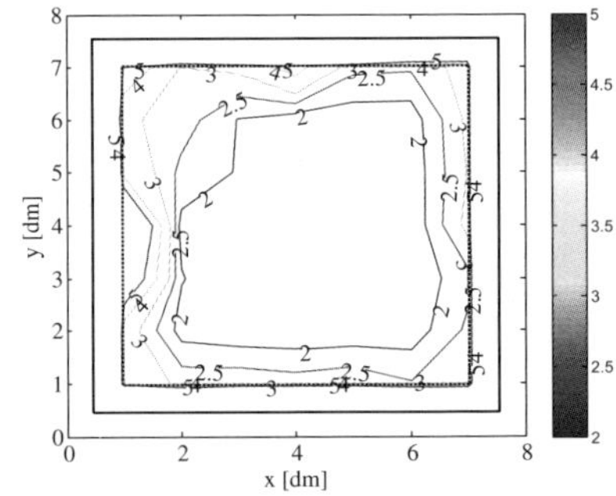

(d) Turbulence intensity contours at $V_\infty = 11.1$ m/s

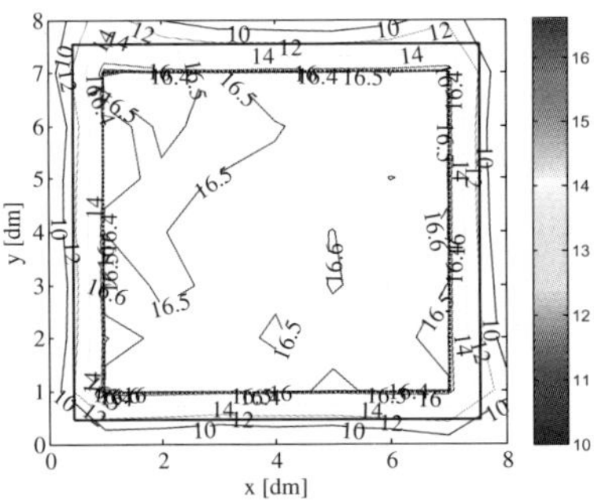

(e) Mean flow speed contours at $V_\infty = 16.5$ m/s

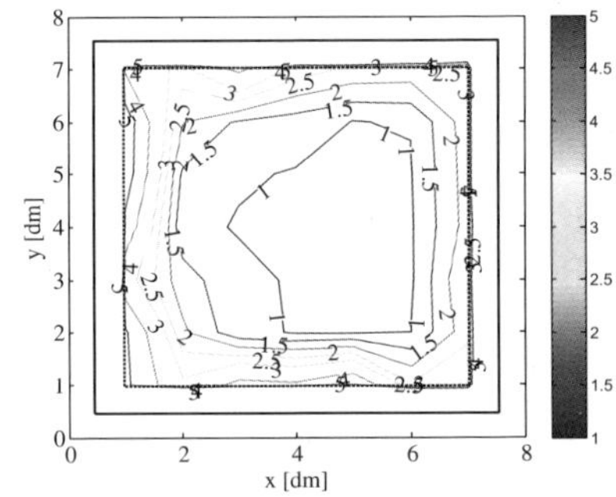

(f) Turbulence intensity contours at $V_\infty = 16.5$ m/s

Fig. 12.7. Flow uniformity and fluctuation levels at 33% of the test section length from the contraction exit. Data are shown for motor settings of 10, 20, and 30 Hz, or mean free-stream speeds (V_∞) of 5.4, 11.1, and 16.5 m/s, respectively. Flow speed is given in m/s and turbulence intensity in %. The area of interest in the design optimization study is shown here with dashed lines (85% value of the contraction exit size). The contraction exit boundary is shown with solid lines.

located too close to the upper-left corner of the inlet. This will be rectified by clearing these objects.

The turbulence intensity contours indicate that the flow unsteadiness is rather high, ranging from 1 to 5% within the region of interest. A desired level of turbulence intensity is less than 1% in the whole cross-section. The turbulence intensity levels are controlled by the turbulence screens; with each new screen the level can be halved (Barlow *et al.*, 1999). Currently, the wind tunnel has three turbulence screens; therefore, three additional screens are needed to reach the desired intensity levels.

12.2 Design Optimization of Trawl-Doors using Local Surrogate Models

Trawling gear is a major contributor to the fuel expenditure of many fishing vessels. As most of the fuel is spent during trawling (often over 60%), there is a great incentive to reduce the drag of the fishing gear. The main parts of a fishing gear assembly are the net, a pair of trawl-doors, and a cable extending from the trawl-doors to the boat and the net (Fig. 12.8(a)). The role of the trawl-doors is to keep the net open during the trawling operation. Typically, their span is 6–8 m with a chord of 2–3 m, while the cables are a few hundred meters long and the net tens of meters. Figure 12.8(b) shows a typical trawl-door. Although the trawl-doors are small compared to the other fishing-gear components, they may be responsible for roughly 10–30% of the total drag of the entire assembly (Garner, 1967). During trawling, the drag of the fishing gear is much greater than the drag of the ship. Therefore, significant fuel consumption savings may be possible by careful design and optimization of trawl-doors.

Trawl-doors function much like aircraft wings, i.e. they maintain certain lift for a given operating condition. Most trawl-doors developed over the years are fundamentally the same and only minor design changes have been made during that time. The reason for this is that their designs are based on time-consuming and expensive physical experiments in tow or flume tanks. Although CFD is widely used in the design of a variety of engineering devices, such as aircraft,

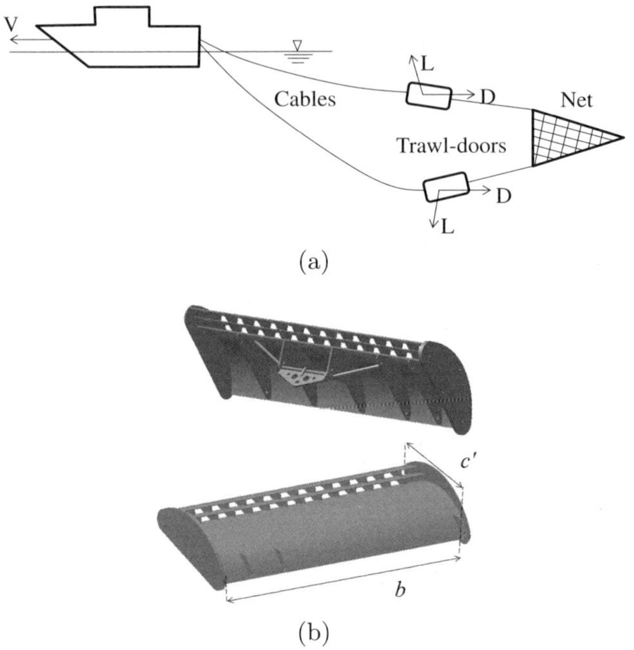

Fig. 12.8. A sketch of a fishing vessel. (a) The main parts of the fishing gear (not drawn to scale); (b) a typical trawl-door with two slats.

ships, and cars, very few applications are reported for trawl-doors in the literature (Haraldsson, 1996).

In this section, we illustrate the use of sequential approximate optimization to CFD-based design of trawl-door shapes. The methodology exploits an iterative scheme with local response surface approximation (RSA) models of the expensive CFD trawl-door model constructed in each iteration (Koziel and Ogurtsov, 2012). The RSA models are constructed using CFD data sparsely sampled in the vicinity of the current design and a low-order polynomial approximation (to reduce the influence of the CFD model numerical noise on the optimization, as well as minimizing the number of required data samples). The size of the vicinity (i.e. the RSA model domain) is automatically adjusted in each iteration based on the performance of the model; more specifically, on the quality of the prediction given by the model compared to the actual changes of the trawl-door performance metrics verified by the CFD simulation. The adjustment scheme is governed by a conventional trust-region framework (Conn *et al.*, 2000).

12.2.1 *CFD modeling*

The flow is assumed to be steady, incompressible, and viscous. The RANS equations are taken to be the governing equations with Menter's k-ω SST turbulence model (see, for example, c *et al.* (1997)).

The farfield is configured in a box-topology in which the trawl-door geometry is placed in the center of the box. The main element leading edge (LE) is placed as the origin, with the farfield extending 100 chord lengths away from the origin in every direction. A sketch of the setup for two dimensions is shown in Fig. 12.9. The grid is an unstructured triangular grid in which the elements are clustered around the trawl-door geometry, growing in size as they move away from the origin. The maximum element size on the geometry is set to 0.1% of the chord length. The maximum element size in domain

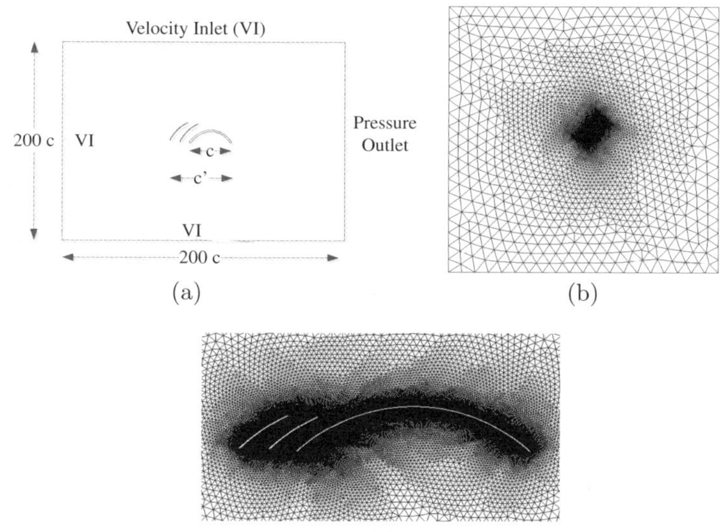

Fig. 12.9. Two-dimensional CFD model. (a) The solution domain and boundary conditions, and an example two-dimensional computational mesh; (b) the farfield; (c) a close-up of the trawl-door surface.

away from the trawl-door is 10 times the chord length. In order to capture the viscous boundary layer well, a prismatic inflation layer is extruded from all surfaces. The initial layer height is chosen so that $y^+ < 1$. The mesh is generated with ICEM CFD (2013). Example mesh is shown in Fig. 12.9.

The flow solver is coupled velocity–pressure-based formulation. The spatial discretization schemes are second order for all flow variables and the gradient information is found using the Green–Gauss node-based method. Additionally, due to the difficult flow condition at high angles of attack, the pseudo-transient option and high-order relaxation terms are used in order to get a stable converged solution. The residuals, which are the sum of the L^2 norms of all governing equations in each cell, are monitored and checked for convergence. The convergence criterion is such that a solution is considered to be converged if the residuals have dropped by six orders of magnitude, or the total number of iterations has reached 1,000. Numerical fluid flow simulations are performed using the commercial computer code FLUENT (2012).

12.2.2 *Design methodology*

The design problem considered in this work can be formulated as a nonlinear minimization problem of the form:

$$\boldsymbol{x}_f^* = \arg\min_{\boldsymbol{x}} U(f(\boldsymbol{x})), \tag{12.5}$$

where $f(\boldsymbol{x})$ is the function representing performance parameters of the trawl-door under design (specifically, the lift and drag coefficients), whereas $\boldsymbol{x}$ is the vector of adjustable geometry parameters; U is a given objective function. For example, if the goal is to improve the trawl-door efficiency, the objective function will be formulated as the ratio of the drag and lift coefficients (so that the minimization of that ratio corresponds to maximization of the efficiency).

The methodology used for trawl-door optimization exploits local RSA models. The procedure is iterative. In each iteration, a local model is constructed using sparse samples of f data and low-order polynomial approximation.

Let $\boldsymbol{x}^{(j)} = [x_1^{(j)} x_2^{(j)} \ldots x_n^{(j)}]^T$ be a design obtained as a result of iteration $j-1$ of the algorithm. Let $\boldsymbol{d}^{(j)} = [d_1^{(j)} d_2^{(j)} \ldots d_n^{(j)}]^T$ be the size parameter that is used to define the vicinity of the vector $\boldsymbol{x}^{(j)}$. The local RSA model is created in the interval $[\boldsymbol{x}^{(j.i)} - \boldsymbol{d}^{(j)}, \boldsymbol{x}^{(j.i)} + \boldsymbol{d}^{(j)}]$. We denote by $X_T^{(j)} = \{\boldsymbol{x}_t^{(j.1)}, \ldots, \boldsymbol{x}_t^{(j.N)}\}$ the training set obtained by sampling the aforementioned vicinity. The RSA model is obtained by approximating the data pairs $\{\boldsymbol{x}_t^{(j.k)}, f(\boldsymbol{x}_t^{(j.k)})\}$, $k = 1, \ldots, N$. In this work, a second-order polynomial model, $q^{(j)}$, is utilized as follows:

$$q^{(j)}(\boldsymbol{x}) = \lambda_0^{(j)} + \sum_{k=1}^{n} \lambda_k^{(j)} x_k + \sum_{k=1}^{n} \sum_{s=k}^{n} \lambda_{k.s}^{(j)} x_k \, x_s. \tag{12.6}$$

The unknown coefficients $\boldsymbol{\lambda} = [\lambda_0 \lambda_1 \ldots \lambda_n \lambda_{11} \lambda_{12} \ldots \lambda_{1n} \lambda_{22} \ldots \lambda_{nn}]$ are found by solving the following the linear regression problems: $q(\boldsymbol{x}_t^{(j.k)}) = f(\boldsymbol{x}_t^{(j.k)})\}$, $k = 1, \ldots, N$. The unique solution to this problem exists and can be found analytically, assuming that the number of training points is equal to or larger than the number of unknown coefficients.

The optimization algorithm workflow is the following (here, $\boldsymbol{x}^{(0)}$ is the initial design and $\boldsymbol{d}^{(0)}$ is the initial vicinity size, usually a fraction of the design space size).

1. Set $j = 0$.
2. Sample the interval $[\boldsymbol{x}^{(j.i)} - \boldsymbol{d}^{(j)}, \boldsymbol{x}^{(j.i)} + \boldsymbol{d}^{(j)}]$ to obtain the training set, $X_T^{(j)}$.
3. Evaluate the function f at $X_T^{(j)}$.
4. Identify the RSA model, $q^{(j)}$.
5. Find a candidate design $\boldsymbol{x}^{tmp}$ as $\boldsymbol{x}^{tmp} = \arg \min_{\boldsymbol{x}^{(j)} - \boldsymbol{d}^{(j)} \le \boldsymbol{x} \le \boldsymbol{x}^{(j)} + \boldsymbol{d}^{(j)}} U(q^{(j)}(\boldsymbol{x}))$.
6. Calculate the gain ratio, $r = \frac{U(f(\boldsymbol{x}^{tmp})) - U(f(\boldsymbol{x}^{(j)}))}{U(q^{(j)}(\boldsymbol{x}^{tmp})) - U(q^{(j)}(\boldsymbol{x}^{(j)}))}$.
7. If $r > r_{incr}$, set $\boldsymbol{d}^{(j+1)} = \boldsymbol{d}^{(j)} \cdot m_{incr}$.
8. If $r < r_{decr}$, set $\boldsymbol{d}^{(j+1)} = \boldsymbol{d}^{(j)} / m_{decr}$.
9. If $r > 0$, set $\boldsymbol{x}^{(j+1)} = \boldsymbol{x}^{tmp}$; otherwise $\boldsymbol{x}^{(j+1)} = \boldsymbol{x}^{(j)}$.

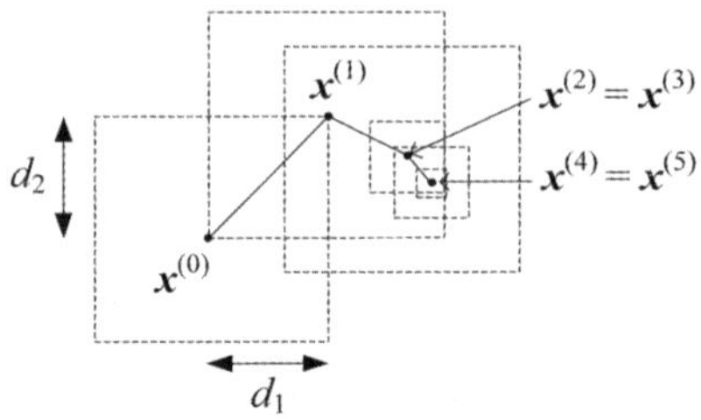

Fig. 12.10. A conceptual illustration of the design optimization procedure ($n = 2$).

10. If the termination condition is satisfied, END; or else $j = j + 1$ and go to Step 2.

In the above procedure, the updating parameters for the trust region size, i.e. r_{incr}, m_{incr}, r_{decr}, and m_{decr}, are set by the user. In this work, we set $r_{incr} = 0.75$, $m_{incr} = 1.5$, $r_{decr} = 0.25$, and $m_{decr} = 2$. The algorithm is terminated if either $\|\boldsymbol{x}^{(j+1)} - \boldsymbol{x}^{(j)}\|$ or $\|\boldsymbol{d}^{(j)}\|$ are smaller than a user prescribed threshold. Figure 12.10 illustrates the operation of the optimization procedure for $n = 2$.

12.2.3 *Design example*

The performance and operation of our optimization procedure are demonstrated through examples of a trawl-door shape design using both two- and three-dimensional CFD models. The trawl-door to be optimized has fixed-element shapes (a slat and a main element). The optimizer improves the performance by adjusting location and orientation of the slat, as well as the angle of attack of the entire assembly. The optimization engine is the algorithm described in Section 12.2.3. Data sampling is realized using the Latin hypercube sampling scheme (Beachkofski and Grandhi, 2002). The local response surface approximation is implemented with a second-order polynomial, and an off-the-shelf gradient-based routine is employed to find the candidate designs (cf. Step 5 of the algorithm). The main element is set as the NACA (National Advisory Committee for Aeronautics) 2412 profile and the slat as the NACA 3210 profile. The chord lengths of both elements are fixed, with the chord length of the slat, c_s, being 20% of

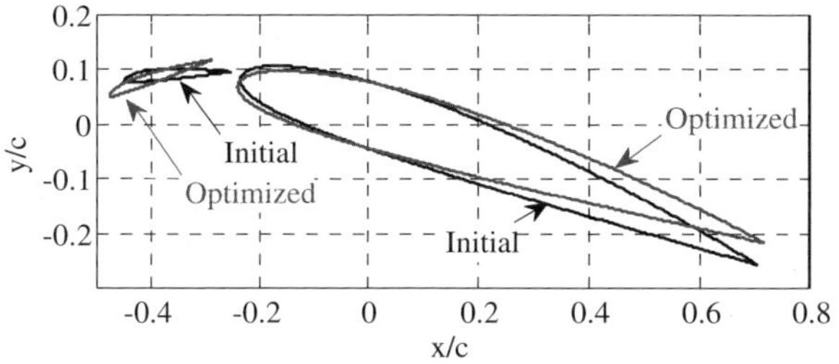

Fig. 12.11. Initial and optimized trawl-door shapes. The flow is parallel to the x/c-axis.

the main element chord length, c. The operating speed is $V_\infty = 2\,\mathrm{m/s}$ (which is a typical speed during trawling).

In our numerical study, the objective is to minimize the section drag coefficient, C_d, subject to a section lift constraint, $C_l \geq 1.2$. The design variables are the x/c and y/c locations of the slat LE with respect to the main element LE, the angle incidence of the slat, θ, with respect to the main element chord line, and the angle of attack, α, of the assembly. The initial search domain is set at $-0.35 \leq x/c \leq -0.17$, $-0.08 \leq y/c \leq 0.07$, $10 \leq \theta \leq 40$ degrees, $5 \leq \alpha \leq 35$ degrees. The initial design is $x/c = -0.2$, $y/c = -0.08$, $\theta = 25$ degrees, and $\alpha = 20$ degrees. At this design, $C_l = 1.61$ (thus, the lift constraint is valid initially) and $C_d = 0.435$.

The initial and optimized trawl-door shapes are shown in Fig. 12.11. The optimized design has $x/c = -0.219$, $y/c = -0.0871$, $\theta = 36.2$ degrees, and $\alpha = 16.8$ degrees, with $C_l = 2.02$ and $C_d = 0.0303$. The optimizer has moved the slat down (by $0.0071c$) and away (by $0.019c$) from the main element, as well as increasing its angle of incidence (by 11.2 degrees). The angle of attack is reduced by 3.2 degrees. This result represents a significant performance improvement: the lift-to-drag ratio increases from 3.7 to 66.7. The lift and drag characteristics for the optimized design are shown in Fig. 12.12. Stalling occurs abruptly, with a stall angle of attack around 20 degrees and maximum lift coefficient around 2.2. An analysis of the optimized shape using the three-dimensional model is also shown in Fig. 12.12. As expected, the three-dimensional model has a significantly lower lift coefficient, although the overall dependence of

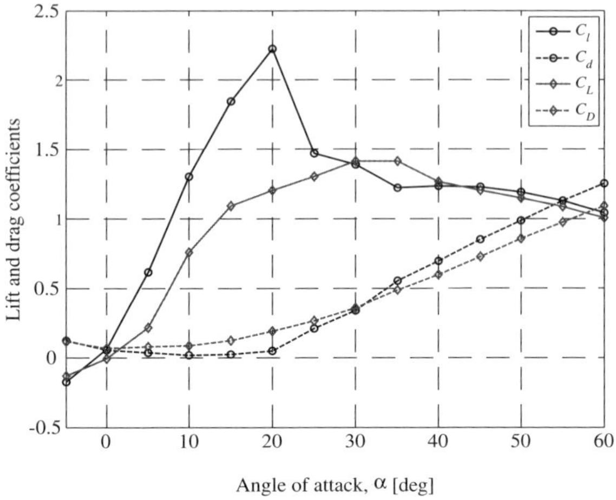

Fig. 12.12. Lift and drag characteristics for the optimized trawl-door design obtained by the two-dimensional (C_l, C_d) and the three-dimensional (C_L, C_D) CFD models.

both lift and drag coefficients as a function of the angle of attack is similar for both two- and three-dimensional models. This indicates that the two-dimensional model can be used as a basis for possible physics-based SBO design of the three-dimensional case.

12.3 Hydrodynamic Shape Optimization of Axisymmetric Bodies using Multi-Fidelity Models

Autonomous underwater vehicles (AUV) are becoming increasingly important in various marine applications, such as oceanography, pipeline inspection, and mine countermeasures (Yamamoto, 2007). Endurance (speed and range) is one of the more important attribute of AUV (Allen *et al.*, 2000). Vehicle drag reduction and/or an increase in the propulsion system efficiency will translate to a longer range for a given speed (or the same distance in a reduced time). A careful hydrodynamic design of the AUV, including hull shape, protrutions, and propulsion system, is therefore essential.

The fluid flow around an underwater vehicle with appendages is characterized by flow features such as thick boundary layers, vortices, and turbulent wakes, generated due to the hull and the appendages

(Huang *et al.*, 1992). These flow features can have adverse effects on, for example, the performance of the propulsion system and the control planes. Moreover, the drag depends highly on the vehicle shape, as well as on the aforementioned flow features. Consequently, it is important to account for these effects during the design of the AUV. The prediction of the flow past the full three-dimensional configuration of the AUV requires the use of CFD. Numerous applications of CFD methods to the flow past AUV and other underwater vehicles are in the literature (see, for example, Yang and Löhener (2003), de Barros *et al.* (2008), Jagadeesh *et al.* (2009)). The purpose of these investigations is to predict properties such as added masses, pressure and friction distributions, drag, normal force and moment coefficients, wake field, and stability derivatives. Comparison with experimental measurements shows that CFD is reliable and can yield accurate results.

Numerous studies on underwater vehicle design and optimization that focus on the clean hull only have been reported, i.e. the appendages and the propulsion system are neglected and the flow is taken to be past an axisymmetric body at a zero angle of attack. Examples of such numerical studies can be found in the literature (e.g. Goldschmied, 1966; Myring, 1976; Lutz and Wagner, 1998; Alvarez *et al.*, 2009; Solov'ev, 2009).

In this section, we present a hydrodynamic shape-optimization methodology based on the SBO concept for AUV. In particular, we adopt the multi-fidelity approach with the high-fidelity model based on the RANS equations, and the low-fidelity model based on the same equations, but with coarse discretization and relaxed convergence criteria. We use a simple response correction to create the surrogate. Here we choose to focus on the clean hull design, which is a convenient case study to implement and test our design approach.

12.3.1 *Problem formulation and computational models*

The goal of hydrodynamic shape optimization is to find an optimal — with respect to given objectives — hull shape, so that given design constraints are satisfied. The general design problem can be formulated as a nonlinear minimization problem, $\text{argmin}\{\boldsymbol{x} : f(\boldsymbol{x})\}$, with appropriate constraints (see Chapter 2 for a detailed formulation). Here, f denotes the high-fidelity objective function.

In this section we consider both direct design (adjustment of the geometrical shape to maximize performance, e.g. minimize drag) and inverse design (obtaining a predefined flow behavior; see, for example, Dalton and Zedan (1980)). In the case of the direct approach that involves the minimization of the drag coefficient, C_D, we set $f(\boldsymbol{x}) = C_D(\boldsymbol{x})$. In the inverse approach, a target pressure distribution, $C_{p,t}$, is prescribed a priori and the function $f(\boldsymbol{x}) = \|C_p(\boldsymbol{x}) - C_{p.t}\|^2$, where $C_p(\boldsymbol{x})$ is the hull surface pressure distribution, is minimized. No constraints are considered in this work, aside from the design variable bounds.

The hull shapes are constrained to the most common AUV hull shape, namely the torpedo shape, i.e. a three-section axisymmetric body with a nose, a cylindrical midsection, and a tail. Figure 12.13(a) shows a typical torpedo-shaped hull with a nose of length a, midsection of length b, overall length L, and maximum diameter of D. The nose and the tail are parameterized using Bezier curves as defined in Equation (12.1). We use five control points for the nose and four for the tail, as shown in Figs. 12.13(b) and 12.13(c). The third and eighth control points are free (x- and y-coordinates), while the other points are fixed. We, therefore, have two design variables for the nose and tail curves, a total of four design variables, aside from the hull dimensions a, b, L, and D.

The flow past the hull is considered to be steady and incompressible. The RANS equations are assumed as the governing flow equations with the two-equation k-ε turbulence model with standard wall functions (Tannehill *et al.*, 1997).

The solution domain is axisymmetric around the hull centreline axis and extends two body lengths in front of the hull, five body lengths behind it, and two body lengths above it (Fig. 12.14). At the inlet, there is a velocity boundary condition, where the velocity is set parallel to the hull axis, i.e. zero angle of attack. Pressure is prescribed at the outlet (zero gauge pressure).

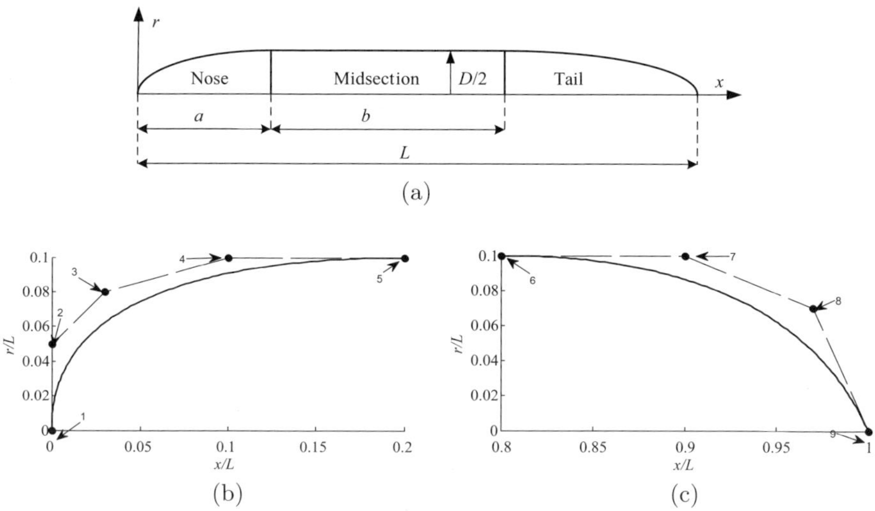

Fig. 12.13. (a) A sketch of a typical torpedo-shaped hull form (axisymmetric with three sections: nose, middle, tail); typically, equipment such as the computer, sensors, electronics, batteries, and payload are housed in the nose and the midsection, whereas the propulsion system is in the tail; (b) Bezier curve representing the nose (five control points); (c) Bezier curve representing the tail (four control points). Control points 3 and 8 are free, while the other points are essentially fixed (depending on L, a, b, and D).

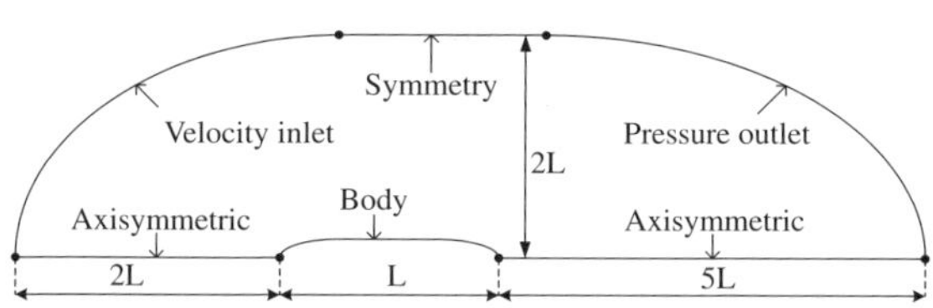

Fig. 12.14. The computational solution domain and the boundary conditions.

The CFD computer code FLUENT (2012) is used for numerical simulations of the fluid flow. The computational grid is structured with quadrilateral elements. The elements are clustered around the body and grow in size with distance from the body. The grids are generated using ICEM CFD (2013). A grid-convergence study was performed to determine the necessary grid density (Fig. 12.15). A torpedo-shaped body with $L/D = 5$ was used in the study. The inlet speed was 2 m/s and the Reynolds number was 2 million. Clearly, the drag coefficient value has converged at the finest grids

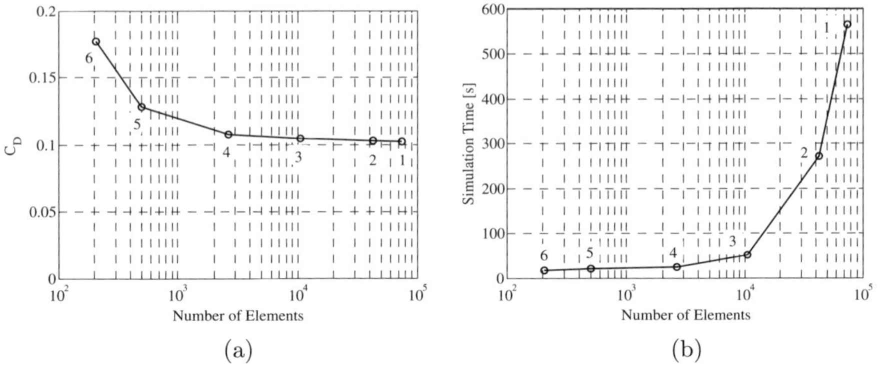

Fig. 12.15. Grid convergence study for a torpedo-shaped hull with length-to-diameter ratio $L/D = 5$ at a speed of 2 m/s and Reynolds number of 2 million. (a) The change in the drag coefficient, C_D, with the number of elements; (b) the variation in the simulation time with number of elements.

(number 1 and 2; Fig. 12.15(a)). There is, however, a large difference in the simulation time between the two finest grids (Fig. 12.15(b)). Therefore, we selected grid number 2, with 42,763 elements, to use for the high-fidelity CFD model in the optimization process. The velocity contours and the pressure and skin friction distributions are shown in Fig. 12.16 for illustration purposes.

The critical component of the design optimization methodology described in this work is a low-fidelity model, which is a simplified representation of the high-fidelity one. The low-fidelity model is based on the same CFD model as the high-fidelity one. The simulation time is substantially reduced by making the grid coarser (Fig. 12.15(b)). Grid number 6 needs the lowest simulation time and is the least accurate. A closer look at that grid reveals that it is too coarse (the responses were too "grainy"). Consequently, we selected grid number 5, with 504 elements, to be used for the low-fidelity model. The simulation time can be reduced further by reducing the number of iterations. Figure 12.17 shows how the drag coefficient reaches a converged value after approximately 50 iterations. We therefore relax the convergence criteria for the low-fidelity model by setting it to 50 iterations. The ratio of simulation time of the high-fidelity model to the low-fidelity model is around 15. A comparison of the high- and low-fidelity responses is given in Fig. 12.16.

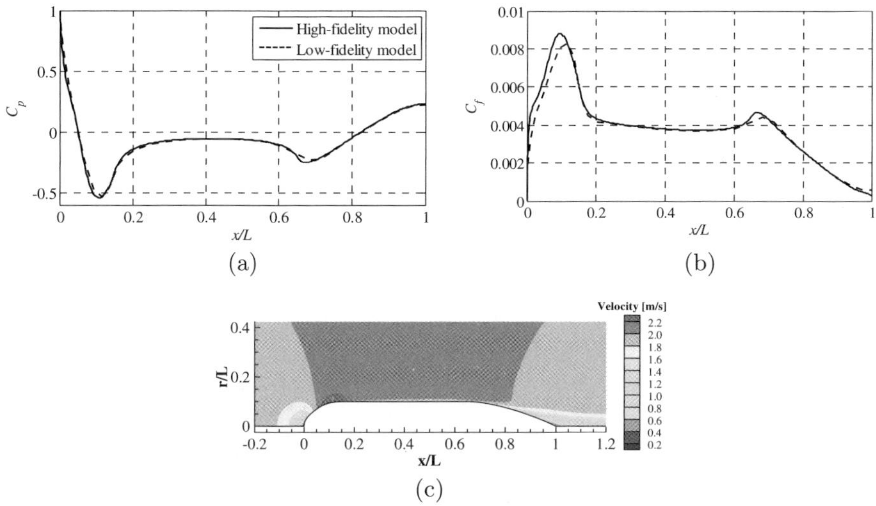

Fig. 12.16. Comparison of the high- and low-fidelity model responses of the axisymmetric hull of diameter ratio $L/D = 5$ at a speed of 2 m/s and Reynolds number of 2 million. (a) Pressure distributions; (b) skin friction distributions; (c) velocity contours of the flow past an axisymmetric torpedo shape obtained by the high-fidelity CFD model.

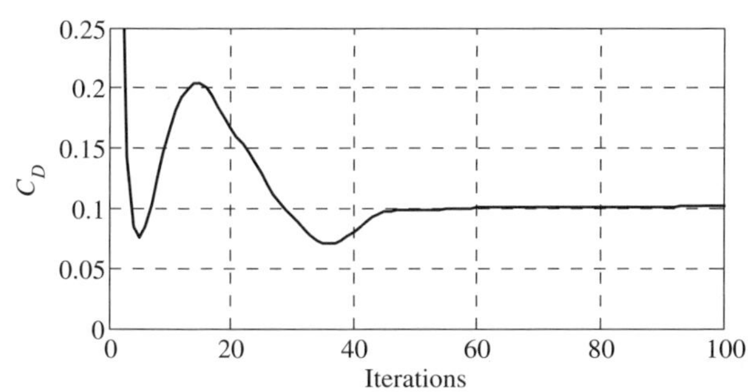

Fig. 12.17. Variation of the drag coefficient with number of iterations for the case shown in Fig. 12.16.

For a body in incompressible flow, the total drag is due to pressure and friction forces, which are calculated by integrating the pressure, C_p, and skin friction, C_f, distributions over the hull surface. The pressure coefficient is defined as $C_p \equiv (p - p_\infty)/q_\infty$, where p is the local static pressure, p_∞ is free-stream static pressure, and $q_\infty = (1/2\rho_\infty V_\infty^2)$ is the dynamic pressure, with ρ_∞ as the free-stream

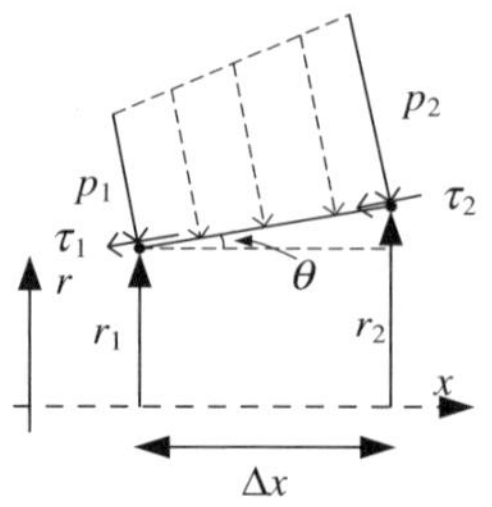

Fig. 12.18. Edge of an element on the hull surface at radius r. The element length is Δx and it makes an angle θ to the x-axis. Pressure, p, acts normal to the hull surface. Shear stress, τ, acts parallel to the surface.

density, and V_∞ the free-stream velocity. The skin-friction coefficient is defined as $C_f \equiv \tau/q_\infty$, where τ is the shear stress. Typical C_p and C_f distributions are shown in Fig. 12.16.

The total drag coefficient is defined as $C_D \equiv d/(q_\infty S)$, where d is the total drag force and S is the reference area. Here we use the frontal area of the hull as the reference area. The drag coefficient is the sum of the pressure and friction drag, or

$$C_D = C_{Dp} + C_{Df}, \tag{12.7}$$

where C_{Dp} is the pressure drag coefficient and C_{Df} is the skin-friction drag coefficient. The CFD analysis yields static pressure and wall shear stress values (which are non-dimensionalized to give C_p and C_f) at the element nodes (Fig. 12.18). The pressure acts normal to the surface and the shear stress parallel to it. The pressure-drag coefficient is calculated by integrating from the LE of the nose to the trailing edge (TE) of the tail:

$$C_{Dp} = 2\pi \int_0^L C_p(x) \sin\theta(x) r(x) dx, \tag{12.8}$$

where $C_p(x)$ is assumed to vary linearly between the element nodes, $\theta(x)$ is the angle of each element relative to the x-axis, and L is the length of the hull.

Similarly, the skin-friction drag coefficient is calculated as:

$$C_{Df} = 2\pi \int_0^L C_f(x) \cos\theta(x)\, r(x) dx. \tag{12.9}$$

12.3.2 *Design optimization methodology*

To solve our design problem, we utilize the standard SBO scheme (Equation (4.2)). The surrogate is constructed using a simple multiplicative response correction of the low-fidelity model, which turns out to be sufficient for our purposes. Let $C_{p.f}(\boldsymbol{x})$ and $C_{f.f}(\boldsymbol{x})$ denote the pressure and skin friction distributions of the high-fidelity model. The respective distributions of the low-fidelity model are denoted as $C_{p.c}(\boldsymbol{x})$ and $C_{f.c}(\boldsymbol{x})$. We will use the notation $C_{p.f}(\boldsymbol{x}) = [C_{p.f.1}(\boldsymbol{x})C_{p.f.2}(\boldsymbol{x})\ldots C_{p.f.m}(\boldsymbol{x})]^T$, where $C_{p.f.j}(\boldsymbol{x})$ is the jth component of $C_{p.f}(\boldsymbol{x})$, with the components corresponding to different coordinates along the x/L axis.

At iteration i, the surrogate model $C_{p.s}^{(i)}$ of the pressure distribution $C_{p.f}$ is constructed using the multiplicative response correction of the form:

$$C_{p.s}^{(i)}(\boldsymbol{x}) = [C_{p.s.1}^{(i)}(\boldsymbol{x})\ C_{p.s.2}^{(i)}(\boldsymbol{x})\ \ldots\ C_{p.s.m}^{(i)}(\boldsymbol{x})]^T, \tag{12.10}$$

$$C_{p.s.j}^{(i)}(\boldsymbol{x}) = A_{p.j}^{(i)} \cdot C_{p.c.j}(\boldsymbol{x}), \tag{12.11}$$

where $j = 1, 2, \ldots, m$ and:

$$A_{p.j}^{(i)} = \frac{C_{p.f.j}^{(i)}(\boldsymbol{x}^{(i)})}{C_{p.c.j}^{(i)}(\boldsymbol{x}^{(i)})}. \tag{12.12}$$

A similar definition holds for the skin-friction distribution model, $C_{f.s}^{(i)}$. Note that Equations (12.10)–(12.12) ensure zero-order consistency (Alexandrov and Lewis, 2001) between the surrogate and the high-fidelity model, i.e. $C_{p.f}(\boldsymbol{x}^{(i)}) = C_{p.s}^{(i)}(\boldsymbol{x}^{(i)})$. Rigorously speaking, this is not sufficient to ensure the convergence of the surrogate-based scheme (Equation (4.2)) to the high-fidelity model optimum. However, because it is constructed from the physically based low-fidelity model, the surrogate exhibits quite good generalization capability.

One of the issues of Equations (12.10)–(12.12) is that Equation (12.12) is not defined whenever $C_{p.c.j}(\boldsymbol{x}^{(i)})$ equals zero and the values of $A_{p.j}^{(i)}$ are very large when $C_{p.c.j}(\boldsymbol{x}^{(i)})$ is close to zero. This may be a source of substantial distortion of the surrogate model response,

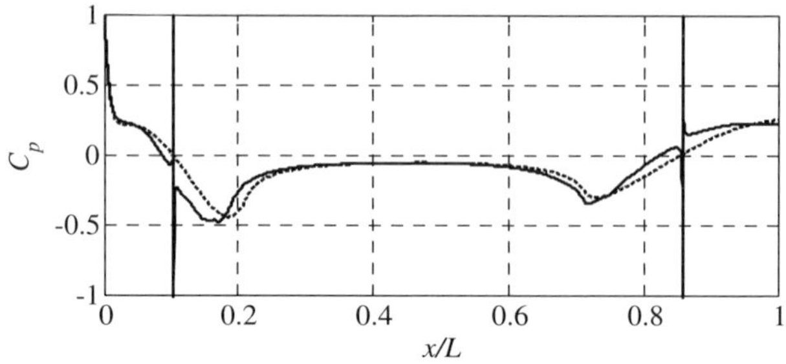

Fig. 12.19. Surrogate model $C_{p.s}^{(i)}$ (Equations (12.9)–(12.11)) at $x^{(i)}$ (- - -), and at some other design, x (—). By definition, $C_{p.s}^{(i)}(x^{(i)}) = C_{p.f}(x^{(i)})$. Note that $C_{p.s}^{(i)}(x)$ has large spikes around the points where $C_{p.s}^{(i)}(x^{(i)})$ is close to zero.

as illustrated in Fig. 12.19. In order to alleviate this problem, the original surrogate model response is "smoothened" in the vicinity of the regions in which $A_{p.j}^{(i)}$ is large (which indicates the problems mentioned above). Let $j_{\max}$ be such that $|A_{p.j\max}^{(i)}| \gg 1$ assumes (locally) the largest value. Let Δj be the user-defined index range (typically, $\Delta j = 0.01 \cdot \mathrm{m}$). The original values of $A_{p.j}^{(i)}$ are replaced, for $j = j_{\max} - \Delta j, \ldots, j_{\max} - 1, j_{\max}, j_{\max} + 1, \ldots, j_{\max} + \Delta j$, by the interpolated values:

$$\begin{aligned}\bar{A}_{p.j}^{(i)} = I(\{&[j_{\max} - 2\Delta j \ldots j_{\max} - \Delta j - 1] \\ &\cup [j_{\max} + \Delta j - 1 \ldots j_{\max} + 2\Delta j]\}, \\ &\{[A_{p.j_{\max}-2\Delta j}^{(i)} \ldots A_{p.j_{\max}-\Delta j-1}^{(i)}] \\ &\cup [A_{p.j_{\max}-2\Delta j}^{(i)} \ldots A_{p.j_{\max}-\Delta j-1}^{(i)}]\}, j),\end{aligned} \tag{12.13}$$

where $I(X, Y, Z)$ is a function that interpolates the function values Y defined over the domain X onto the set Z. Here, we use cubic splines. In other words, the values of $A_{p.j}^{(i)}$ in the neighborhood of $j_{\max}$ are "restored" using the values of $A_{p.j}^{(i)}$ from the surrounding of $j = j_{\max} - \Delta j, \ldots, j_{\max} + \Delta j$. Figure 12.20(a) shows the "smoothened" surrogate model response corresponding to that of Fig. 12.19. Figure 12.20 shows the surrogate and the high-fidelity model responses, both C_p and C_f, at $\boldsymbol{x}^{(i)}$ and at some other design, $\boldsymbol{x}$.

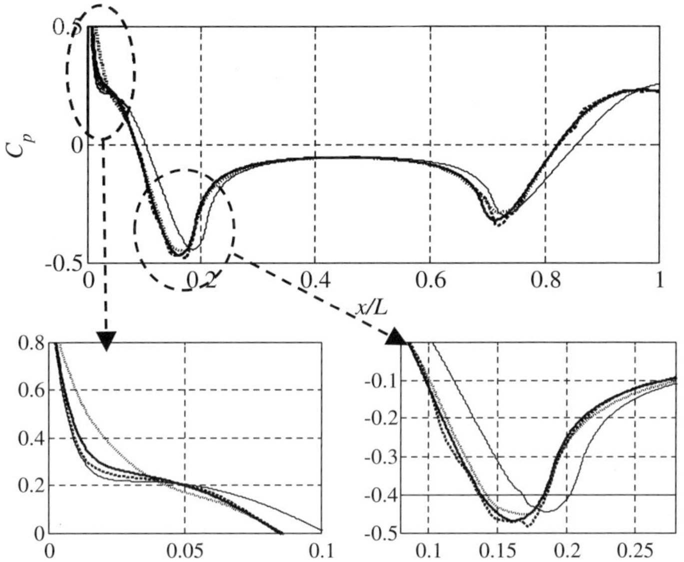

(a)

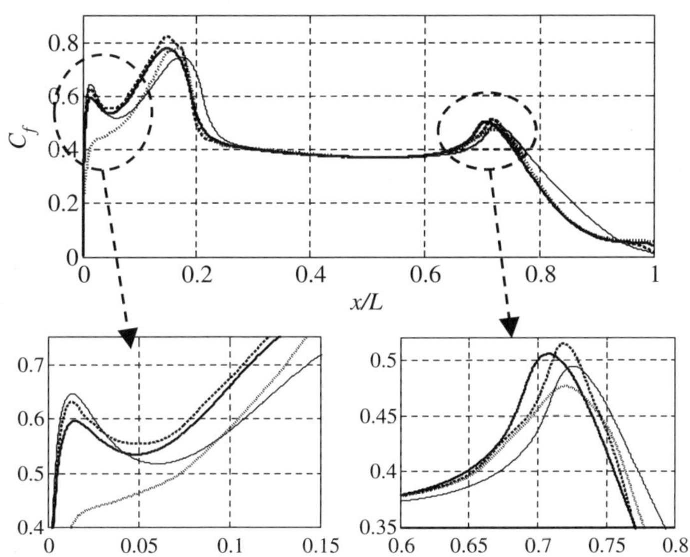

(b)

Fig. 12.20. (a) Smoothened surrogate model (Equations (12.9)–(12.11)) $C_{p.s}^{(i)}(\boldsymbol{x}^{(i)}) = C_{p.f}(\boldsymbol{x}^{(i)})$ (—), $C_{p.s}^{(i)}(\boldsymbol{x})$ (- - -), $C_{p.c}(\boldsymbol{x})$ ($\cdots$), and $C_{p.s}(\boldsymbol{x})$ (—); (b) smoothened responses $C_{f.s}^{(i)}(\boldsymbol{x}^{(i)}) = C_{f.f}(\boldsymbol{x}^{(i)})$ (—), $C_{f.s}^{(i)}(\boldsymbol{x})$ (- - -), $C_{f.c}(\boldsymbol{x})$ ($\cdots$), and $C_{f.s}(\boldsymbol{x})$ (—).

12.3.3 *Results*

The methodology of Section 12.3.2 is applied to the hydrodynamic shape optimization of torpedo-type hulls, involving both the direct and inverse design approaches. Designs are obtained with the surrogate model, optimized using the pattern-search algorithm (Koziel, 2010a). For comparison purposes, designs obtained with direct optimization of the high-fidelity model using the pattern-search method are also presented.

For both the direct and the inverse design approaches, the design-variable vector is $\boldsymbol{x} = [a\ x_n\ y_n\ x_t\ y_t]^T$, where a is the nose length and (x_n, y_n) and (x_t, y_t) are the coordinates of the free control points on the nose and tail Bezier curves, respectively, i.e. points 3 and 8 in Fig. 12.13. The lower and upper bounds of design variables are $\boldsymbol{l} = [0\ 0\ 0\ 80\ 0]^T$ cm and $\boldsymbol{u} = [30\ 30\ 10\ 100\ 10]^T$ cm, respectively. Other geometrical shape parameters are, for both cases, $L = 100$ cm, $d = 20$ cm, and $b = 50$ cm. The flow speed is 2 m/s and the Reynolds number is 2 million.

Numerical results for a direct design case are presented in Table 12.2. The hull drag coefficient is minimized by finding the

Table 12.2 Numerical results for design Cases 1 and 2*.

	Case 1 (drag minimization)			Case 2 (inverse design)		
Variable	Initial	Pattern-search	SBO	Initial	Pattern-search	SBO
A	15.0000	21.8611	20.9945	18.000	24.7407	24.7667
x_n	5.0000	5.6758	5.6676	7.0000	7.3704	6.8333
y_n	5.0000	2.7022	2.7531	8.0000	4.7407	4.5667
x_t	90.0000	98.000	96.6701	85.0000	88.1111	88.6333
y_t	5.0000	0.8214	3.0290	7.0000	5.5926	5.3000
F	—	—	—	0.0204	$1.64 \cdot 10^{-5}$	$1.93 \cdot 10^{-5}$
C_D	0.0915	0.0853	0.0857	0.0925	0.0894	0.0893
N_c	—	0	300	—	0	500
N_f	—	282	3	—	401	5
Total cost	—	282	13	—	401	<22

*Design goals: axisymmetric hull direct-drag minimization (Case 1) and inverse design with a target-pressure distribution (Case 2). In both cases, the flow speed is 2 m/s and the Reynolds number is $2 \cdot 10^6$. All the numerical values are from the high-fidelity model. N_c and N_f are the number of low- and high-fidelity model evaluations, respectively. F is the norm of the difference between the target and the design shapes.

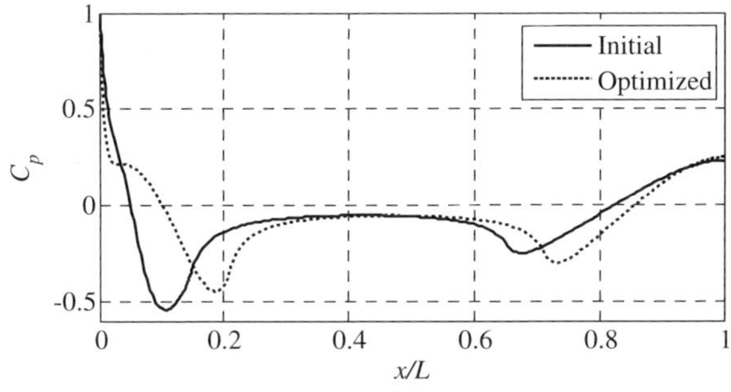

(a)

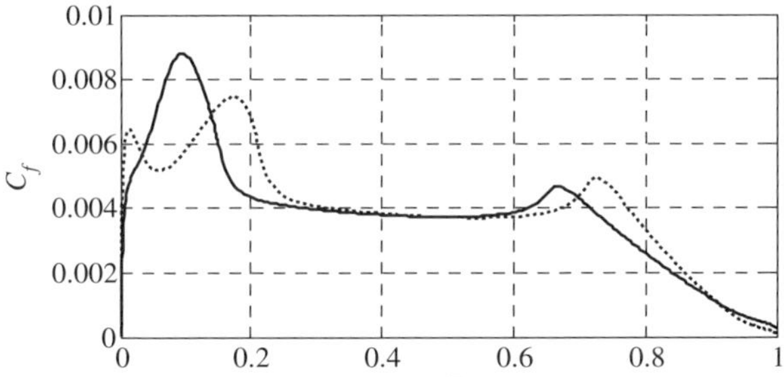

(b)

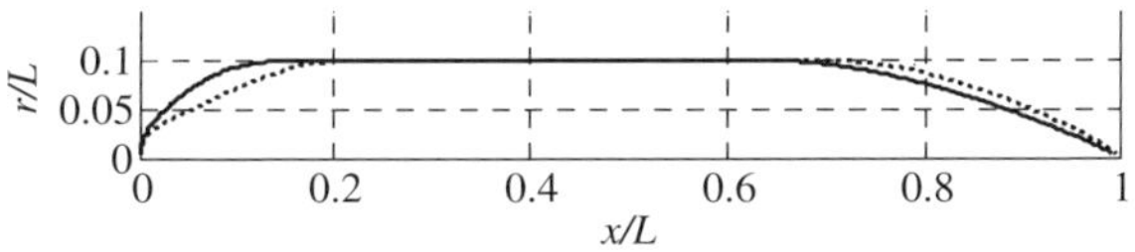

(c)

Fig. 12.21. Case 1 results for the direct hull drag minimization. (a) Initial and optimized pressure distributions; (b) initial and optimized skin friction distributions; (c) initial and optimized hull shapes.

appropriate shape and length of the nose and tail sections for a given hull length, diameter, and cylindrical section length. In this case, the drag coefficient is reduced by 6.3%. This drag reduction comes from a reduction in skin friction and a lower pressure peak, where the nose and tail connect with the midsection (Fig. 12.21). These changes are due to a more streamlined nose (longer by 6 cm) and a fuller tail, when compared to the initial design. SBO requires three high-fidelity and 300 low-fidelity model evaluations. The ratio of the high-fidelity model evaluation time to the corrected low-fidelity model evaluation time varies between 11 and 45, depending on whether the flow solver converges the residuals by six orders of magnitude or hits the maximum iteration limit of 1,000. We express the total optimization cost of the presented method in the equivalent number of high-fidelity model evaluations. For the sake of simplicity, we use a fixed value of 30 as the high-to-low-fidelity model evaluation time ratio. The results show that the total optimization cost of the presented approach is around 13 equivalent high-fidelity model evaluations. The direct optimization method yields a very similar design, but at the substantially higher computational cost of 282 high-fidelity model evaluations.

Inverse design of the hull shape was performed by prescribing a target-pressure distribution. The objective is to minimize the norm of the difference between the pressure distribution of the hull design and the target-pressure distribution. The numerical results of the inverse design are presented in Table 12.2; see also Fig. 12.22. The SBO algorithm matched the target-pressure distribution (the norm of the distributions is less than 2×10^{-5}), using less than 22 equivalent high-fidelity model evaluations. The direct optimization of the high-fidelity model using the pattern-search algorithm required 401 function calls to yield a comparable match with the target.

12.4 Design of High-Lift Single-Element Airfoil Shapes

High-lift systems have a major effect on the sizing and economics of most transport aircraft. A 1% improvement in performance during takeoff of a generic large twin-engine transport translates, for a given range, into an increase in payload of 14 passengers or 2,800 lbs (van Dam, 2002). The design of high-lift systems has traditionally been carried out using detailed wind-tunnel testing, which is both challenging and expensive due to complicated flow features. The improvement of CFD capabilities to accurately predict the flow past airfoils at high-lift conditions has motivated the development of numerical design methods (Rumsey and Ying, 2002).

Early work on numerical optimization of high-lift systems focused on using high-fidelity CFD simulations (the Navier–Stokes equations)

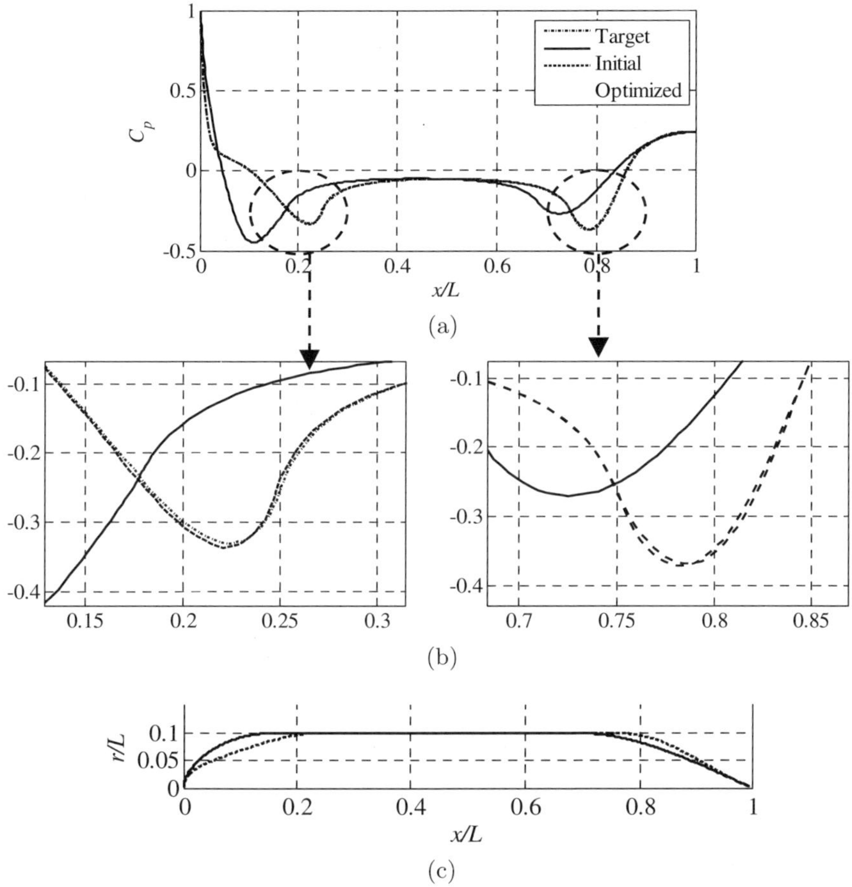

Fig. 12.22. Case 2 results of the inverse design optimization with a prescribed target-pressure distribution. (a) Target, initial, and optimized pressure distributions; (b) zoom on two regions of Fig. 12.22(a); (c) the initial and optimized hull shapes.

with a gradient-based optimizer, in which the necessary gradients are obtained by finite-difference methods (Eyi *et al.*, 1996). Kim *et al.* (2000, 2002) applied a viscous adjoint method to two-dimensional high-lift system design. The adjoint method is gradient-based, but it is very efficient as the computational expense incurred in the calculation of the gradient is effectively independent of the number of design variables. Nemec and Zingg (2003) apply a gradient-based Newton–Krylov algorithm to high-lift system design, in which the gradient of the objective function is computed using the discrete-adjoint approach. Heuristic methods have also been applied to high-lift system design (Giannakoglou, 2002).

High-fidelity CFD simulation is computationally far too expensive to be used in a direct, simulation-based design optimization, especially when using traditional, gradient-based techniques, such as the methods mentioned in the previous paragraph. SBO methods use computationally cheap surrogate functions in lieu of the computationally more expensive high-fidelity models (Queipo *et al.*, 2005; Forrester and Keane, 2009). SBO has been successfully applied to cruise (transonic) design of airfoils and wings (Alexandrov *et al.*, 2000; Forrester *et al.*, 2006; Toal *et al.*, 2008). SBO has not been as extensively applied in high-lift system design. Examples of such studies include those by Greenman (1998), Kanazaki *et al.* (2006), and Meunier (2007). In all these studies the surrogates are constructed by approximating the high-fidelity model data using nonlinear regression, e.g. neural networks and kriging.

This section describes the use of shape-preserving response prediction (SPRP)-based surrogate models for the design optimization of single-element airfoil shapes at high-lift conditions. The high-fidelity CFD model solves the RANS equations. SPRP surrogates are constructed for both the airfoil surface pressure distribution, C_p, and the skin friction distribution, C_f. The computational efficiency of the methodology is demonstrated by several examples of airfoil-shape optimization problems.

12.4.1 *Problem formulation*

The objective is to maximize the lift coefficient without increasing the drag coefficient. In particular, we solve:

$$\min_{x} -C_l(\mathbf{x}), \tag{12.14}$$

subject to:

$$g_1 = C_d(\mathbf{x}) - C_d^* \leq 0, \tag{12.15}$$

where C_d^* is the drag coefficient of the baseline design.

The constraint is handled through a penalty function. More specifically, the objective function is defined as:

$$H(C_p(\boldsymbol{x}), C_f(\boldsymbol{x})) = -C_{l.s}(C_p(\boldsymbol{x}), C_f(\boldsymbol{x})) + \boldsymbol{\beta}[\Delta C_{d.s}(C_p(\boldsymbol{x}), C_f(\boldsymbol{x}))]^2, \tag{12.16}$$

where $\Delta C_{d.s} = 0$ if $C_{d.s} \leq C_{d.s}^*$ and $\Delta C_{d.s} = C_{d.s} - C_{d.s}^*$ otherwise. In our numerical experiments we use $\beta = 1{,}000$. Here C_p and C_f are the pressure and skin friction distributions ($C_p = C_{p.s}$ for the surrogate model and $C_p = C_{p.f}$ for the high-fidelity model; similarly for C_f) and $C_{l.s}$ and $C_{d.s}$ denote the lift and drag coefficients (both being functions of the pressure and skin friction distributions).

12.4.2 *Aerodynamic model*

The airfoil shapes are described by the NACA four-digit parameterization method, in which the airfoil shape is defined by only three parameters: m (the maximum ordinate of the mean camberline as a fraction of chord), p (the chordwise position of the maximum ordinate), and t/c (the thickness-to-chord ratio). A detailed description of the method is given in Section 7.6.

The flow past the airfoil is taken to be steady, incompressible, and fully turbulent. The RANS equations are taken as the governing flow equations. For steady incompressible flow, the RANS equations can be written (in Cartesian tensor form) as (Tannehill *et al.*, 1997):

$$\frac{\partial \bar{u}_j}{\partial x_j} = 0, \tag{12.17}$$

and:

$$\frac{\partial}{\partial x_j}(\rho\, \bar{u}_i\, \bar{u}_j) = -\frac{\partial \bar{p}}{\partial x_i} + \frac{\partial}{\partial x_j}(\bar{\tau}_{ij} - \rho\, \overline{u_i' u_j'}), \tag{12.18}$$

where $\bar{u}_i$ is the time-averaged velocity component ($i = 1, 2, 3$), ρ is the fluid density, $\bar{p}$ is the time-averaged pressure, and $\bar{\tau}_{ij}$ is the time-averaged viscous stress tensor, defined as:

$$\bar{\tau}_{ij} = \mu \left(\frac{\partial \bar{u}_i}{\partial x_j} + \frac{\partial \bar{u}_j}{\partial x_i} \right), \tag{12.19}$$

where μ is the fluid dynamic viscosity. The last component of Equation (12.18), $-\rho\overline{u_i' u_j'}$, commonly called the Reynolds stress, represents the effects of turbulence and needs to be modeled in order to close this equation. There are numerous turbulence models in the literature (Tannehill *et al.*, 1997). We selected one equation model by Spalart and Allmaras (1992) which solves a single partial differential equation for the Reynolds stress. The model was developed for aerospace applications and is considered to be accurate for attached wall-bounded flows and flows with mild separation and recirculation. The Spalart–Allmaras model employs the Boussinesq approach, which provides an equation for the Reynolds stress as:

$$-\rho\,\overline{u_i'u_j'} = \mu_T\left(\frac{\partial u_i}{\partial x_j} + \frac{\partial u_j}{\partial x_i}\right) - \frac{2}{3}\delta_{ij}\left(\mu_T\frac{\partial u_k}{\partial x_k} + \rho\bar{k}\right), \qquad (12.20)$$

where δ_{ij} is Kronecker delta, $\bar{k}$ the kinetic energy of turbulence, $\bar{k} = \overline{u_i'u_i'}/2$, and μ_T is the turbulent viscosity. The Spalart–Allmaras turbulence model defines the turbulent viscosity as:

$$\mu_T = \rho\tilde{\nu}f_{\nu 1}, \qquad (12.21)$$

where the parameter $\tilde{\nu}$ is obtained from the solution of the transport equation:

$$u_j\frac{\partial\tilde{\nu}}{\partial x_j} = \frac{1}{\sigma}\frac{\partial}{\partial x_k}\left[(\nu+\tilde{\nu})\frac{\partial\tilde{\nu}}{\partial x_k}\right] + c_{b1}(1-f_{\nu 2})\tilde{S}\tilde{\nu}$$
$$- c_{w1}f_w\left(\frac{\tilde{\nu}}{d}\right)^2 + \frac{c_{b2}}{\sigma}\frac{\partial\tilde{\nu}}{\partial x_k}\frac{\partial\tilde{\nu}}{\partial x_k}, \qquad (12.22)$$

where the closure coefficients and functions are given by:

$$c_{b1} = 0.1355 \quad c_{b2} = 0.622 \quad c_{\nu 1} = 7.1 \quad \sigma = 2/3$$

$$c_{w1} = \frac{c_{b1}}{\kappa^2} + \frac{(1+c_{b2})}{\sigma} \quad c_{w2} = 0.3 \quad c_{w3} = 2 \quad \kappa = 0.41$$

$$f_{\nu 1} = \frac{\chi^3}{\chi^3 + c_{\nu 1}^3} \quad f_{\nu 2} = 1 - \frac{\chi}{1+\chi f_{\nu 1}} \quad f_w = g\left(\frac{1+c_{w3}^6}{g^6+c_{w3}^6}\right) \quad \chi = \frac{\tilde{\nu}}{\nu},$$

$$g = r + c_{w2}(r^6 - r) \quad r = \frac{\tilde{\nu}}{\tilde{S}\kappa^2 d^2} \quad \tilde{S} = S + \frac{\tilde{\nu}}{\kappa^2 d^2}f_{\nu 2}$$

$$S = \sqrt{2\Omega_{ij}\Omega_{ij}} \quad \Omega_{ij} = \frac{1}{2}\left(\frac{\partial u_i}{\partial x_j} - \frac{\partial u_j}{\partial x_i}\right)$$

and d is the distance from the closest surface.

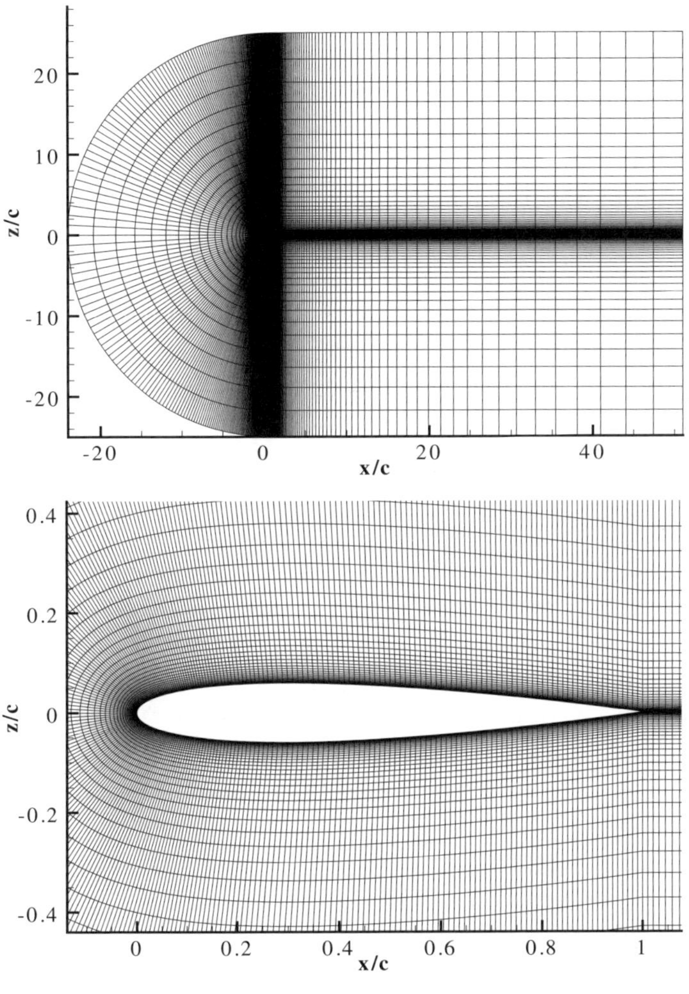

Fig. 12.23. An example computational grid with a C-topology and 27,600 quadrilateral cells.

Computational grids are developed using the computer code ICEM CFD (2013). All grids used in this study are structured curvilinear body-fitted grids of C-topology, i.e. at the inlet the grid surrounds the LE of the airfoil but is open at the other end (a topology that can be associated to the letter C). An example computational grid is shown in Fig. 12.23. The computational region is set such that there are 24 chord lengths in front of the airfoil, 50 chord lengths behind it, and 25 chord lengths above and below it. At the boundaries in front of and above and below the airfoil the velocity is taken as constant. A constant pressure boundary condition is applied to the boundary behind the airfoil.

The CFD computer code FLUENT (2012) is used for numerical simulations of the fluid flow. Asymptotic convergence to a steady state solution is obtained for each case. The iterative convergence of each solution is examined by monitoring the overall residual, which is the sum (over all the cells in the computational domain) of the L^2 norm of all the governing equations solved in each cell. In addition to this, the lift and drag forces (defined below) are monitored for convergence.

The aerodynamic forces are calculated by integrating the pressure and the viscous shear stress over the surface of the airfoil, as described in Section 8.1.4.

Example RANS simulation results are given in Figs. 12.24 and 12.25, showing the flow past NACA 0012 at a free-stream Mach number $M_\infty = 0.2$ and a Reynolds number (based on chord length) $Re = 2.3$ million. The velocity field contours are given in Fig. 12.24 for several angles of attack. As the angle of attack increases, the flow separates from the upper surface, starting at the TE, and at $\alpha = 18°$ there is a massive flow separation over most of the upper surface. The lift coefficient, shown in Fig. 12.25, increases linearly with an angle of attack from 0° up to about 12°, when flow separation on the upper surface begins, and then it falls sharply at around 16°. The drag coefficient increases quadratically with the increase of the angle of attack, as can be seen in Figure 12.25. Example pressure and skin friction distributions are given in Fig. 12.26. The pressure is highest at the LE on the lower surface and lowest on the upper surface near the LE.

12.4.3 *Surrogate model exploiting SPRP*

We adopt the SPRP technique described in Section 6.3. As SPRP uses a low-fidelity model of the structure under consideration, we have to begin with building such a model. As the figures of interest in an aerodynamic analysis (e.g. the lift, drag, and pitching-moment

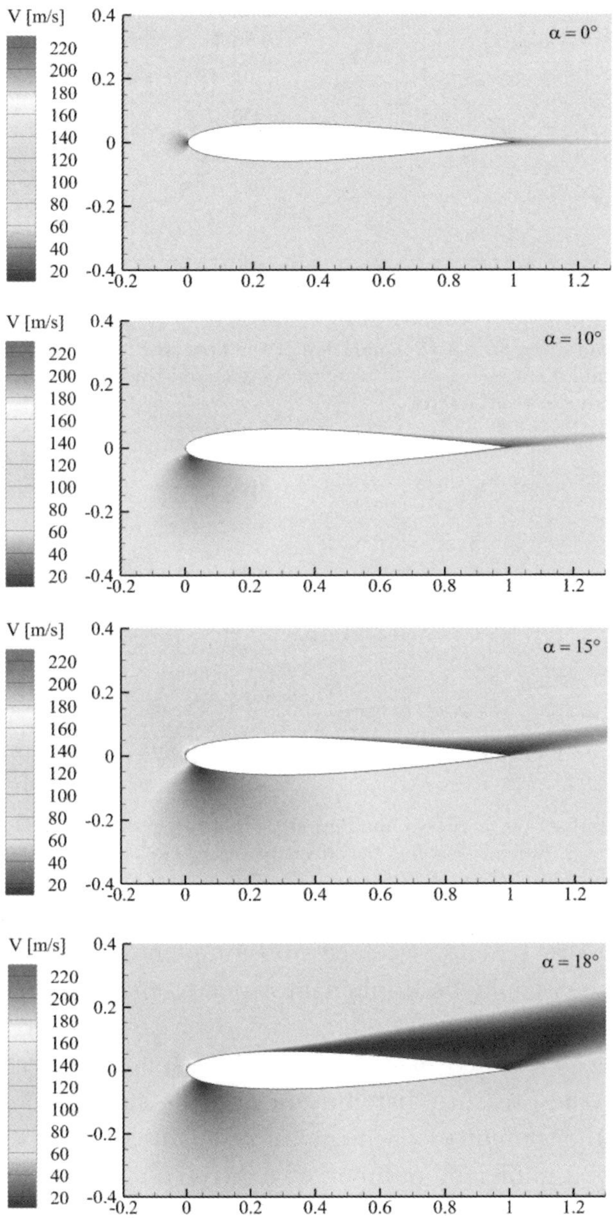

Fig. 12.24. Velocity field contours of the flow past NACA 0012 at $M_\infty = 0.2$, $Re = 2.3$ million, and four different angles of attack.

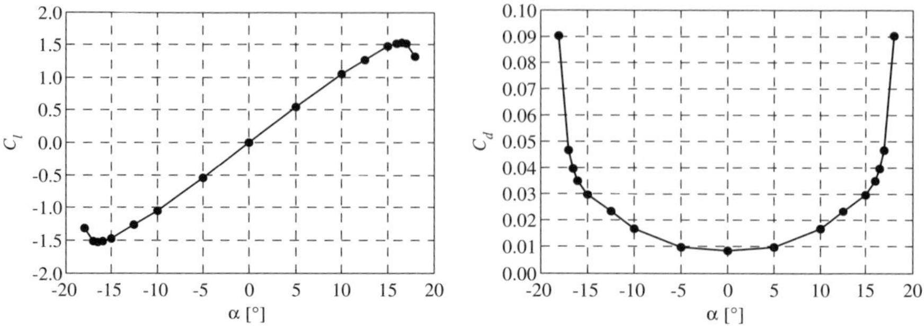

Fig. 12.25. Results from a RANS simulation of the flow past NACA 0012 at $M_\infty = 0.2$ and $Re = 2.3$ million, showing the lift coefficient (C_l) and the drag coefficient (C_d) as a function of the angle of attack (α).

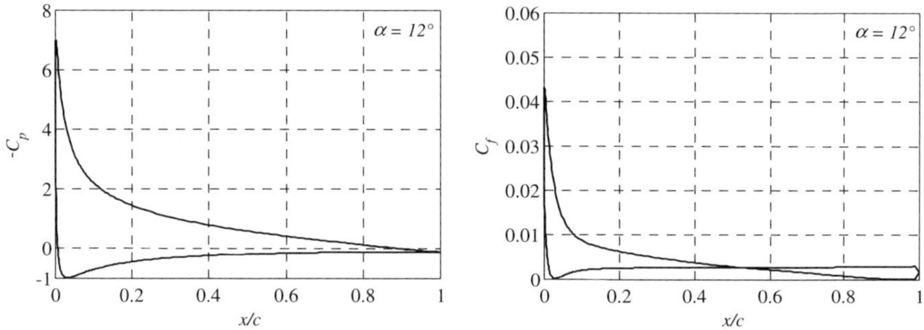

Fig. 12.26. Results from a RANS simulation of the flow past NACA 0012 at $M_\infty = 0.2$ and $Re = 2.3$ million, showing the pressure coefficient (C_p) and the skin friction coefficient (C_f) as a function of x/c at $\alpha = 12°$.

coefficients) are scalars, we face non-uniqueness of any alignment procedure that could be applied in order to match the low-fidelity model with the high-fidelity one at any given design. Here the model alignment is performed using intermediate simulation results; more specifically, the pressure distribution and the skin friction distribution, the dimensionality of which can be made as large as necessary by selecting a sufficient number of control points along the airfoil chord. As the objectives and constraints are uniquely determined by the pressure and skin friction distributions, the alignment of corresponding distributions for the low- and high-fidelity models will result in (unique) alignment of the figures of interest.

The low-fidelity model, c, is constructed as a low-order polynomial approximation of the high-fidelity model data. The low-fidelity model is established in the entire design space (here, $0.0 \leq m \leq 0.08$, $0.3 \leq p \leq 0.6$, and $0.08 \leq t/c \leq 0.14$), using evaluations of f at the following seven designs: $\boldsymbol{x}^0 = [0.04\ 0.45\ 0.11]^T$ (center point), and $\boldsymbol{x}^1 = [0.0\ 0.45\ 0.11]^T$, $\boldsymbol{x}^2 = [0.08\ 0.45\ 0.11]^T$, $\boldsymbol{x}^3 = [0.04\ 0.3\ 0.11]^T$, $\boldsymbol{x}^4 = [0.04\ 0.6\ 0.11]^T$, $\boldsymbol{x}^5 = [0.04\ 0.45\ 0.08]^T$, $\boldsymbol{x}^6 = [0.044\ 0.45\ 0.14]^T$ (single-variable perturbations for all design variables).

The low-fidelity model is defined as a reduced quadratic model (no mixed terms):

$$c(\boldsymbol{x}) = c([m\,p\,t]^T) = \lambda_0 + \lambda_1 m + \lambda_2 p + \lambda_3 t + \lambda_4 m^2 + \lambda_5 p^2 + \lambda_6 t^2. \tag{12.23}$$

The coefficients λ can be uniquely found by solving the linear system $c(\boldsymbol{x}^j) = f(\boldsymbol{x}^j)$, $j = 0, 1, \ldots, 6$.

The reason for choosing the approximation-based model c is that the pressure distribution does not change significantly for the operating conditions considered in this paper (i.e. high-lift, low Mach number, small design space) so the model is sufficiently accurate to be used by SPRP. Still, the low-fidelity model has to be corrected in order to become a reliable representation of the high-fidelity one.

As mentioned above, the surrogate model is constructed based on the pressure distribution, C_p, and skin friction distribution, C_f, rather than directly on the figures of interest such as lift, C_l, or drag, C_d. The reason for this is that these figures are not uniquely determined by the design variable vector, $\boldsymbol{x}$. The pressure and skin friction distributions, on the other hand, are uniquely determined by $\boldsymbol{x}$, and uniquely determine lift and drag at the same time. The pressure distribution for the high- and low-fidelity models will be denoted as $C_{p.f}$ and $C_{p.c}$, respectively ($C_{f.f}$ and $C_{f.c}$ for skin friction distribution).

For the sake of brevity, the construction of the surrogate model is explained for the pressure distribution only; the skin friction part is analogous. The surrogate model is constructed assuming that the change of $C_{p.f}$ due to the adjustment of the design variables, $\boldsymbol{x}$, can be predicted using the actual changes of $C_{p.c}$. The change of $C_{p.c}$ is described by the translation vectors corresponding to a certain

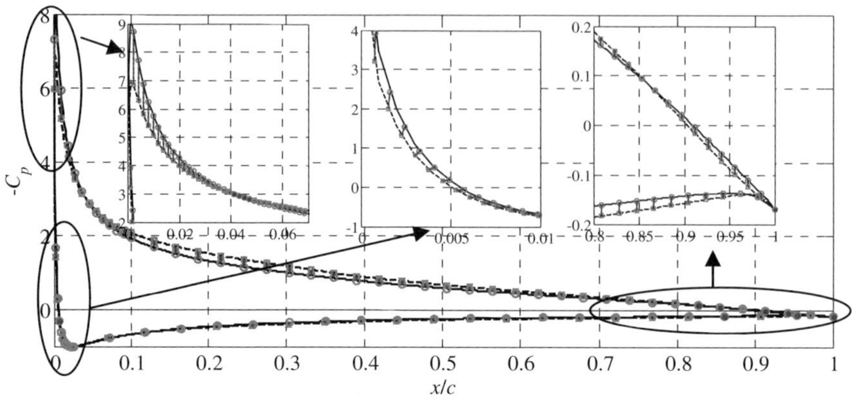

Fig. 12.27. Example low fidelity model pressure distribution at the design $\boldsymbol{x}^{(i)}$, $C_{p.c}(\boldsymbol{x}^{(i)})$ (solid line); the low-fidelity model pressure distribution at other design $\boldsymbol{x}$, $C_{p.c}(\boldsymbol{x})$ (dotted line); characteristic points of $C_{p.c}(\boldsymbol{x}^{(i)})$ (circles) and $C_{p.c}(\boldsymbol{x})$ (squares); and the translation vectors (short lines). Only selected points and vectors are shown for the sake of clarity. Selected parts of the distributions are magnified.

(finite) number of its characteristic points. These translation vectors are subsequently used to predict the change of $C_{p.f}$, whereas the actual $C_{p.f}$ at the current design, $C_{p.f}(\boldsymbol{x}^{(i)})$, is treated as a reference.

Figure 12.27 shows the pressure distribution, $C_{p.c}$, of the low-fidelity model at $\boldsymbol{x}^{(i)} = [0.01\ 0.40\ 0.09]^T$ for $M_\infty = 0.2$ and $\alpha = 10°$, as well as $C_{p.c}$ at $\boldsymbol{x} = [0.02\ 0.35\ 0.10]^T$; $\boldsymbol{x}^{(i)}$ will denote a current design (at the ith iteration of the optimization algorithm; the initial design will be denoted as $\boldsymbol{x}^{(0)}$ accordingly). Circles denote characteristic points of $C_{p.c}(\boldsymbol{x}^{(i)})$, here representing, among others, x/c equal to 0 and 1 (airfoil LE and TE, respectively), the minima/maxima of $C_{p.c}$ for the lower and upper airfoil surfaces (if there are any), as well as other points corresponding, for example, to specific values of x/c or specific values (absolute or relative to minimal/maximal ones) of $C_{p.c}$. Squares denote corresponding characteristic points for $C_{p.c}(\boldsymbol{x})$, while small line segments represent the translation vectors that determine the "shift" of the characteristic points of $C_{p.c}$ when changing the design variables from $\boldsymbol{x}^{(i)}$ to $\boldsymbol{x}$. For clarity, Fig. 12.27 only shows selected characteristic points.

The pressure distribution of the high-fidelity model at the given design, here $\boldsymbol{x}$, can be predicted using the translation vectors applied

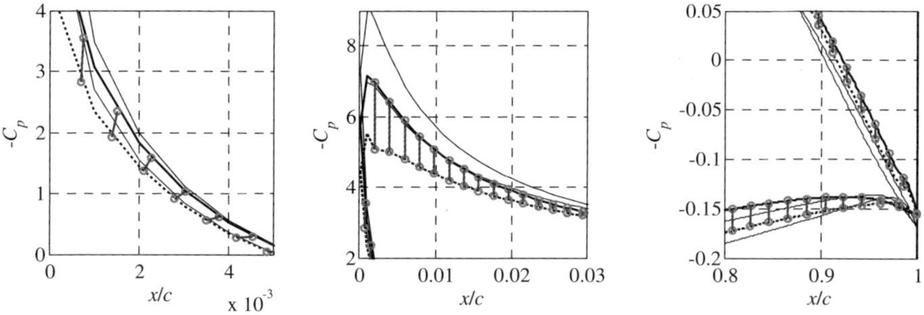

Fig. 12.28. High-fidelity model pressure distribution at $\boldsymbol{x}^{(i)}$, $C_{p.f}(\boldsymbol{x}^{(i)})$ (solid line) and the predicted high-fidelity model C_p at $\boldsymbol{x}$ (dotted line), obtained using SPRP based on characteristic points of Fig. 12.27. Characteristic points of $C_{p.f}(\boldsymbol{x}^{(i)})$ (circles) and the translation vectors (short lines) were used to find the characteristic points (squares) of the predicted high-fidelity model pressure distribution.

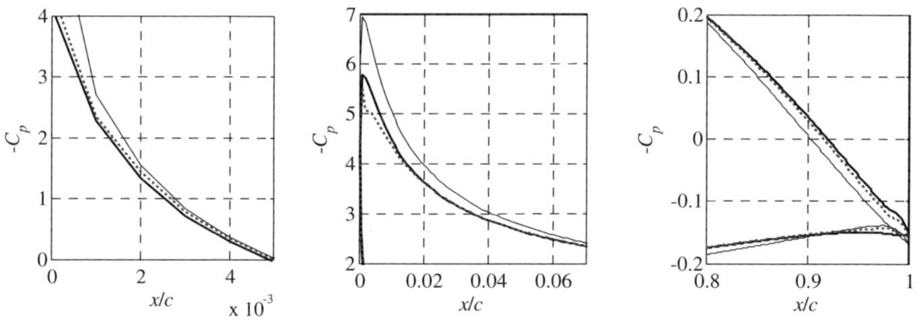

Fig. 12.29. High-fidelity model pressure distribution at $\boldsymbol{x}$, $C_{p.f}(\boldsymbol{x})$ (solid line), and the predicted high-fidelity model pressure distribution at $\boldsymbol{x}$, obtained using SPRP (thick dotted line). The low-fidelity model pressure distribution at $\boldsymbol{x}$ is shown using a thin dashed line. The SPRP model ensures better accuracy than the low-fidelity model.

to the corresponding characteristic points of the pressure distribution of the high-fidelity model at $\boldsymbol{x}^{(i)}$, $C_{p.f}(\boldsymbol{x}^{(i)})$. This is illustrated in Fig. 12.28, in which the magnified parts of the distributions are shown for clarity. Figure 12.29 shows the predicted pressure distribution (magnified parts only) of the high-fidelity model at $\boldsymbol{x}$ as well as the actual $C_{p.f}(\boldsymbol{x})$. The agreement between both curves is very good.

Rigorous formulation of the SPRP method can be found in Section 6.3. It should be noted that our prediction method assumes that the high- and low-fidelity model pressure and skin friction distributions have corresponding sets of characteristic points. This assumption is satisfied for the entire design space and for the operating conditions considered in this study.

12.4.4 *Optimization algorithm*

Our optimization algorithm exploits the SPRP-based surrogate model and a trust-region convergence safeguard (Conn *et al.*, 2000). It can be summarized as follows.

1. Set $i = 0$; select λ (initial trust region radius); evaluate $C_{p.f}(\boldsymbol{x}^{(0)})$ and $C_{f.f}(\boldsymbol{x}^{(0)})$.
2. Set up SPRP model.
3. Obtain $\boldsymbol{x}^{(i+1)} = \text{argmin}\{l \leq \boldsymbol{x} \leq u,\ \|\boldsymbol{x} - \boldsymbol{x}^{(i)}\| \leq \lambda\colon H(C_{p.s}^{(i)}(\boldsymbol{x}), C_{f.s}^{(i)}(\boldsymbol{x}))\}$.
4. Evaluate high-fidelity model to get $C_{p.f}(\boldsymbol{x}^{(i)})$ and $C_{f.f}(\boldsymbol{x}^{(i)})$.
5. If $H(C_{p.f}^{(i)}(\boldsymbol{x}^{(i+1)}), C_{f.f}^{(i)}(\boldsymbol{x}^{(i+1)})) < H(C_{p.f}^{(i)}(\boldsymbol{x}^{(i)}), C_{f.f}^{(i)}(\boldsymbol{x}^{(i)}))$, accept $\boldsymbol{x}^{(i+1)}$; otherwise $\boldsymbol{x}^{(i+1)} = \boldsymbol{x}^{(i)}$.
6. Update λ.
7. Set $i = i + 1$.
8. If termination condition is not satisfied, go to Step 2.

The SPRP surrogate model is updated before each iteration of the optimization algorithm using the high-fidelity model data at the design obtained in the previous iteration. The trust-region parameter, λ, is updated after each iteration, i.e. decreased if the new design was rejected or the improvement of the high-fidelity model objective function was too small compared to the prediction given by the SPRP surrogate, or increased otherwise. We use classical updating rules (see, for example, Koziel *et al.* (2006)). The algorithm is terminated if $\|\boldsymbol{x}^{(i+1)} - \boldsymbol{x}^{(i)}\| < 0.001$ or $\lambda < 0.001$.

12.4.5 *Numerical examples*

Six design studies are presented for $M_\infty = 0.2$ and $Re = 2.3$ million. Initial designs considered are NACA 0012 at $\alpha = 12°$, $14°$, and $16°$, NACA 2412 at $\alpha = 14°$, NACA 6608 at $\alpha = 16°$, and NACA 8314 at

$\alpha = 16°$. The objective is to maximize the lift coefficient at a constant angle of attack, subject to a constraint on the drag coefficient (see Section 12.4.1 for details). The airfoil shapes are parameterized using the NACA four-digit airfoil and the design variable vector is $\boldsymbol{x} = [m\ p\ t]^T$. Note that the chord length is set to 1. The design variable bounds are set to $0 \leq m \leq 0.08$, $0.3 \leq p \leq 0.6$, and $0.08 \leq t \leq 0.14$. For each case the constraint value for drag coefficient constraint is the value of the initial design. Details of the test cases and optimization results are given in Table 12.3. Results are presented for the SPRP

Table 12.3 Results from the design optimization of six NACA airfoil sections at $M_\infty = 0.2$ and $Re = 2.3$ million using direct optimization and the SPRP method. All the numerical values presented are from the high-fidelity model.

Case	Initial designs			
	α [deg]	$\boldsymbol{x}$	C_l	C_d
1	12	[0.0 0.45 0.12]	1.235	0.0212
2	14	[0.0 0.45 0.12]	1.408	0.0297
3	16	[0.0 0.45 0.12]	1.536	0.0429
4	14	[0.02 0.4 0.12]	1.597	0.0312
5	16	[0.06 0.6 0.08]	2.089	0.0893
6	16	[0.08 0.3 0.14]	1.930	0.0695

Case	Method	Optimized designs				Design cost	ΔC_l [%]
		α [deg]	$\boldsymbol{x}$	C_l	C_d		
1	Direct*	12	[0.0150 0.4840 0.1247]	1.392	0.0212	65	11.3
	SPRP$		[0.0234 0.6000 0.1400]	1.491	0.0210	11	17.2
2	Direct*	14	[0.0209 0.4457 0.1255]	1.603	0.0295	103	12.2
	SPRP$		[0.0298 0.6000 0.1400]	1.702	0.0298	8	17.3
3	Direct*	16	[0.0169 0.4415 0.1252]	1.684	0.0422	103	8.8
	SPRP$		[0.0317 0.6000 0.1375]	1.826	0.0430	11	15.9
4	Direct*	14	[0.0280 0.3787 0.1221]	1.660	0.0313	103	3.8
	SPRP$		[0.0376 0.5274 0.1398]	1.744	0.0311	9	8.4
5	Direct*	16	[0.0800 0.6000 0.0862]	2.205	0.0893	67	5.3
	SPRP$		[0.0800 0.6000 0.0863]	2.205	0.0892	4	5.3
6	Direct*	16	[0.0733 0.3000 0.0951]	2.044	0.0648	84	5.6
	SPRP$		[0.0684 0.5179 0.1024]	2.079	0.0694	9	7.2

*Design obtained through direct optimization of the high-fidelity CFD model using the pattern-search algorithm (Section 3.3.1).

$Design obtained using the SPRP method described in Section 12.4.4; surrogate model optimization performed using the pattern-search algorithm (Section 3.3.1).

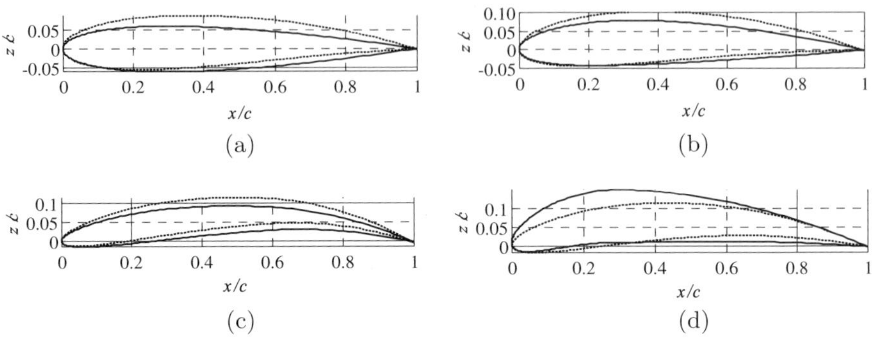

Fig. 12.30. Comparison of initial (solid lines) and optimized (dashed lines) designs for Cases 1, 4, 5 and 6 (plots (a) to (d), respectively).

method and, for comparison, direct optimization of the high-fidelity model using the pattern-search algorithm (see Section 3.3.1).

In the first design case, the SPRP method improves the lift coefficient from 1.235 to 1.491 (+17.2%) by increasing camber by 2.34% and moving the location of maximum camber more aft, from 0.45 to 0.60 (which is the upper bound). The thickness is increased from 12% to 14% (which is the upper bound). A comparison of the initial and optimized airfoil shapes is given in Figure 12.30. The proposed method achieves this by using only 11 high-fidelity CFD evaluations. The direct optimization method requires 65 high-fidelity CFD evaluations and improves the lift by only 11.3%. Figures 12.31 and 12.32 show the optimization history for Case 1. In particular, one can observe a convergence plot, as well as the evolution of the objective function, in Fig. 12.31, and the evolution of the lift and drag coefficients in Fig. 12.32. It follows that the algorithm exhibits a good convergence pattern and that the mechanisms introduced in the algorithm (i.e. the trust-region approach and the penalty function) enforce the drag limitation to be satisfied while increasing the lift coefficient as much as possible.

Figure 12.33 shows contours of velocity around the initial and optimized airfoil shapes. Effectively, the optimizer has done three things. Firstly, the increased camber opens up the pressure distribution over the whole airfoil, as can be seen in Fig. 12.34, and thus the lift increases. Secondly, the aft camber opens the

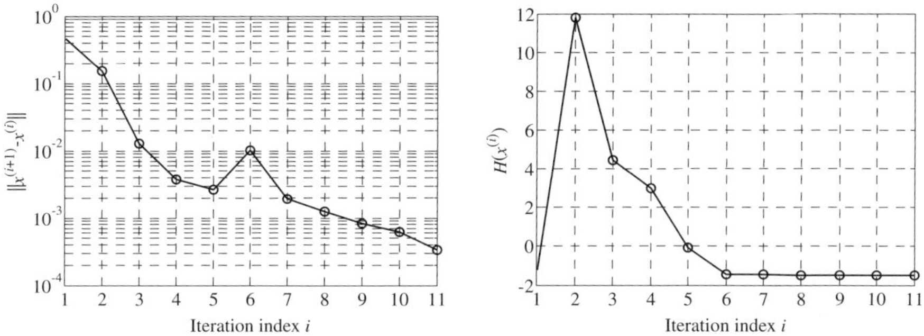

Fig. 12.31. Optimization history for Case 1 showing the convergence plot (top) and the evolution of the objective function (bottom).

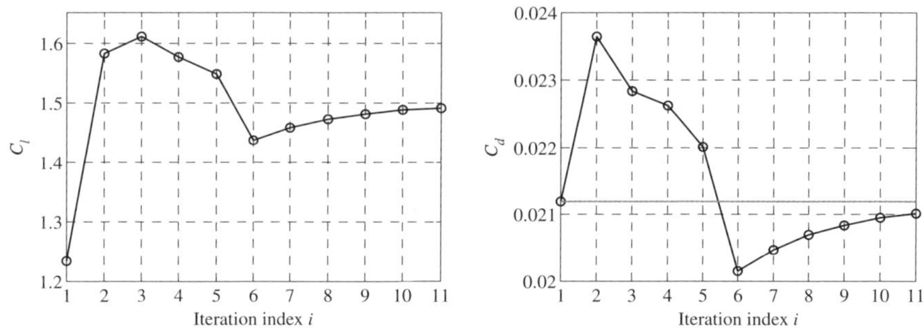

Fig. 12.32. Evolution of the lift coefficient (top) and the drag coefficient (bottom) in Case 1. The drag constraint is marked using a solid horizontal line.

pressure distribution up near the TE, also increasing lift. Thirdly, the increased thickness reduces the pressure peak near the LE, thus creating a milder expansion around the LE and thereby reducing pressure drag. The result is an optimized airfoil with improved lift coefficient at the same drag coefficient.

A comparison of the lift and drag curves is given in Fig. 12.35. Although the airfoil was optimized at $\alpha = 12°$, the entire lift curve is shifted upwards. In this case, the angle of attack at maximum lift increases slightly (by approximately $1°$). Design Cases 2 and 3 also use NACA 0012 as an initial design, but the optimization is performed at higher angles of attack, i.e. $\alpha = 14°$ and $16°$,

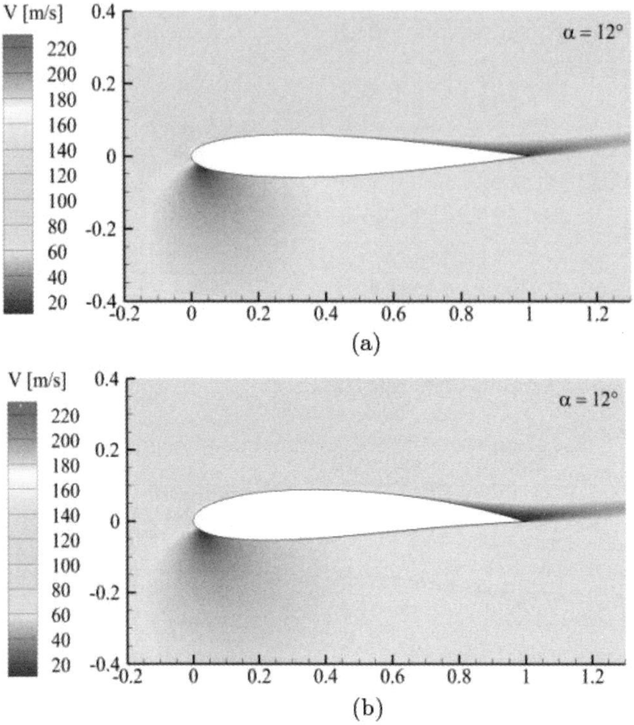

Fig. 12.33. Velocity contours of Case 1. (a) initial design; (b) optimized design.

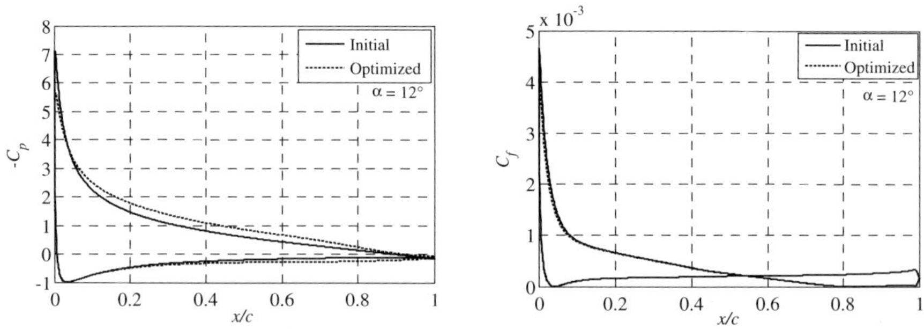

Fig. 12.34. Case 1 initial and optimized pressure distributions (top) and skin friction distributions (bottom).

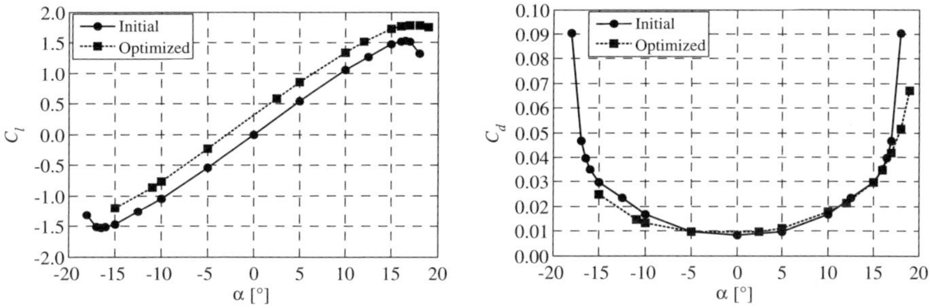

Fig. 12.35. Case 1 lift (above) and drag (below) curves of the initial and optimized designs.

respectively. Similar results are obtained in both these cases as for the first one.

Design Cases 4 and 5 use NACA 2412 and NACA 6608 as initial designs, respectively, and are optimized at 14° and 16°, respectively. Similar airfoil shape changes are obtained for these two cases as for the first three cases, i.e. both the camber and thickness increases and location of maximum camber is shifted aft. The SPRP method requires only nine and four high-fidelity CFD evaluations for Cases 4 and 5, respectively.

In the sixth design case the initial shape used is NACA 8314, which is both thick and with high camber. The optimizer now reduces both the camber, from 8% to 6.84%, and the thickness, from 14% to 10.24%. The location of maximum camber is moved more aft from 30% to 52%. As a result the lift coefficient improves by 7.2% and the drag coefficient stays constant. The SPRP method achieved this result in nine high-fidelity CFD evaluations. Direct optimization fails to achieve a similar result and improves the lift by 5.6% using 84 high-fidelity CFD evaluations.

12.5 Conclusion

SBO — particularly in variable-fidelity models — may be an attractive alternative for conventional numerical optimization techniques. As demonstrated in this chapter, using relatively simple means, such as basic multiplicative response correction, may be sufficient to carry out the design process at low computational cost. At the same time, the choice of a particular surrogate modeling approach is usually problem-dependent and requires some experience and engineering insight. Full automation of the SBO process, including selection of the low-fidelity model and a particular algorithmic approach, is still an open research problem.

Chapter 13

Selected Applications of Surrogate-Based Optimization in Other Areas

Surrogate-based optimization (SBO) techniques are attractive alternatives for conventional methods whenever the computational cost of the design process is of major concern. Nowadays, simulation tools are used across most disciplines of engineering and science and reliable software packages (both commercial and open source) are in abundance. The high cost of accurate simulation using fine discretization of the structure of interest is a common problem for most types of numerical solvers, not only computational fluid dynamics (CFD) ones. As a consequence, many SBO techniques can cover various areas and provide computational speed-up of the design optimization processes.

In this chapter, we illustrate the applications of several physics-based SBO methods described in Chapter 6 for solving computationally expensive problems in microwave engineering, antenna design, and climate modeling. The specific methods considered here include space mapping (SM), shape-preserving response prediction (SPRP), adaptive response correction (ARC), and multi-fidelity design optimization. To illustrate problem-specific SBO methods, we also discuss simulation-based tuning (or tuning SM), which is a dedicated approach for the design of planar microwave components and structures.

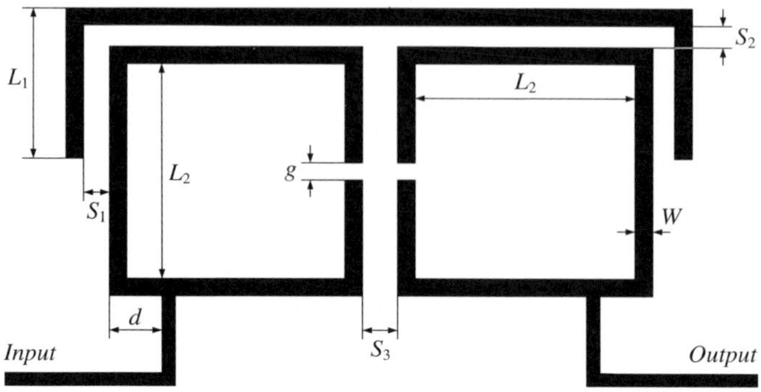

Fig. 13.1. Dual-band bandpass filter: geometry (Guan *et al.*, 2008).

13.1 Microwave Filter Design using SPRP, ARC, and SM

In this section, we illustrate the use of SM (Section 6.2), SPRP (Section 6.3), and ARC (Section 6.4) in the design optimization of a dual-band microstrip filter (Guan *et al.*, 2008), shown in Fig. 13.1. The filter consists of metal elements, so-called microstrip lines (Pozar, 2012), allocated on a dielectric substrate. The filter performance is determined by its geometry. The designable parameters are $\boldsymbol{x} = [L_1\ L_2\ S_1\ S_2\ S_3\ d\ g\ W]^T$, all in millimeters. The goal is to adjust $\boldsymbol{x}$ so that the transmission coefficient, $|S_{21}|$, of the filter (i.e. the relative power of the signal transmitted from the input to the output of the device) meets given design specifications. Here, the specifications are: $|S_{21}| \geq -3\,\text{dB}$ for $0.85\,\text{GHz} \leq \omega \leq 0.95\,\text{GHz}$ and $1.75\,\text{GHz} \leq \omega \leq 1.85\,\text{GHz}$, and $|S_{21}| \leq -20\,\text{dB}$ for $0.5\,\text{GHz} \leq \omega \leq 0.7\,\text{GHz}$, $1.1\,\text{GHz} \leq \omega \leq 1.6\,\text{GHz}$ and $2.0\,\text{GHz} \leq \omega \leq 2.2\,\text{GHz}$. In other words, the filter is supposed to transmit the signals in two frequency bands, 0.85–0.95 GHz and 1.75–1.85 GHz, and reject it at other frequencies. For the purpose of optimization, we use a minimax objective function (or specification error), which determines the maximum violation (in dB) of the above specifications by the actual $|S_{21}|$ response of the filter. In particular, the negative value of the objective function refers to the situation in which the design criteria are met in all frequency bands of interest.

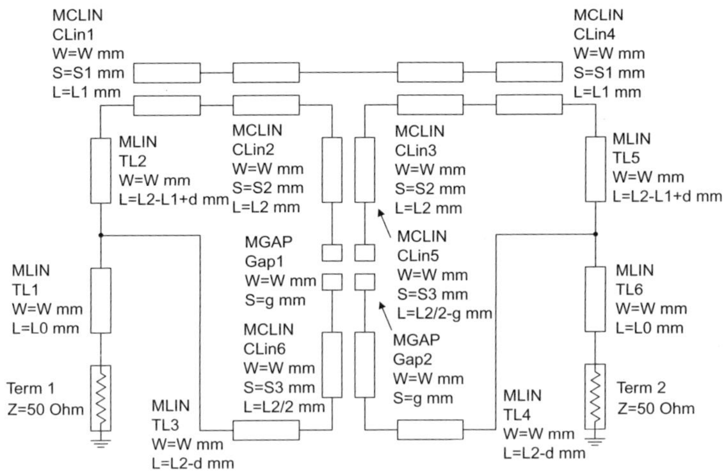

Fig. 13.2. Dual-band bandpass filter: low-fidelity model (Agilent ADS, 2011).

The high-fidelity model of the filter is simulated using a commercial full-wave electromagnetic (EM) solver Sonnet, ***em*** (2012). Simulation time is around 20 minutes per design. The low-fidelity model is an equivalent circuit implemented in Agilent ADS (2011) (Fig. 13.2). The low-fidelity model is a simplified representation of the filter, where the microstrip line models are connected together using circuit theory rules. The low-fidelity model is very fast (simulation time is in the range of milliseconds), so all the operations on the low-fidelity model can be neglected in the design process.

The initial design is $\boldsymbol{x}^{(0)} = [16.14\ 17.28\ 1.16\ 0.38\ 1.18\ 0.98\ 0.98\ 0.20]^T$ mm, which is the optimum design of the low-fidelity model. The main optimization engine is SPRP, which is then compared with ARC and SM. To set up the SPRP surrogate model, the following characteristic points are selected: four points for which $|S_{21}| = -20\,\text{dB}$, four points with $|S_{21}| = -5\,\text{dB}$, and six additional points located between the points corresponding to the $-5\,\text{dB}$ level (according to the generalized SPRP procedure described in Koziel (2010a)).

For the purpose of optimization, the low-fidelity model was preconditioned (enhanced) by tuning the dielectric constants and the substrate heights of the microstrip models corresponding to the

Table 13.1 Optimization results for dual-band bandpass filter.

Algorithm	Final specification error [dB]	Number of high-fidelity model evaluations[1]
Output SM[2]	−1.4	8
Output SM[2] + ISM[3]	−1.9	8
Adaptive response correction [6]	−1.8	3
Adaptive response correction [6] + ISM[3]	−1.5	4
Shape-preserving response prediction	−2.0[4]	3
Shape-preserving response prediction+ ISM[3]	−1.9[5]	2

[1]Excludes the high-fidelity model evaluation at the starting point.
[2]The surrogate model is of the form of $s^{(i)}(\boldsymbol{x}) = c(\boldsymbol{x}) + [f(\boldsymbol{x}^{(i)}) - c(\boldsymbol{x}^{(i)})]$.
[3]The surrogate model is of the form of $s^{(i)}(\boldsymbol{x}) = c(\boldsymbol{x} + \boldsymbol{c}^{(i)})$; $\boldsymbol{c}^{(i)}$ is found using parameter extraction (Koziel *et al.*, 2006).
[4]Design specifications satisfied after the first iteration (specification error −1.2 dB).
[5]Design specifications satisfied after the first iteration (specification error −1.0 dB).

design variables L_1, L_2, d, and g (original values of ε_r and H were 10.2 and 0.635 mm, respectively), so that improved alignment with the high-fidelity model is obtained at the initial design.

Table 13.1 shows the optimization results. We can observe that all of the considered SBO methods were able to yield solutions satisfying the design specifications. However, the quality of the designs produced by SPRP is slightly better than the quality of the designs yielded by the other methods.

Also, the computational cost of the SPRP technique is lower than that of the other approaches, particularly SM. This indicates that SRPR is capable of fully exploiting the knowledge embedded in the low-fidelity model, which is only partially done by ARC and SM. Figure 13.3 shows the initial high-fidelity model response as well as the response at the design obtained using SPRP.

13.2 Microwave Filter Design using Simulation-Based Tuning

In this section, we illustrate the use of so-called tuning space mapping (TSM) for the design of microwave filters. TSM (Koziel *et al.*, 2009b) is a unique technology that combines SM and the concept of tuning, widely used in microwave engineering (Swanson and Wenzel, 2001;

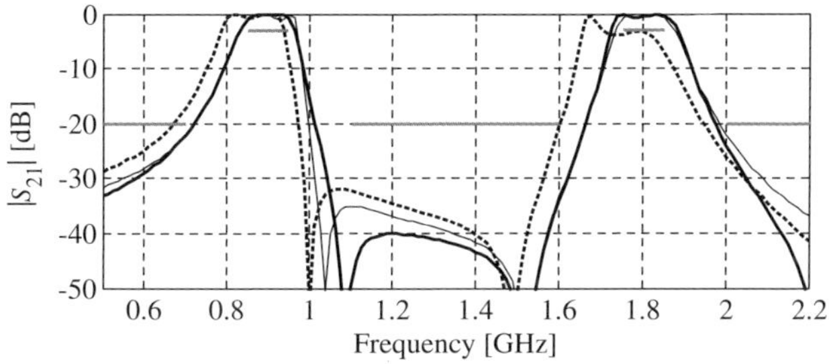

Fig. 13.3. Dual-band bandpass filter: high-fidelity model (dashed line) and low-fidelity model (thin dashed line) response at $\boldsymbol{x}^{(0)}$, and the optimized high-fidelity model response (solid line) at the design obtained using SPRP.

Rautio, 2006; Swanson and Macchiarella, 2007). Formally speaking, it is a physics-based SBO technique in which the surrogate model's role is taken by a so-called tuning model, which is constructed by introducing circuit-theory-based tuning elements (e.g. capacitors, inductors or coupled-line models) into the EM-simulated high-fidelity model structure. The tuning model is updated and optimized with respect to the tuning parameters. Their optimal values are subsequently transformed — in a calibration process — into an appropriate modification of the design variables, which are then assigned to the high-fidelity model. Because the tuning model is based on an "image" of the high-fidelity one (normally stored in the form of numerical data), its generalization capability is good, which results in a small number of high-fidelity model evaluations required to find a satisfactory design.

13.2.1 *TSM concept*

Here we present one of the TSM variations, also-referred to as "Type-0" TSM (Cheng *et al.*, 2010b). Other versions can be found in the literature (Koziel *et al.*, 2009b; Cheng *et al.*, 2010a; Rasmita *et al.*, 2009). TSM is an iterative optimization procedure that assumes the existence of two surrogate models. The first model is a so-called tuning model, c_t, that contains relevant high-fidelity data (typically a high-fidelity model response) at the current iteration point and tuning parameters (typically implemented through circuit

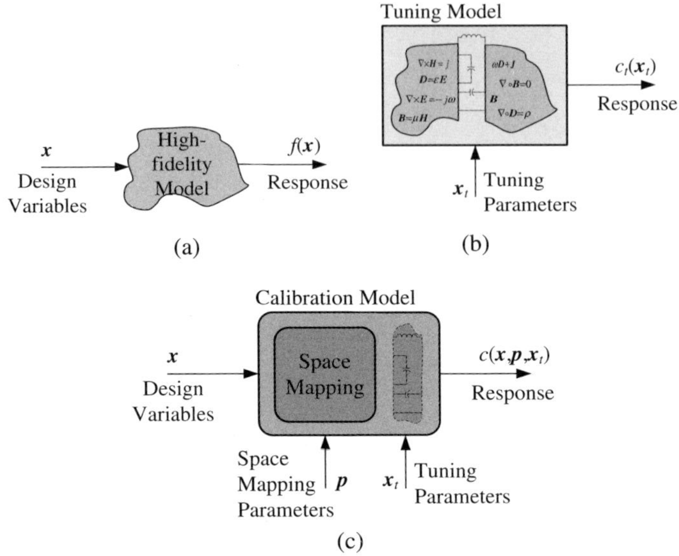

Fig. 13.4. Conceptual illustrations of the high-fidelity model, the tuning model, and the calibration model. (a) The high-fidelity model is typically based on full-wave electromagnetic simulation; (b) the tuning model exploits the high-fidelity model "image" (e.g. in the form of S-parameters corresponding to the current design imported to the tuning model using suitable data components) and a number of circuit-theory-based tuning elements; (c) the calibration model is usually a circuit equivalent dependent on the same design variables as the high-fidelity model, the same tuning parameters as the tuning model, and, additionally, a set of SM parameters used to align the calibration model with both the high-fidelity and the tuning model during the calibration process (Koziel *et al.*, 2011a).

elements inserted into so-called tuning ports; Koziel *et al.* (2009b)). These data are obtained from a so-called "cut" high-fidelity model which is the same as the original microwave structure under consideration, except that it contains the tuning ports (see, for example, Figs. 13.4 and 13.5). The tuning ports have to be perfectly calibrated so that inserting them does not affect the simulation results. This kind of capability is available through the co-calibrated port technology of Sonnet ***em*** (2012). The tunable parameters are adjusted so that the model, c_t, satisfies the design specifications. The second model, c, is used for calibration purposes: it allows us to translate the change of the tuning parameters into relevant changes of the actual design variables. The calibration model is dependent on three sets of

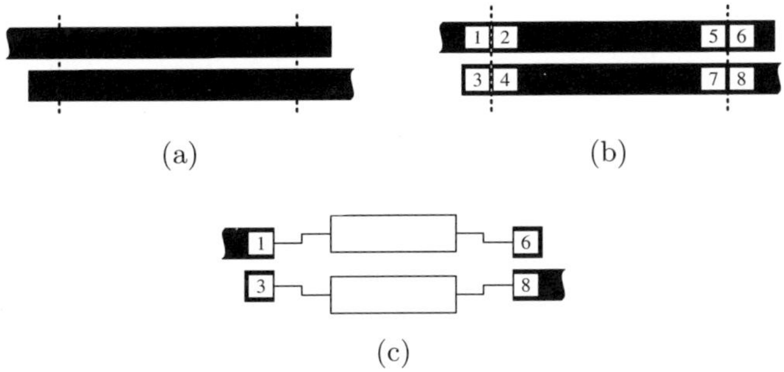

Fig. 13.5. TSM concept. (a) A coupled microstrip line; (b) coupled line with co-calibrated ports inserted; (c) tuning model of the coupled line; middle section is replaced by the microstrip model (tuning element).

variables: design parameters, tuning parameters (which are actually the same parameters as the ones used in c_t), and SM parameters that are adjusted using the usual parameter extraction process (Koziel *et al.*, 2006) in order to have the model c meet certain matching conditions. Typically, the model c is a standard SM surrogate (i.e. a low-fidelity model composed with suitable transformations) enhanced by the same or corresponding tuning elements as the model t. The conceptual illustrations of the high-fidelity model, the tuning model, and the calibration model are shown in Fig. 13.4. Figure 13.5 shows an example of tuning port and tuning component insertion.

13.2.2 *TSM algorithm*

The Type-0 TSM algorithm produces a sequence of points (design variable vectors), $\boldsymbol{x}^{(i)}$, $i = 0, 1, \ldots$. The iteration of the algorithm consists of two steps: optimization of the tuning model and a calibration procedure. First, the current tuning model, $c_t^{(i)}$, is built using high-fidelity model data at point $\boldsymbol{x}^{(i)}$. In general, because the high-fidelity model with inserted tuning ports is not identical to the original structure, the tuning model response may not agree with the response of the high-fidelity model at $\boldsymbol{x}^{(i)}$ even if the values of the tuning parameters, $\boldsymbol{x}_t$, are zero, so these values must be adjusted to, say, $\boldsymbol{x}_{t.0}^{(i)}$, in order to obtain alignment (Koziel *et al.*, 2009b):

$$\boldsymbol{x}_{t.0}^{(i)} = \arg\min_{\boldsymbol{x}_t} \|f(\boldsymbol{x}^{(i)}) - c_t^{(i)}(\boldsymbol{x}_t)\|. \tag{13.1}$$

In the next step, the tuning model, $c_t^{(i)}$, is optimized to have it meet the design specifications. The optimal values of the tuning parameters, $\boldsymbol{x}_{t.1}^{(i)}$, are obtained as:

$$\boldsymbol{x}_{t.1}^{(i)} = \arg\min_{\boldsymbol{x}_t} U(c_t^{(i)}(\boldsymbol{x}_t)). \tag{13.2}$$

Having $\boldsymbol{x}_{t.1}^{(i)}$, we perform the calibration procedure to determine changes in the design variables that yield the same change in the calibration model response as that caused $\boldsymbol{x}_{t.1}^{(i)} - \boldsymbol{x}_{t.0}^{(i)}$ (Koziel *et al.*, 2009b). Calibration is a two-step process. First, the SM parameters, $\boldsymbol{p}^{(i)}$, of the calibration model are adjusted to obtain a match with the high-fidelity model response at $\boldsymbol{x}^{(i)}$:

$$\boldsymbol{p}^{(i)} = \arg\min_{\boldsymbol{p}} \|f(\boldsymbol{x}^{(i)}) - c(\boldsymbol{x}^{(i)}, \boldsymbol{p}, \boldsymbol{x}_{t.0}^{(i)})\|. \tag{13.3}$$

In the second step, the low-fidelity model, c, is optimized with respect to the design variables in order to obtain the next iteration point $\boldsymbol{x}^{(i+1)}$:

$$\boldsymbol{x}^{(i+1)} = \arg\min_{\boldsymbol{x}} \|c_t^{(i)}(\boldsymbol{x}_{t.1}^{(i)}) - c(\boldsymbol{x}, \boldsymbol{p}^{(i)}, \boldsymbol{x}_{t.0}^{(i)})\|. \tag{13.4}$$

Note that we use $\boldsymbol{x}_{t.0}^{(i)}$ in Equation (13.3), which corresponds to the state of the tuning model after performing the alignment procedure in Equation (13.1), and $\boldsymbol{x}_{t.1}^{(i)}$ in Equation (13.4), which corresponds to the optimized tuning model (cf. Equation (13.2)). In other words, Equations (13.3) and (13.4) allow us to find the change of design variable values, $\boldsymbol{x}^{(i+1)} - \boldsymbol{x}^{(i)}$, necessary to compensate the effect of changing the tuning parameters from $\boldsymbol{x}_{t.0}^{(i)}$ to $\boldsymbol{x}_{t.1}^{(i)}$.

The calibration procedure described here represents the most generic approach. In some cases, there might be an explicit analytical relation between the design variables and the tuning parameters so that the updated design can be found simply by applying an appropriate formula. In particular, the calibration formula may be just a linear function so that $\boldsymbol{x}^{(i+1)} = \boldsymbol{x}^{(i)} + \boldsymbol{s}^{(i)} * (\boldsymbol{x}_{t.1}^{(i)} - \boldsymbol{x}_{t.0}^{(i)})$, where $\boldsymbol{s}^{(i)}$ is a

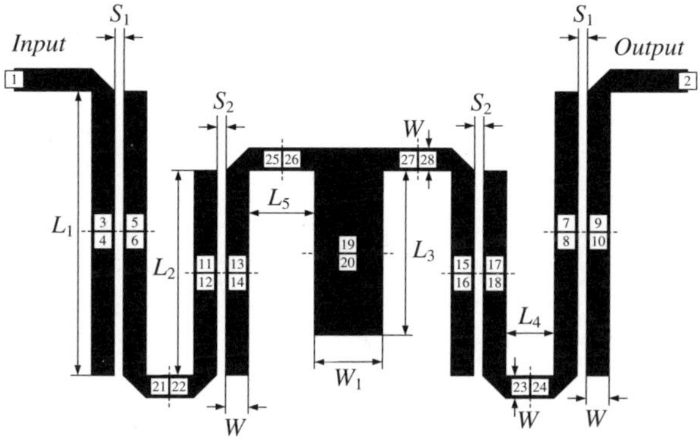

Fig. 13.6. Box-section Chebyshev bandpass filter: geometry [50], and the tuning port insertion points.

real vector and $*$ denotes a Hadamard product (i.e. component-wise multiplication). Obviously, if the analytical calibration is possible, there is no need to use the calibration model. Other approaches to the calibration process can be found in the literature (Koziel *et al.*, 2009b; Cheng *et al.*, 2010b).

13.2.3 *Illustration example: microstrip filter design using TSM*

TSM design optimization procedure is illustrated using the box-section Chebyshev microstrip bandpass filter (Liao *et al.*, 2007), shown in Fig. 13.6. The design parameters are $\boldsymbol{x} = [L_1\ L_2\ L_3\ L_4\ L_5\ S_1\ S_2]^T$. The high-fidelity model is simulated in Sonnet ***em*** (2012), with a grid of 1 mil $\times$ 2 mil (1 mil $= 0.0254$ mm). The width parameters are $W = 40$ mil and $W_1 = 150$ mil. Substrate parameters are relative permittivity $\varepsilon_r = 3.63$ and height $H = 20$ mil. The design specifications are $|S_{21}| \leq -20$ dB for 1.8 GHz $\leq \omega \leq$ 2.15 GHz and 2.65 GHz $\leq \omega \leq$ 3.0 GHz, and $|S_{21}| \geq -3$ dB for 2.4 GHz $\leq \omega \leq$ 2.5 GHz.

The tuning model is constructed by dividing the polygons corresponding to parameters L_1–L_5 in the middle and inserting the tuning ports at the new cut edges. Its S28P data file (i.e. the file generated by the EM solver and containing the high-fidelity model

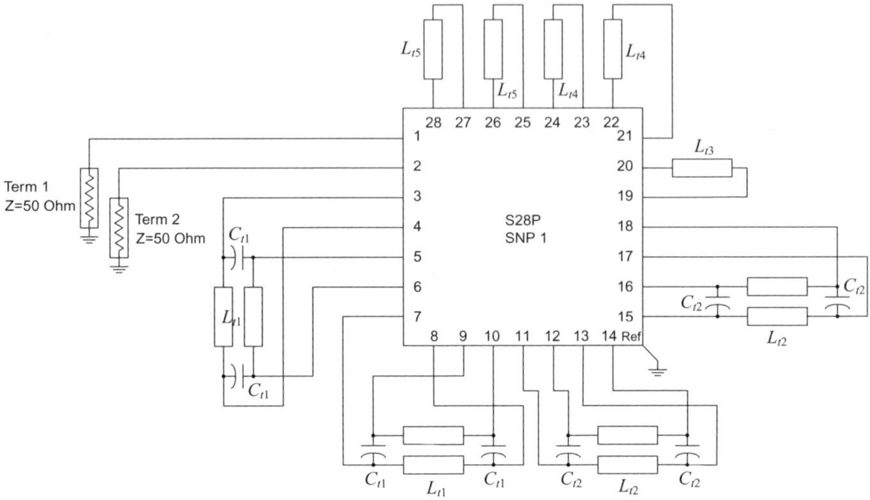

Fig. 13.7. Box-section Chebyshev bandpass filter: tuning model (Agilent ADS, 2011).

S-parameter data) is then loaded into the S-parameter component in Agilent ADS (2011). The circuit-theory coupled-line components and capacitor components are chosen to be the tuning elements and are inserted into each pair of tuning ports (Fig. 13.7). The lengths of the imposed coupled lines and the capacitances of the capacitors are assigned to be the tuning parameters, so that we have $\boldsymbol{x}_t = [L_{t1}\ L_{t2}\ L_{t3}\ L_{t4}\ L_{t5}\ C_{t1}\ C_{t2}]^T$ (L_{tk} in mil, C_{tk} in pF).

The calibration model is implemented in Agilent ADS and shown in Fig. 13.8. It contains the same tuning elements as the tuning model. It basically mimics the division of the coupled lines performed while preparing c_t. The calibration model also contains six (implicit) SM parameters that will be used as parameters, $\boldsymbol{p}$, in the calibration process (Equations (13.3) and (13.4)). These parameters are $\boldsymbol{p} = [\varepsilon_{r1}\ \varepsilon_{r2}\ \varepsilon_{r3}\ \varepsilon_{r4}\ \varepsilon_{r5}\ H]^T$, where ε_{rk} is the dielectric constant of the microstrip line segment of length L_k (cf. Fig. 13.5), and H is the substrate height of the filter. Initial values of these parameters are $[3.63\ 3.63\ 3.63\ 3.63\ 3.63\ 20]^T$.

The initial design, $\boldsymbol{x}^{(0)} = [928\ 508\ 50\ 50\ 201\ 5\ 19]^T$ mil, is the optimal solution of the low-fidelity model, i.e. the calibration model with zero values of the tuning parameters. The minimax specification

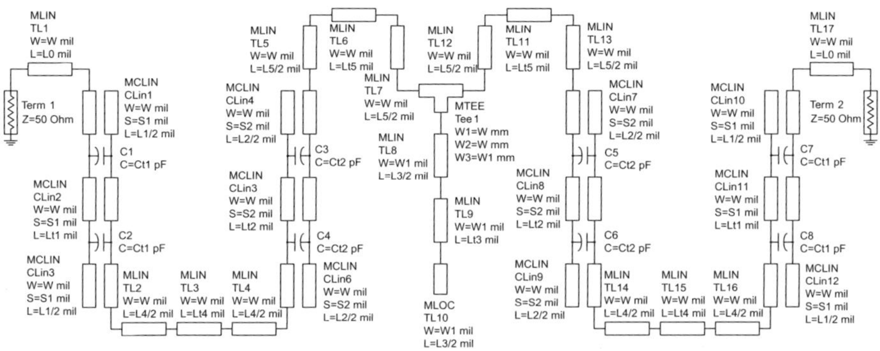

Fig. 13.8. Box-section Chebyshev bandpass filter: calibration model (Agilent ADS, 2011).

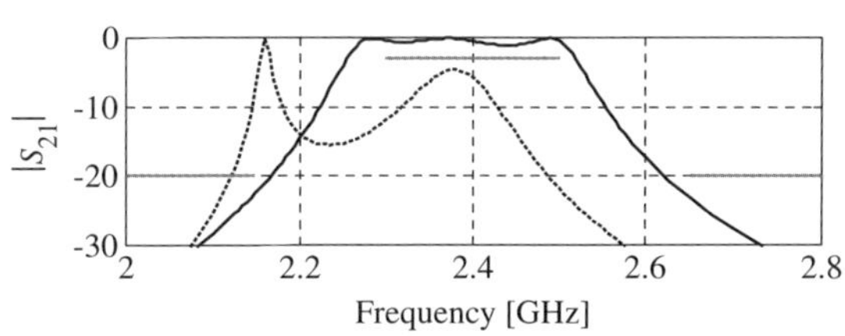

Fig. 13.9. Box-section Chebyshev bandpass filter: high-fidelity model response at the initial design (dashed line) and after one iteration of the TSM algorithm (solid line).

error is +19 dB. The misalignment between the high-fidelity and the tuning model response with the tuning elements set to zero is negligible (thanks to the co-calibrated port feature in Sonnet ***em*** (2012), so that $\boldsymbol{x}_{t.0}^{(0)} = [0\ 0\ 0\ 0\ 0\ 0\ 0]^T$ was used throughout. The values of the tuning parameters at the optimal design of the tuning model are $\boldsymbol{x}_{t.1}^{(0)} = [-85.2\ 132.5\ 5.24\ 1.13\ -15.24\ 0.169\ -0.290]^T$. Note that some of the parameters take negative values, which is permitted in Agilent ADS. The values of pre-assigned parameters obtained in the first calibration phase (Equation (13.3)) are $\boldsymbol{p}^{(0)} = [3.10\ 6.98\ 4.29\ 7.00\ 6.05\ 17.41]^T$.

Figure 13.9 shows the low- and high-fidelity model response at the initial design, as well as the high-fidelity model response after just one TSM iteration with $\boldsymbol{x}^{(1)} = [1022\ 398\ 46\ 56\ 235\ 4\ 10]^T$ mil (specification error −1.8 dB). It should be emphasized that the evaluation time of both the tuning and the calibration model is very low (a fraction of a second), and it is negligible compared to the evaluation time of the low-fidelity model. Consequently, the computational cost of each TSM iteration corresponds to only two electromagnetic simulations (one for the high-fidelity model and one for the "cut" high-fidelity model).

13.3 Antenna Design using Variable-Fidelity Electromagnetic Models

In this section, we discuss the design of a wideband microstrip antenna (Chen, 2008) using a multi-fidelity design optimization algorithm (see Section 7.5 for the method details). The antenna structure is shown in Figure 13.10. The design variables are $\boldsymbol{x} = [l_1\ l_2\ l_3\ l_4\ w_2\ w_3\ d_1\ s]^T$. The multilayer substrate is $l_s \times l_s$ ($l_s = 30$ mm). The stack (from the bottom to the top) is: (i) ground, (i) the RO4003 substrate, (iii) the signal trace, (iv) the RO3006 substrate (with a through via trace to patch), (v) the driven patch, (vi) the RO4003 substrate, and (vii) four patches. Feeding is with a 50 ohm SMA (SubMiniature version A) connector. The high-fidelity

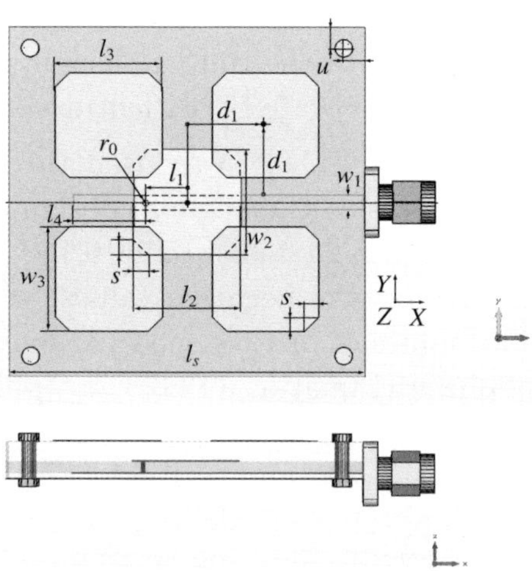

Fig. 13.10. Microstrip antenna: top/side views, substrates shown transparent.

antenna model, f, is simulated using CST Microwave Studio (2012) with ~2,300,000 mesh cells (evaluation time 160 min per design).

The design objective is to minimize the reflection coefficient, $|S_{11}|$, of the antenna in the frequency band 3.1–4.8 GHz so that we have $|S_{11}| \leq -10$ dB for the aforementioned frequencies. Reduction of the reflection coefficient improves the antenna performance in the sense of having a larger portion of the incoming signal to be radiated (and not reflected from the antenna input). At the same time, we want the IEEE (Institute of Electrical and Electronics Engineers) antenna gain to be more than 5 dB for the zero elevation angle over the same frequency band. The gain requirements are handled as optimization constraint.

The initial design is $\boldsymbol{x}^{(0)} = [15\ 15\ 15\ 15\ 20\ {-4}\ 2\ 2]^T$ mm. We use the MFDO (multi-fidelity design optimization) algorithm with two coarse-discretization models (also simulated in CST Microwave Studio): c_1 (~120,000 mesh cells at $\boldsymbol{x}^{(0)}$, simulation time 3 min) and c_2 (~780,000 mesh cells, simulation time 18 min).

The antenna was optimized using the MFDO algorithm described in Section 7.5. Figure 13.11 shows the responses of c_1 at $\boldsymbol{x}^{(0)}$ and at its optimal design, $\boldsymbol{x}^{(1)}$. Figure 13.12 shows the responses of c_2 at $\boldsymbol{x}^{(1)}$ and at its optimized design, $\boldsymbol{x}^{(2)}$. Figure 13.13 shows the responses of f at $\boldsymbol{x}^{(0)}$, at $\boldsymbol{x}^{(2)}$, and at the refined design, $\boldsymbol{x}^* = [14.87\ 13.95\ 15.4\ 13.13\ 20.87\ {-5.90}\ 2.88\ 0.68]^T$ mm ($|S_{11}| \leq -11.5$ dB for 3.1 GHz to 4.8 GHz), obtained in two iterations of the refinement step (cf. Equation (7.16)). The design cost (Table 13.2) corresponds to about 12 runs of the high-fidelity model f. Antenna gain at the final design is shown in Fig. 13.14. For comparison, direct optimization of the

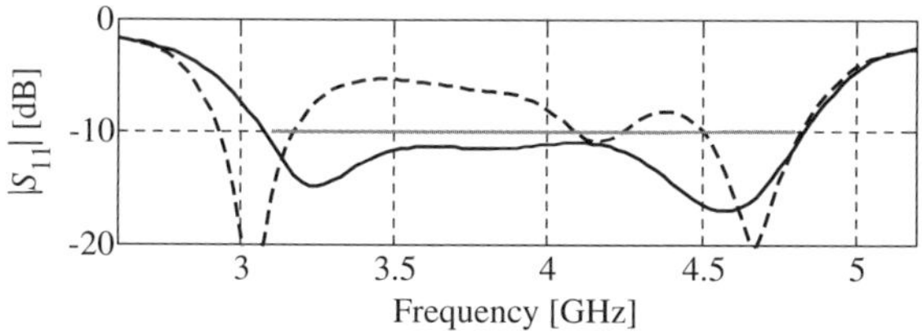

Fig. 13.11. Microstrip antenna: responses of the coarse-discretization model, c_1, at the initial design, $\boldsymbol{x}^{(0)}$ (- - -), and at the optimized design, $\boldsymbol{x}^{(1)}$ (—).

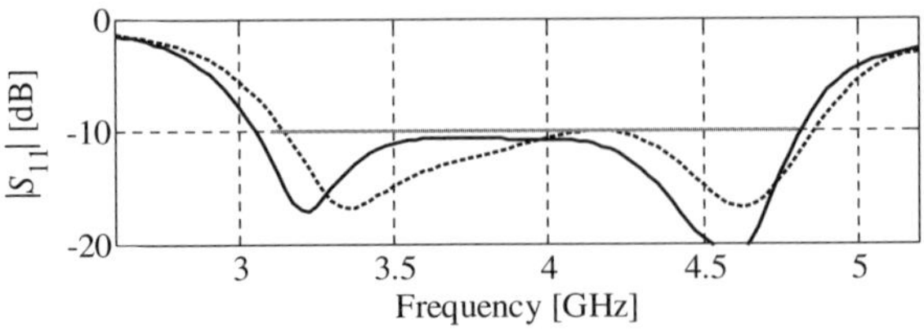

Fig. 13.12. Microstrip antenna: responses of the coarse-discretization model, c_2, at $\boldsymbol{x}^{(1)}$ (- - -), and at its optimized design, $\boldsymbol{x}^{(2)}$ (—).

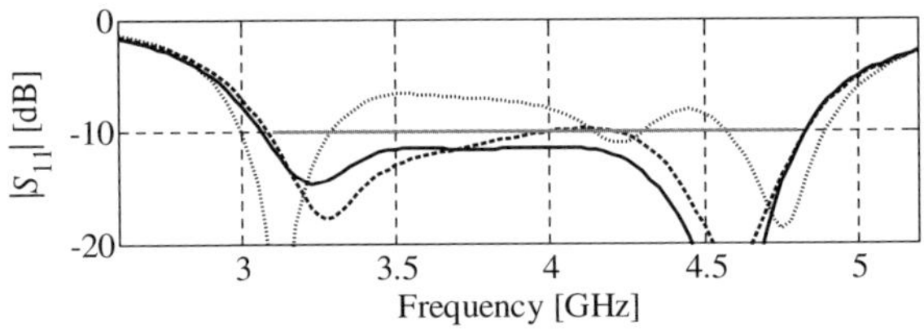

Fig. 13.13. Microstrip antenna: responses of the original, high-fidelity model, f, at $\boldsymbol{x}^{(0)}$ (· · ··), at $\boldsymbol{x}^{(2)}$ (- - -), and at the refined final design, $\boldsymbol{x}^{*}$ (—).

Table 13.2 Microstrip antenna: design cost.

Design step	Model evaluations	Computational cost*	
		Absolute [h]	Relative to f
Optimization of c_1	$125 \times c_1$	6.3	2.6
Optimization of c_2	$48 \times c_2$	14.4	5.4
Setup of model q	$17 \times c_2$	5.1	1.9
Evaluation of f	$2 \times f$	5.3	2.0
Total design time	N/A	31.1	11.9

*Excludes f evaluation at the initial design.

high-fidelity model using a pattern search algorithm (Koziel, 2010a) requires 200 evaluations of the model.

13.4 Antenna Design using Space Mapping and Response Surface Approximations

One of the challenges in SBO of antenna structures is that the underlying low-fidelity model is usually relatively expensive. As opposed to equivalent circuit models that are commonly used for filter structures and are very fast (evaluation times are in the range

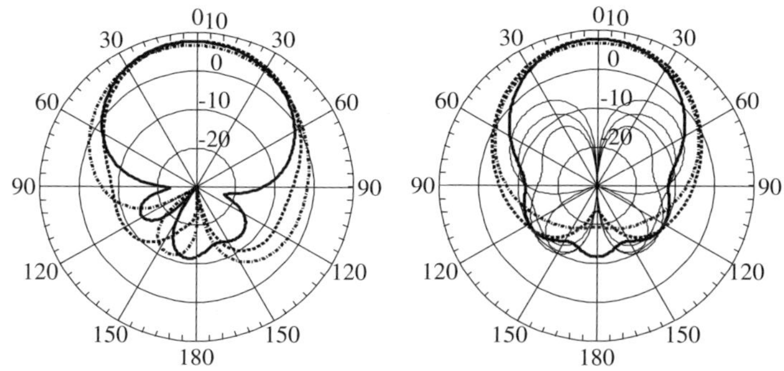

Fig. 13.14. Microstrip antenna, gain [dBi] of the final design at 3.5 GHz (· – ·), 4.0 GHz (– –), and 4.5 GHz (—). (a) Co-polarization in the E-plane (XOZ) and connector is at 90° on the right; (b) x-polarization, primary (thick lines), and co-polarization (thin lines) in the H-plane (YOZ).

of milliseconds), low-fidelity antenna models are normally obtained from coarse-discretization EM simulations. The practical time evaluation ratios between the high- and low-fidelity models are therefore relatively small, typically between 10 and 50. This poses a challenge for SBO antenna design because, without proper handling, the cost of multiple low-fidelity model evaluations may be the major contributor to the overall cost of the SBO process (Koziel and Ogurtsov, 2013c).

The above problem is particularly visible in SM, where multiple evaluations of the low-fidelity model are required not only to optimize the surrogate model but also to carry out the parameter extraction process (cf. Section 6.2). This difficulty can be alleviated to some extent through replacing the low-fidelity antenna model (here referred to as c_{cd}) — before executing the SM algorithm — with its response surface approximation (RSA) model (referred to as c). In practice, this is done in the small vicinity of the low-fidelity model optimum, which allows for the reduction of the computational overhead of creating the RSA model. The design optimization procedure implementing this concept, also outlined in a flowchart shown in Fig. 13.15, can be summarized as follows.

1. Take an initial design $\boldsymbol{x}^{init}$.
2. Find the starting point, $\boldsymbol{x}^{(0)}$, for the SM algorithm by optimizing the coarse-discretization model, c_{cd}.
3. Allocate N base (training) designs, $X_B = \{\boldsymbol{x}^1, \ldots, \boldsymbol{x}^N\}$.

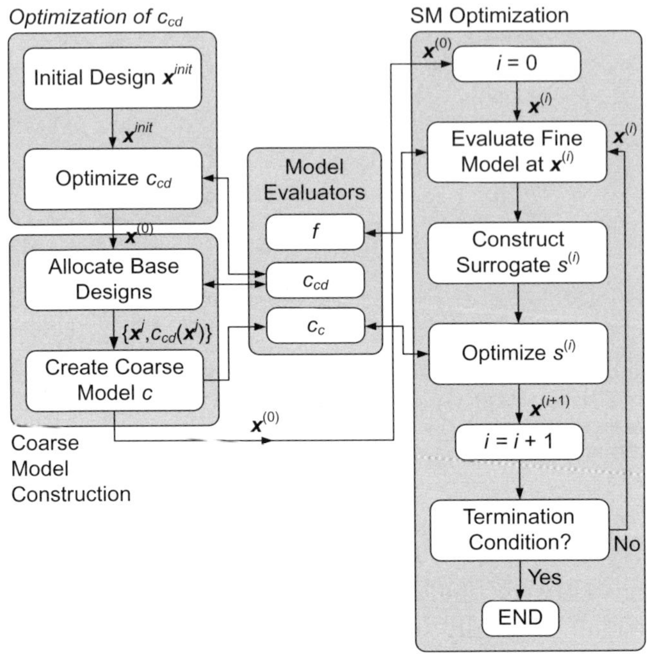

Fig. 13.15. Flowchart of the design optimization procedure exploiting an RSA-based coarse model and SM (Koziel and Ogurtsov, 2011).

4. Evaluate c_{cd} at each design $\boldsymbol{x}^j$, $j = 1, 2, \ldots, N$.
5. Build the coarse model, c, as an RSA of the data set $\{(\boldsymbol{x}^j, c_{cd}(\boldsymbol{x}^j))\}_{j=1,\ldots,N}$.
6. Set $i = 0$.
7. Evaluate the high-fidelity model f at $\boldsymbol{x}^{(i)}$.
8. Construct the surrogate model, $s^{(i)}$, from c, using suitable SM correction.
9. Find a new design, $\boldsymbol{x}^{(i+1)}$, as the surrogate model optimum: $\boldsymbol{x}^{(i+1)} = \text{argmin}\{\boldsymbol{x}\colon U(s^{(i)}(\boldsymbol{x}))\}$.
10. Set $i = i + 1$.
11. If the termination condition is not satisfied go to Step 7.
12. END.

The first phase of the design process is to find an optimized design of the coarse-discretization model, c_{cd}. The optimum of c_{cd} is usually the best design we can get at a reasonably low computational cost. This cost can be further reduced by relaxing tolerance requirements while searching for $\boldsymbol{x}^{(0)}$: due to a limited accuracy of c_{cd}, it is sufficient to find only a rough approximation of its optimum. Steps 3–5 describe the construction of the RSA coarse model c. In the example below, we use kriging interpolation (Queipo *et al.*, 2005), with the base points allocated using Latin hypercube sampling (LHS) (Beachkofski and Grandhi, 2002). Steps 6–12 describe the flow of the SM algorithm with particular operations detailed (see Section 6.2 for more details).

For the sake of illustration, consider a rectangular dielectric resonator antenna (DRA), shown in Fig. 13.16. The DRA comprises: a suspended rectangular dielectric resonator (DR), estimated to operate at the perturbed $TE_{\delta 11}$ mode (Petosa, 2007); supporting RO4003C slabs; and polycarbonate housing. The housing is fixed to the circuit board with four M1 bolts. The DRA is energized with a

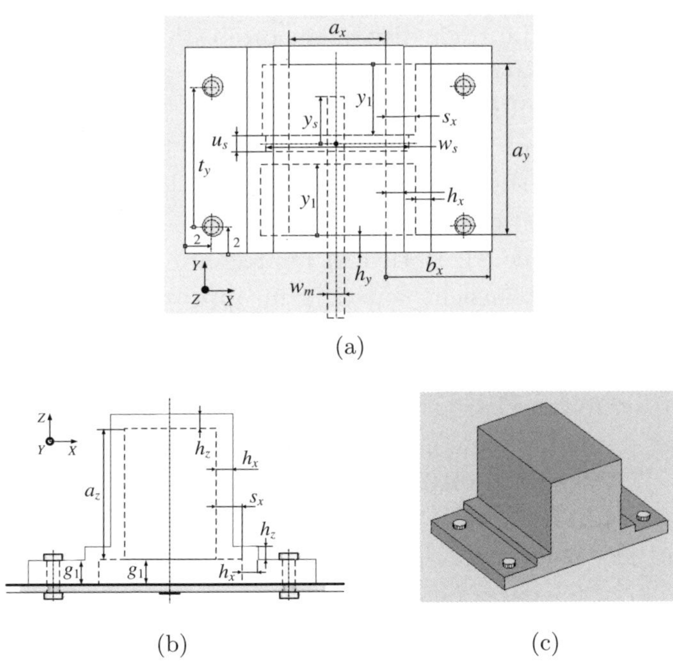

Fig. 13.16. DRA. (a) Top view; (b) front view; (c) 3D view (Koziel and Ogurtsov, 2011).

50 ohm microstrip through a slot made in the metal ground. The substrate is 0.5 mm-thick RO4003C.

Design specifications are $|S_{11}| \leq -15\,\text{dB}$ for 5.1–5.9 GHz; also the DRA is required to have antenna gain better than 5 dBi for the zero zenith angle over the bandwidth of interest.

There are nine design variables: $\boldsymbol{x} = [a_x\ a_y\ a_z\ a_{y0}\ u_s\ w_s\ y_s\ g_1\ y_1]^T$, where a_x, a_y, and a_z are dimensions of the DR brick; a_{y0} stands for the offset of the DR center relative to the slot center, marked by a black dot in Fig. 13.16(a); u_s and w_s are the slot dimensions; y_s is the length of the open-ended microstrip stub; and g_1 and y_1 are dimensions of the supporting slabs. Relative permittivity and loss tangent of the DR are 10 and 10^{-4} respectively at 6.5 GHz. The width of the microstrip signal trace is 1.15 mm. Metallization of the trace and ground is with 50 μm copper. Relative permittivity and loss tangent of the polycarbonate housing are 2.8 and 0.01 at 6.5 GHz, respectively. DRA models are defined with the CST Microwave Studio (2012), and the built in single-pole Debye model is used for all dielectrics to describe their dispersion properties. Other dimensions are fixed as follows: $h_x = h_y = h_z = 1$, $b_x = 7.5$, $s_x = 2$, and $t_y = a_y - a_{y0} - 1$, all in mm.

The initial design is $\boldsymbol{x}^{init} = [8.00\ 14.00\ 9.00\ 0\ 1.75\ 10.00\ 3.00\ 1.500\ 6.00]^T$ mm. Here we use two EM models: the high-fidelity model f (~1,100,000 mesh cells) and the coarse-discretization model c_{cd} (~27,000 mesh cells). Both EM models are evaluated with CST Microwave Studio transient solver; f in approximately 35 min and $\boldsymbol{R}_{cd}$ in about 42 seconds.

In the optimization process, the antenna reflection $|\text{S}_{11}|$ is treated as the objective function and gain requirements are handled through a constraint function. In the first step, the optimum of c_{cd} is found at $\boldsymbol{x}^{(0)} = [7.444\ 13.556\ 9.167\ 0.250\ 1.750\ 10.500\ 2.500\ 1.500\ 6.000]^T$ mm. Figure 13.17 shows the high-fidelity model reflection response at the initial design, $\boldsymbol{x}^{init}$, as well as the responses of f and c_{cd} at $\boldsymbol{x}^{(0)}$. The kriging coarse model, c, is set up using 200 samples of c_{cd}, allocated in the vicinity of $\boldsymbol{x}^{(0)}$ of the size $[0.5\ 0.5\ 0.5\ 0.25\ 0.5\ 0.25\ 0.25\ 0.25\ 0.5]^T$ mm.

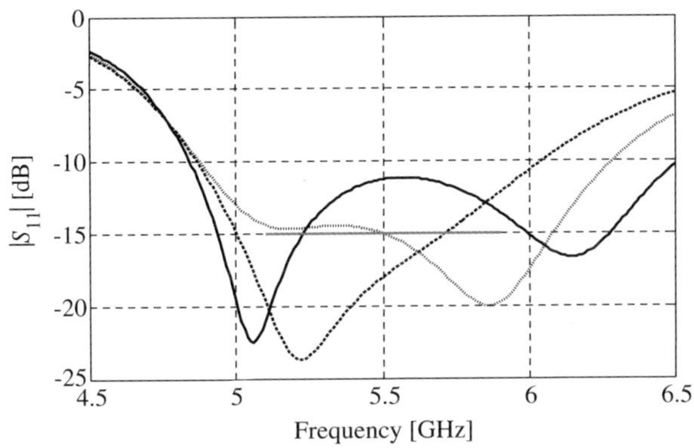

Fig. 13.17. DRA, $|S_{11}|$, versus frequency: fine model $\boldsymbol{R}_f$ at the initial design (- - -), optimized coarse-discretization model $\boldsymbol{R}_{cd}$ (· · · ·), and $\boldsymbol{R}_f$ at the optimum of $\boldsymbol{R}_{cd}$ (—) (Koziel and Ogurtsov, 2011).

Table 13.3 DRA: optimization cost (Koziel and Ogurtsov, 2011).

Algorithm component	The number of model evaluations	CPU time	
		Absolute	Relative to f
Optimization of c_{cd}	$150 \times c_{cd}$	105 min	2.9
Setting up c	$200 \times c_{cd}$	140 min	3.9
Evaluation of f	$4 \times f$	145 min	4.0
Total cost	N/A	390 min	**10.8**

The final design, $\boldsymbol{x}^{(4)} = [7.556\ 13.278\ 9.630\ 0.472\ 1.287\ 10.593\ 2.667\ 1.722\ 6.482]^T$ mm, is obtained after four SM iterations. The reflection response is in Fig. 13.18. The farfield response of the final design is shown in Fig. 13.19. For the bandwidth of interest, the peak gain is above 5 dBi, and the back radiation level is below -14 dB (relative to the maximum). All responses shown include the effect of 25 mm input microstrip. The surrogate model used by the optimization algorithm exploited input and output SM of the form $\boldsymbol{R}_s(\boldsymbol{x}) = \boldsymbol{R}_c(\boldsymbol{x} + \boldsymbol{c}) + \boldsymbol{d}$. Optimization costs are summarized in Table 13.3. The total design time corresponds to about 11 evaluations of the high-fidelity model.

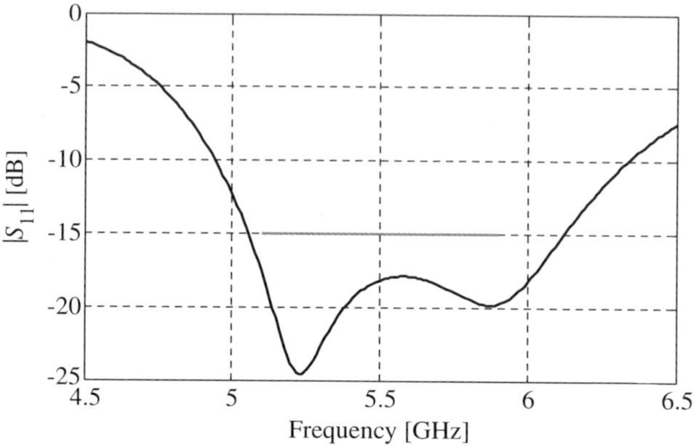

Fig. 13.18. DRA, $|S_{11}|$, versus frequency: $\boldsymbol{R}_f$ at the final design. The design specification is shown with the solid horizontal line (Koziel and Ogurtsov, 2011).

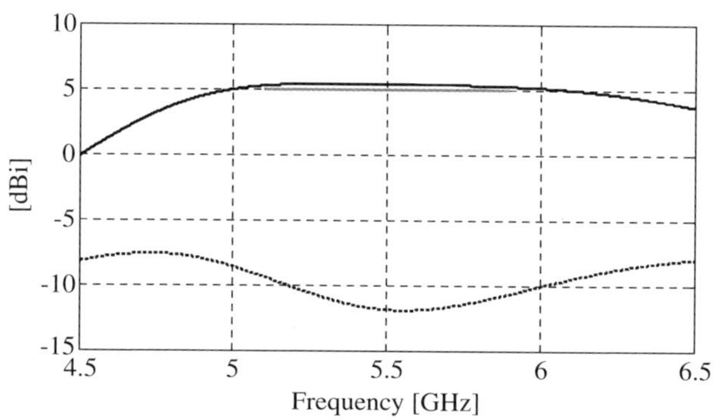

Fig. 13.19. DRA, realized gain versus frequency: (—) is for the zero zenith angle ($\theta = 0°$); (- - -) is back radiation for $\theta = 180°$ (Koziel and Ogurtsov, 2011). Here, only θ-polarization ($\phi = 90°$) contributes to the gain for the listed directions. The design constrain is shown with the solid horizontal line.

13.5 Multi-Objective Optimization of Antennas using Physics-Based and Approximation Surrogates

Multi-objective optimization is usually carried out in order to find a set of alternative designs representing possible trade-offs between conflicting objectives (referred to as a Pareto front; Coello Coello (2013)). Perhaps the most popular solution approaches are multi-objective evolutionary algorithms (MOEA; Deb (2009)), as they allow the identification of the entire representation of the Pareto front in one run of the algorithm. However, MOEA are computationally very expensive, with thousands of objective function evaluations necessary to yield a reasonable Pareto front approximation. This is particularly problematic for problems in which the objectives are evaluated using time-consuming computer simulation. In this section, we discuss the use of surrogate models to speed up the multi-objective design of antenna structures.

13.5.1 *Multi-objective antenna design problem formulation*

Let $f(\boldsymbol{x})$ be a response of an accurate model of the antenna under consideration. The response $f(\boldsymbol{x})$ is computed using high-fidelity EM simulation, and it may represent, for example, an antenna gain or reflection coefficient. Here $\boldsymbol{x}$ is a vector of designable parameters, i.e. antenna dimensions.

Let $F_k(\boldsymbol{x})$, $k = 1, \ldots, N_{obj}$ be a kth design objective. A typical performance objective would be to minimize antenna reflection over a certain frequency band of interest, and to ensure that $|S_{11}| < -10\,\text{dB}$ over that band. There might also be geometrical objectives, such as to minimize $F_k(\boldsymbol{x}) = A(\boldsymbol{x})$ — the antenna size, defined in a convenient way (e.g. maximal lateral size, height, the maximal dimension, area of the footprint, antenna volume). Similar objectives can be formulated with respect to antenna gain, radiation pattern, efficiency, etc.

If $N_{obj} > 1$ then any two designs, $\boldsymbol{x}^{(1)}$ and $\boldsymbol{x}^{(2)}$, for which $F_k(\boldsymbol{x}^{(1)}) < F_k(\boldsymbol{x}^{(2)})$ and $F_l(\boldsymbol{x}^{(2)}) < F_l(\boldsymbol{x}^{(1)})$ for at least one pair $k \neq l$, are not commensurable, i.e. none is better than the other in the multi-objective sense. We define Pareto dominance relation, $\prec$ (Coello Coello, 2013), stating that for the two designs $\boldsymbol{x}$ and $\boldsymbol{y}$, we have $\boldsymbol{x} \prec \boldsymbol{y}$ ($\boldsymbol{x}$ dominates $\boldsymbol{y}$) if $F_k(\boldsymbol{x}) < F_k(\boldsymbol{y})$ for all $k = 1, \ldots, N_{obj}$. The goal of the multi-objective optimization if to find a representation of a so-called Pareto front (of Pareto-optimal set), X_P, of the design space X, such that for any $\boldsymbol{x} \in X_P$, there is no $\boldsymbol{y} \in X$ for which $\boldsymbol{y} \prec \boldsymbol{x}$ (Fonseca, 1995).

13.5.2 *Optimization algorithm*

Because the high-fidelity model, f, is computationally too expensive to be handled directly in the multi-objective optimization, we use a surrogate model constructed as follows. Let c_{cd} be a coarse-discretization EM simulation model of the antenna. Typically, c_{cd} is 10–50 times faster than f, but it is still too expensive for multi-objective optimization. Therefore, we sample the design space using LHS (Beachkofski and Grandhi, 2002) and create a fast surrogate model, s, by approximating the sampled c_{cd} data using kriging interpolation (Queipo *et al.*, 2005). The kriging model s is very fast, smooth, and easy to optimize.

In the next stage, we apply a MOEA to optimize s and find a set of designs representing Pareto-optimal solutions with respect to the objectives, F_k, of interest. Here we use a standard MOEA with fitness sharing, Pareto-dominance tournament selection, and mating restrictions (Fonseca, 1995).

Let $\boldsymbol{x}_s^{(k)}$, $k = 1, \ldots, K$ be the selected elements of the Pareto front found by the MOEA. These solutions have to be refined because they were obtained by optimizing the surrogate model, whereas we are interested in optimizing the high-fidelity model. For simplicity of the notation, the design refinement stage below is defined assuming two objectives, F_1 and F_2; however, it can be generalized for any value of N_{obj}. The refinement stage exploits the output space mapping (OSM) (Koziel *et al.*, 2006) process of the following form:

$$\boldsymbol{x}_f^{(k.i+1)} = \arg \min_{\boldsymbol{x}, F_2(\boldsymbol{x}) \le F_2(\boldsymbol{x}_s^{(k.i)})} F_1(s(\boldsymbol{x}) + [f(\boldsymbol{x}_s^{(k.i)}) - s(\boldsymbol{x}_s^{(k.i)})]). \tag{13.5}$$

The optimization process (Equation (13.5)) is constrained so as not to increase the second objective compared to $\boldsymbol{x}_s^{(k)}$. The surrogate model s is corrected using the OSM term $f(\boldsymbol{x}_s^{(k.i)}) - s(\boldsymbol{x}_s^{(k.i)})$ (here, $\boldsymbol{x}_f^{(k.0)} = \boldsymbol{x}_s^{(k)}$, so that the corrected surrogate model coincides with f at the beginning of each iteration. In practice, two or three iterations of Equation (13.5) are sufficient to find a refined high-fidelity model design, $\boldsymbol{x}_f^{(k)}$. After completing this stage, we create a set of Pareto-optimal high-fidelity model designs. This set is the final outcome of our multi-objective optimization process.

The entire design flow can be summarized as follows.

1. Sample the design space and acquire the c_{cd} data.
2. Construct the kriging interpolation model, s.
3. Obtain the Pareto front by optimizing s using MOEA.
4. Refine selected elements of the Pareto front, $\boldsymbol{x}_s^{(k)}$, to obtain corresponding high-fidelity model designs, $\boldsymbol{x}_f^{(k)}$.

It should be emphasized that the high-fidelity model f is not evaluated until the refinement stage (Step 4 above). It is worth noting that the cost of finding the high-fidelity model Pareto-optimal set is only about three evaluations of the high-fidelity model per design.

13.5.3 *Design example: UWB monocone*

The design approach of Section 13.5.2 is demonstrated using an ultra-wideband (UWB) monocone antenna (Koziel and Ogurtsov, 2013d), where the minimization of the antenna reflection and its size are the two objectives of interest. The antenna structure is shown in Fig. 13.20. The antenna is energized through a 50 ohm coaxial input (Teflon filling, $r_0 = 0.635$ mm). No extra circuitry is used for matching here. The design specification imposed on the reflection response of the monocone is $|S_{11}| \leq -10$ dB from 3 to 10 GHz. Design variables are $\boldsymbol{x} = [z_1\ z_2\ r_1]^T$ (sizes in mm), where z_1 is the extension of the

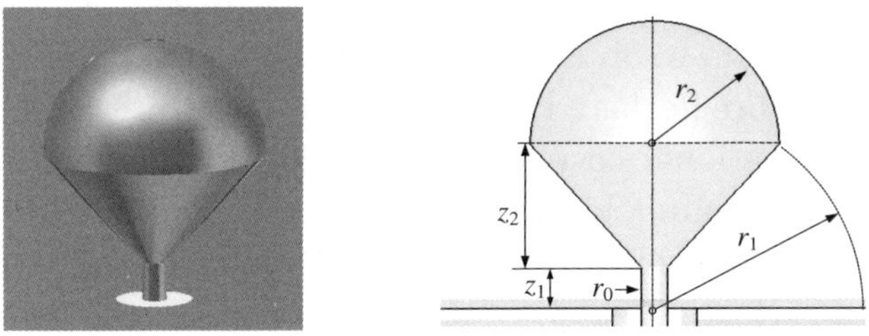

(a) (b)

Fig. 13.20. UWB monocone. (a) 3D view; (b) cut view.

coax pin, z_2 is the length of the cone section, and r_1 is the size of the radial line section, as shown in Fig. 13.20(b).

The high-fidelity model of the antenna is simulated in CST Microwave Studio (2012) (~1,400,000 mesh cells, evaluation time 23 minutes). The coarse-discretization model, c_{cd}, is also simulated in CST Microwave Studio (~33,000 mesh cells, 33 s). The design space is defined by $0 \leq z_1 \leq 4$, $2 \leq z_2 \leq 15$, $4 \leq r_1 \leq 20$, and a linear constraint $z_1 + z_2 \leq r_1 - 0.25$. The antenna size defined here is the maximal dimension out of vertical and lateral ones: $A(\boldsymbol{x}) = \max\{2r_2, z_1 + z_2 + r_2\}$, where $r_2 = (r_1^2 - (z_1 + z_2)^2)^{1/2}$ is the radius of the hemisphere terminating the conical section.

The design objectives for our example are:

- $F_1(\boldsymbol{x}) = \max\{|S_{11}(\boldsymbol{x}, f)|\colon 3\,\text{GHz} \leq f \leq 10\,\text{GHz}\}$ — maximum of $|S_{11}|$ over the frequency band of 3–10 GHz; and
- $F_2(\boldsymbol{x}) = A(\boldsymbol{x})$ — antenna size, as defined in the previous paragraph.

In the definitions above, explicit dependence of antenna reflection on the design variables is indicated.

The kriging surrogate model, s, is created using 600 low-fidelity model samples allocated in the design space using LHS (Beachkofski and Grandhi, 2002). The surrogate was then optimized using MOEA. Figure 13.21 shows the Pareto front obtained by applying MOEA. It indicates that the minimum size of the considered antenna that ensures the satisfaction of the reflection requirements ($|S_{11}| \leq -10\,\text{dB}$) is 18.5 mm. A number of selected solutions of the Pareto front, corresponding to sizes of $18, 19, \ldots, 28$, have been refined using the OSM algorithm to obtain high-fidelity model designs. Those optimum solutions are gathered in Table 13.4.

The computational cost of creating the kriging surrogate corresponds to only around 15 evaluations of the high-fidelity model, f. Multi-objective optimization of s takes about 10 minutes of CPU time as the kriging model is very fast. Design refinement takes, on average, two iterations of the OSM algorithm (i.e. two evaluations of f). Thus, the total design cost (obtaining the Pareto front and 11 antenna designs of various sizes, as in Table 1) corresponds

Table 13.4 UWB monocone optimization results.

Antenna size A [mm]	Design variables [mm] z_1	z_2	r_1	max $\|S_{11}\|$ for 3 GHz to 10 GHz
18	0.994	10.36	13.16	−9.5 dB
19	0.021	12.18	12.97	−10.9 dB
20	0.065	12.48	14.59	−11.6 dB
21	0.051	12.84	15.23	−12.3 dB
22	0.000	12.92	15.79	−13.2 dB
23	0.000	12.06	16.28	−14.7 dB
24	0.008	12.08	16.97	−16.0 dB
25	0.079	12.42	17.68	−17.1 dB
26	0.142	12.99	18.38	−18.1 dB
27	0.169	13.43	19.09	−18.4 dB
28	0.231	13.27	19.45	−19.4 dB

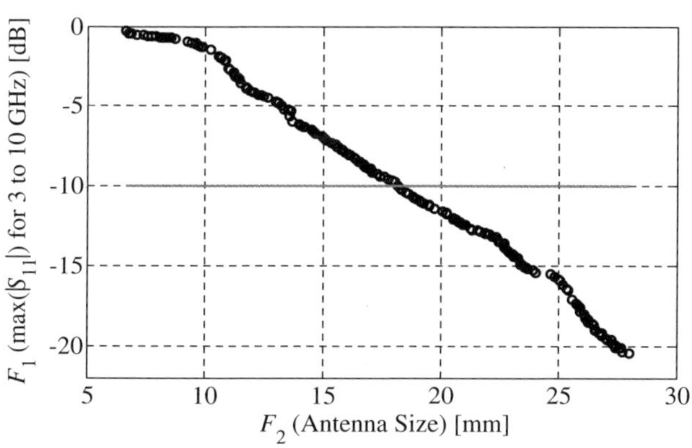

Fig. 13.21. Pareto front for the UWB monocone obtained by optimizing the kriging surrogate model using MOEA.

to about 38 evaluations of the high-fidelity model (∼14 hours). Figure 13.22 shows the Pareto-optimal set of the high-fidelity model designs obtained after the design refinement. Figure 13.23 shows the reflection responses for selected designs of this set.

It should be emphasized that the SBO approach allows, at a low computational cost, the obtaining of comprehensive information about a given UWB antenna, including the trade-offs between the reflection coefficient and the antenna size, as well as the minimum

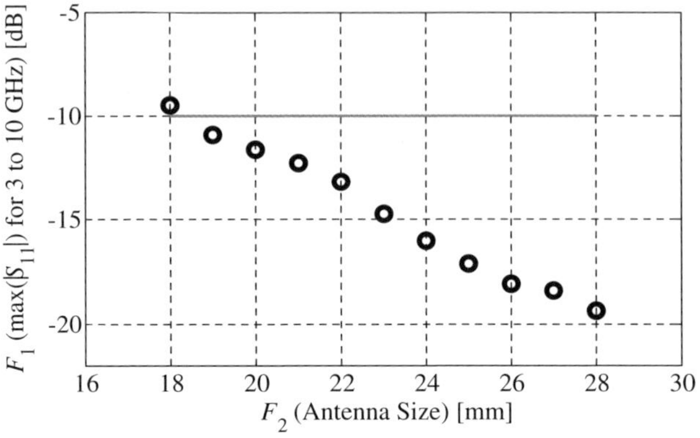

Fig. 13.22. Pareto-optimal high-fidelity model designs obtained in the refinement stage (cf. Table 13.4).

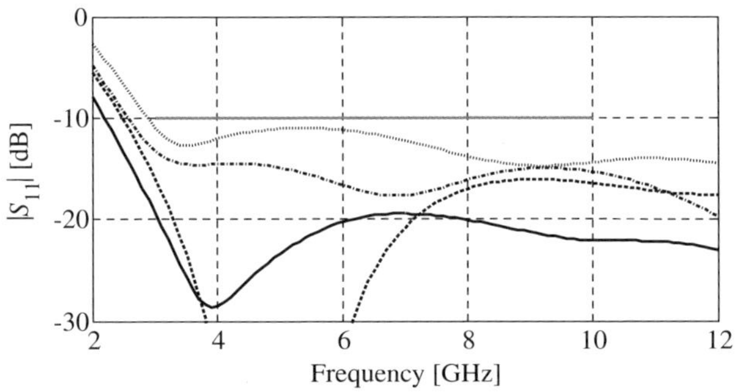

Fig. 13.23. Reflection responses of the UWB monocone corresponding to the optimized high-fidelity model designs of sizes 19 mm (· · ··), 22 mm (- · — ·), 24 mm (- - -), and 28 mm (—).

size, while satisfying given reflection requirements. For the specific example considered here, the minimum size of the antenna still satisfying reflection requirements is 34% smaller than the size of the antenna optimized for reflection only. It is also important that the total cost of obtaining the set of Pareto-optimal high-fidelity model designs using the proposed approach is comparable to the cost of conventional single-objective optimization of the antenna structure (that would yield a single design).

13.6 SBO of the Climate Model Parameters using Response Correction

Climate models are typically formulated as time-dependent partial differential or differential algebraic equations (PDE/DAE) (Majda, 2003; McGuffie and Henderson-Sellers, 2005). A specific example is marine ecosystem models (Fennel and Neumann, 2004; Sarmiento and Gruber, 2006) that describe photosynthesis and other biogeochemical processes in the marine ecosystem. These kinds of models are important, e.g. to compute and predict the oceanic uptake of carbon dioxide as a part of the global carbon cycle. As the number of processes that have to be included and the necessary temporary and special resolution is high, so is the computational cost of model evaluation. In this section, we discuss the use of SBO for the low-cost identification of marine model parameters.

13.6.1 *Marine model description*

Here we consider a one-dimensional marine ecosystem model (Oschlies and Garcon, 1999) driven by precomputed ocean circulation data. In the following, we briefly describe the high-fidelity model and its low-fidelity counterpart, which is a basis to construct a surrogate for further use in the optimization process.

Simulating the marine ecosystem has become a key tool for understanding the ocean carbon cycle and its variability. The marine ecosystem contains several biogeochemical quantities (called tracers), for example nutrients, phytoplankton, and zooplankton, which interact and are transported by the ocean circulation and influenced by temperature and salinity. Consequently, ecosystem simulations require the modeling and computation of both ocean circulation and biogeochemistry. The underlying continuous models are governed by coupled systems of nonlinear, parabolic PDE or DAE for ocean circulation (ocean models, i.e. Navier–Stokes equations with additional temperature and salinity transport equations) and the transport of biogeochemical tracers (marine ecosystem models, i.e. convection- or advection-diffusion-reaction-type equations) (Sarmiento and Gruber, 2006). In ecosystem models, the parameters to be optimized

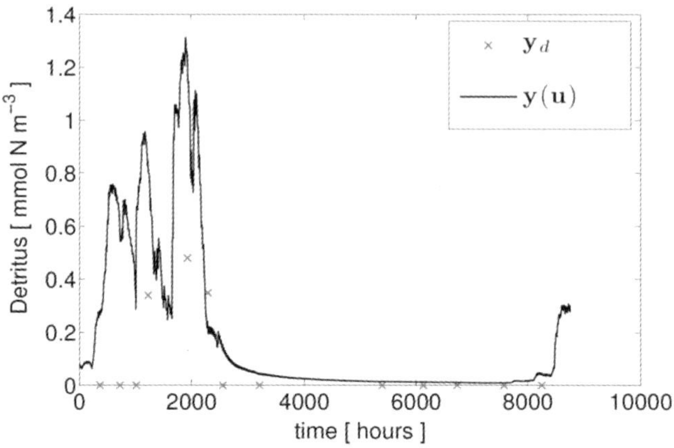

Fig. 13.24. Model response, $\mathbf{y}^{(D)}$ (detritus), and observation data, $\mathbf{y}_d^{(D)}$, for one year at depth $z \simeq -25\,\mathrm{m}$ (Priess *et al.*, 2011b).

(represented by a vector $\boldsymbol{u}$) are, for example, growth and dying rates of the tracers, and thus appear in the usually nonlinear coupling or interaction terms in the model.

Our example ecosystem model simulates the interaction of dissolved inorganic nitrogen, phytoplankton, zooplankton, and detritus (dead material), thus it is also called the NPZD model. One aim was to reproduce observations, $\boldsymbol{y}_d$, at different North Atlantic locations by the optimization of model parameters within credible limits. Figure 13.24 shows the model response and target data, respectively, as an illustration of the tracer detritus for a certain depth and part of the time interval.

The model uses pre-computed ocean circulation and temperature data from an ocean model (in what is sometimes called an off-line modus), i.e. no feedback by the biogeochemistry on the circulation and temperature is noted (Oschlies and Garcon, 1999).

As a test case and since biogeochemistry — except for the sinking processes—mainly happens locally in space, we use here a one-dimensional version of the model. This version simulates one water column at a given horizontal position. Similar SBO methodology can be used for three-dimensional models, as demonstrated in Priess *et al.* (2013).

In the NPZD model, the concentrations (in mmol·N·m^{-3}) of dissolved inorganic nitrogen, N, phytoplankton, P, zooplankton, Z, and detritus, D, are summarized in the vector $\boldsymbol{y} = (y^{(l)})_{l=N,P,Z,D}$ and described by the following coupled PDE system:

$$\left.\begin{aligned} \frac{\partial y^{(l)}}{\partial t} &= \frac{\partial}{\partial z}\left(\kappa \frac{\partial y^{(l)}}{\partial z}\right) + Q^{(l)}(\boldsymbol{y}, u_2, \ldots, u_n), \quad l = N, P, Z \\ \frac{\partial y^{(D)}}{\partial t} &= \frac{\partial}{\partial z}\left(\kappa \frac{\partial y^{(D)}}{\partial z}\right) + Q^{(D)}(\boldsymbol{y}, u_2, \ldots, u_n) - \frac{\partial y^{(D)}}{\partial z} u_1 \end{aligned}\right\}, \tag{13.6}$$

in $(-H, 0) \times (0, T)$, with additional appropriate initial values. Here, z denotes the only remaining, vertical spatial coordinate and H the depth of the water column. The terms $Q^{(l)}$ are the biogeochemical coupling (or source-minus-sink) terms for the four tracers and $\boldsymbol{u} = (u_1, \ldots, u_n)$ is the vector of unknown physical and biological parameters. The sinking term is only apparent in the equation for detritus. In the one-dimensional model no advection term is used, since a reduction to vertical advection would make no sense. Thus the circulation data (taken from an ocean model) are the turbulent mixing coefficient $\kappa = \kappa(z,t)$ and the temperature $\Theta = \Theta(z,t)$, which goes into the nonlinear coupling terms, $Q^{(l,)}$ but is omitted in the notation.

The continuous model (Equation (13.6)) is discretized and solved using an operator splitting method (Marchuk, 1982), an explicit Euler timestepping scheme for the nonlinear coupling term, Q, and the sinking term, while using an implicit euler timestepping scheme for the diffusion term. For further details we refer the reader to Priess *et al.* (2011b). In the original model, the time step denoted by τ is chosen as one hour. The model with this particular time step will be referred to as the high-fidelity or fine one.

We will denote by $\boldsymbol{y}_j \approx y(\cdot, t_j)$ the discrete fine model solution of the continous model (Equation (13.6)) in time step j (containing all tracers, N, P, Z, D), given as $\boldsymbol{y}_j = (y_{ji})_{i=1,\ldots,I}$, $j = 1, \ldots, M, \boldsymbol{y} \in R^{MI}$, where $I = 66 \times 4$ denotes the number of spatial discrete points for all tracers given by 66, the number of spatial discrete points per tracer, times four, the number of tracers (cf. Equation (13.6)), and where $M = 8760$ time steps/year $\times$ 5 years denotes the number of discrete time steps for each tracer.

The low-fidelity (or coarse) model, which is a less accurate but computationally cheap representation of $\boldsymbol{y}$ is obtained by using a coarser time discretization, $\hat{\tau}$, given as $\hat{\tau} = \beta\tau$, with a coarsening factor, β, being a positive integer while keeping the spatial discretization fixed. The state variable for this coarser, discretized model will be denoted by $\hat{\boldsymbol{y}}$, the corresponding number of discrete time steps by $\hat{M} = M/\beta$, i.e. we have $\hat{\boldsymbol{y}}_j = (\hat{y}_{ji})_{i=1,\ldots,I}$, $j = 1, \ldots, \hat{M}$, $\hat{\boldsymbol{y}} \in R^{\hat{M}I}$. Note that the parameters, $\boldsymbol{u}$, for the low-fidelity model are the same as for the high-fidelity one. The surrogate created using the low-fidelity model will be used for low-cost optimization of the high-fidelity model.

13.6.2 *Model identification problem and optimization algorithm*

The goal is to minimize a least-squares-type cost function measuring the misfit between the discrete model output, $\boldsymbol{y} = \boldsymbol{y}(\boldsymbol{u})$, and given observational data $\boldsymbol{y}_d$ (Banks and Kunisch, 1989; Tarantola, 2005). In most cases, the problem is constrained by parameter bounds. Thus the parameter optimization problem can be written as argmin$\{\boldsymbol{u} \in U_{ad}$: $J(\boldsymbol{y}(\boldsymbol{u}))\}$, where $J(\boldsymbol{y}) = 0.5 \cdot \|\boldsymbol{y} - \boldsymbol{y}_d\|^2$, $U_{ad} = \{\boldsymbol{u} \in R^n$: $\boldsymbol{b}_l \leq \boldsymbol{u} \leq \boldsymbol{b}_u\}$, with $\boldsymbol{b}_l$, $\boldsymbol{b}_u \in R^n$ being the lower and upper bounds for the parameters. The functional J may additionally include a regularization term for the parameters, which is not necessary in our case.

To solve the parameter identification problem we use the standard SBO process (Equation (4.2)), here rewritten as $\boldsymbol{u}_{k+1} = $ argmin$\{\boldsymbol{u} \in U_{ad}$: $J(s_k(\boldsymbol{u}))\}$, where s_k is the surrogate model at iteration k, whereas $\boldsymbol{u}_{k+1}$ is the approximated solution of the original problem obtained by optimizing s_k. Here the surrogate is generated by multiplicative correction of the smoothed low-fidelity model response denoted as $\tilde{\hat{\boldsymbol{y}}}$ (smoothing is necessary in order to capture the most important features of the otherwise very complex model response).

We have:

$$\left.\begin{aligned} s_{kji}(\boldsymbol{u}) &= A_{kji}\tilde{\hat{\boldsymbol{y}}}_{ji}(\boldsymbol{u}) \\ A_{kji} &= \frac{\tilde{y}^{\beta}_{ji}(\boldsymbol{u}_k)}{\tilde{\hat{y}}^{\beta}_{ji}(\boldsymbol{u}_k)} \end{aligned}\right\} \begin{aligned} k &= 1,2,\ldots \\ j &= 1,\ldots,\hat{M} \\ i &= 1,\ldots,I \\ \beta &= M/\hat{M} \end{aligned}, \tag{13.7}$$

where A_{kji} denotes the correction factor given as the pointwise division of the smoothed and downsampled high-fidelity model response, denoted by $\tilde{\boldsymbol{y}}^{\beta}$, by the smoothed coarse model response at the point $\boldsymbol{u}_k$. This simple correction scheme is justified by the fact that the overall "shape" of the low-fidelity model response resembles that of the high-fidelity one (Priess *et al.*, 2011).

As indicated in Priess *et al.* (2011b), the basic correction (Equation (13.7)) may exhibit some difficulties, either because of possible non-physical (negative) values of the low-fidelity model responses being the results of coarse time discretization in Equation (13.6) or because of a large value of the correction terms (in case the low-fidelity model output is very small). The following enhancement to the basic correction has been introduced in Priess *et al.* (2011b):

$$\begin{aligned} &(i)\ \ \hat{y}_{ji}(\boldsymbol{u}_k) = \max\{\hat{y}_{ji}(\boldsymbol{u}_k), 10^{-8}\} \\ &(ii)\ \ A_{kji} = \min\{A_{kji}, 10\} \\ &(iii)\ A_{kji} = 1 \quad \text{if } (\tilde{y}^{\beta}_{ji}(\boldsymbol{u}_k) \le \varepsilon \quad \text{and} \quad \tilde{\hat{y}}_{ji}(\boldsymbol{u}_k) \le \varepsilon) \end{aligned} \quad . \tag{13.8}$$

Here we use $\varepsilon = 10^{-10}$. The enhancements (Equation (13.8)) allow for the removal of non-physical artefacts from the surrogate model response, which results in a further improvement of the SBO algorithm performance, as indicated in the next section.

13.6.3 *Results*

Identification of model parameters was performed using SBO with the surrogate model created with both the basic correction (Equation (13.6)) and the enhanced one (Equations (13.6) and (13.7)). For comparison purposes, the optimization was also performed using the high-fidelity model directly, similarly for the low-fidelity model. The obtained sets of parameters are denoted as $\boldsymbol{u}^*_{s,1}$, $\boldsymbol{u}^*_{s,2}$, $\boldsymbol{u}^*$, and

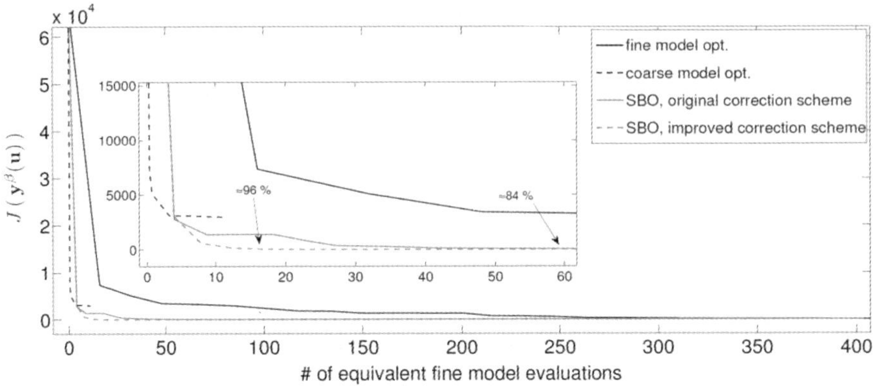

Fig. 13.25. Values of the cost function J versus the equivalent number of high-fidelity model evaluations for the SBO, run using the surrogate model exploiting the original (cf. Equation (13.6)) and the enhanced correction scheme (cf. Equation (13.7)), as well as for high- and low-fidelity model optimization runs. The figure also indicates those points in the SBO runs that correspond to a termination condition of $J(\boldsymbol{y}^{\beta}(\boldsymbol{u}_k)) \leq 50$, ensuring good visual agreement between the fine-model output and the target. By using the enhancements (Equation (13.7)), the number of equivalent high-fidelity model evaluations required to satisfy this termination condition was reduced from 60 to only 17, resulting in an increase of the corresponding time savings, from 84% to about 96%, when compared to the direct high-fidelity model optimization (Priess *et al.*, 2011b).

$\hat{\boldsymbol{u}}^*$, respectively. The termination condition for the SBO algorithm was $J(\boldsymbol{y}^{\beta}(\boldsymbol{u}_k)) \leq 50$, as the value ensures good visual agreement between the model output and the target data. The optimization history is shown in Fig. 13.25 (only two representative tracers at a certain depth level and time interval are included for illustration). Figure 13.26 shows the downsampled high-fidelity model response for the optimal parameter values obtained using all aforementioned methods.

It should be noted that the model parameters obtained by directly optimizing the coarse model result in a cost function value of about 2,960 (optimization cost is 11 equivalent fine-model evaluations). This solution is far from that obtained by the direct fine-model optimization, which indicates that the coarse model is not a reliable prediction tool.

Direct high-fidelity model optimization yields a very low cost function of about 0.01, corresponding to a solution close to the target data. However, the optimization cost is substantially higher: 983

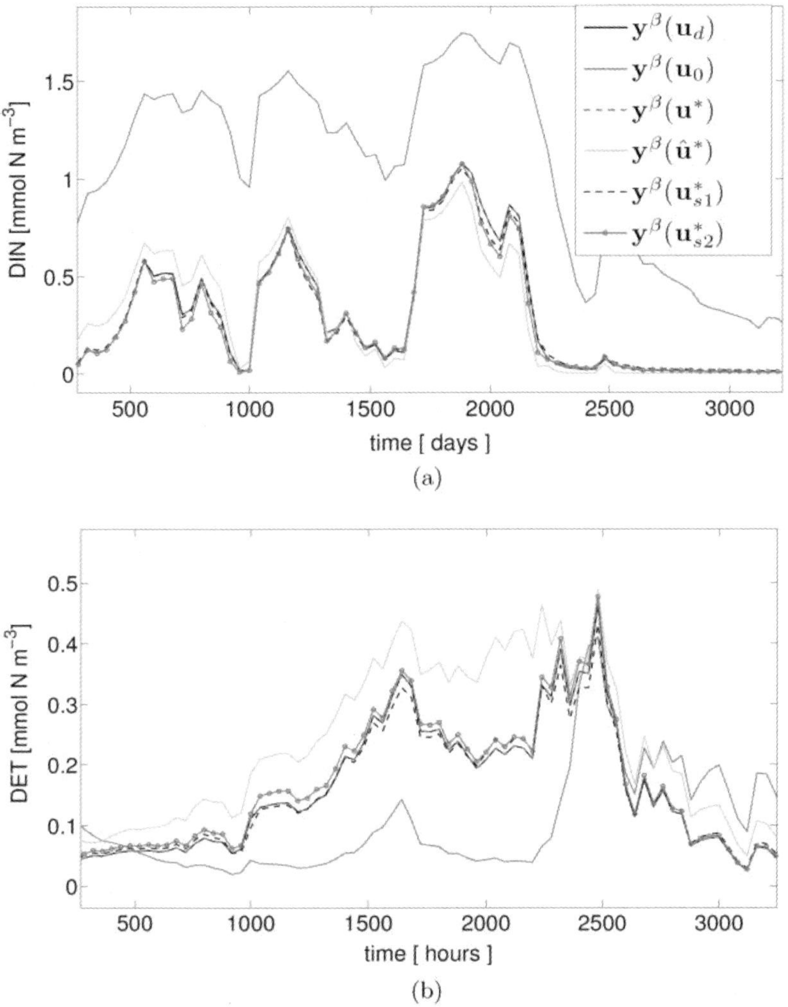

Fig. 13.26. High-fidelity model output, $\boldsymbol{y}^{\beta}$ (down-sampled) at depth $z \approx -2.68\,\text{m}$. (a) State-dissolved inorganic nitrogen; (b) state detritus. Shown are, in the legend from top to bottom: (i) Target $\boldsymbol{y}_d$; (ii) high-fidelity model output at the initial value $\boldsymbol{u}_0$; (iii) at the result of the direct high-fidelity model optimization yielding $\boldsymbol{u}^*$; (iv) at the coarse model optimum $\hat{\boldsymbol{u}}^*$; and (v) at the optima $\mathbf{u}^*_{s1}$, $\mathbf{u}^*_{s2}$, obtained through SBO using the original and the enhanced-response correction schemes, respectively. Solutions (iv) and (v) are both very close to (iii) but solution (v) was obtained at the computational cost three times lower than (iv) (Priess *et al.*, 2011b).

high-fidelity model evaluations (Figure 13.25 only shows the range 0–400 function evaluations for clarity).

The SBO based on the original response correction scheme (Equation (13.6)) yields a solution after approximately 60 equivalent high-fidelity model evaluations (assuming the termination condition discussed above). This results in a reduction of the total optimization cost of about 84% when compared to the high-fidelity model optimization (the direct high-fidelity model optimization required 375 evaluations to reach the same cost function value, cf. Fig. 13.26).

After employing the improvements suggested in Equation (13.7), only 17 equivalent fine-model evaluations were required to satisfy the same termination condition, which is over three times less than the original response correction scheme (Equation (13.6)). The corresponding reduction of the total optimization cost, compared to the direct high-fidelity model optimization, is about 96%. The corresponding solution is close to that obtained by direct high-fidelity model optimization.

13.7 Conclusion

In this chapter, several applications of SBO exploiting physics-based surrogate models have been demonstrated, including design optimization of microwave filters, single- and multi-objective optimization of antenna structures, and parameter identification for marine models. SBO seems to be the right tool to dramatically reduce the computational cost of the design process. On the other hand, there are certain issues that have to be carefully addressed in order to make SBO efficient. One of them is the proper choice and setup of the underlying low-fidelity model. Another is the suitable selection of the surrogate modeling approach, which usually is problem-dependent. Finally, a combination of several SBO-based tools (in particular, the use of auxiliary response surface models) may be advantageous in improving the performance of the optimization algorithm. All of these factors indicate that the successful use of SBO requires certain experience and problem-related expertise. Automation of the SBO procedures is still the aim of the future research in the area.

Chapter 14

Surrogate-Based Optimization with MATLAB

In this chapter, we present basic MATLAB implementations of some of the optimization techniques considered in this book. The primary purpose is to give the reader a flavor of different design optimization approaches and, perhaps, provide a starting point for developing more involved algorithms. It should be emphasized that the particular implementations shown here are very basic versions, with the emphasis on simplicity. They should be viewed as illustrations of the methods rather than fully featured general purpose codes.

All of the algorithms presented here are illustrated using two selected problems concerning transonic airfoil design, in particular single-point drag minimization and lift maximization. We provide the problem formulations, describe the computational fluid dynamics (CFD) models, and discuss a few specific methods including direct gradient-based optimization, sequential approximate optimization (SAO), space mapping (SM), and multi-level design.

In order to run the algorithms described here, any recent release of MATLAB (2012) as well as MATLAB's optimization toolbox are necessary.

14.1 Example Problems and Objectives

In order to illustrate the operation of the algorithm implementations discussed in this chapter, we consider two example problems concerning transonic airfoil design. We assume that the airfoil models are vector-valued functions of the geometry parameters $\boldsymbol{x}$, so we have

$\boldsymbol{f}(\boldsymbol{x}) = [C_{l.f}(\boldsymbol{x})\ C_{d.f}(\boldsymbol{x})\ A_f(\boldsymbol{x})]^T$, where $C_{l.f}$, $C_{d.f}$, and A refer to the lift coefficient, drag coefficient, and cross-sectional area of the airfoil, respectively. Here, f refers to the high-fidelity airfoil model. For variable-fidelity methods we will also use low-fidelity models as described in Section 14.2.

14.1.1 *Design space*

Here, we use four-digit NACA (National Advisory Committee for Aeronautics) airfoil parameterization (cf. Section 7.6). The design space is defined through lower/upper bounds for the design parameters $\boldsymbol{x} = [m\ p\ t/c]^T$ as follows: $0 \leq m \leq 0.03$, $0.4 \leq p \leq 0.8$, and $0.07 \leq t/c \leq 0.12$, with $c = 1$.

14.1.2 *Test problem 1: drag minimization*

The first test problem concerns drag minimization. It is defined as:

$$\boldsymbol{x}^* = \arg\min_{\boldsymbol{x}} C_{d.f}(\boldsymbol{x}), \tag{14.1}$$

with the following constraints:

$$C_{l.f}(\boldsymbol{x}) = 0.5, \tag{14.2}$$

$$A_f(\boldsymbol{x}) \geq 0.075. \tag{14.3}$$

Additionally, the vector $\boldsymbol{x}$ has to be within the design space determined by the lower/upper bounds, as described in Section 14.1.1.

The operating conditions are Mach number $M_\infty = 0.75$ and the angle of attack $\alpha = 0$ degrees.

For simplicity, we handle both constraints using the penalty functions, so that the objective function for our test problem is defined as follows:

$$\begin{aligned} H_1(\boldsymbol{f}(\boldsymbol{x})) = C_{d.f}(\boldsymbol{x}) &+ \beta_1 \cdot [(C_{l.f}(\boldsymbol{x}) - 0.5)/0.5]^2 \\ &+ \beta_2 \cdot [\min\{0, A_f(\boldsymbol{x}) - 0.075\}/0.075]^2. \end{aligned} \tag{14.4}$$

Here, we set $\beta_1 = \beta_2 = 1{,}000$. The original Equation (14.1) is then replaced with the following unconstrained version:

$$\boldsymbol{x}^* = \arg\min_{\boldsymbol{x}} H_1(\boldsymbol{f}(\boldsymbol{x})). \tag{14.5}$$

14.1.3 *Test problem 2: lift maximization*

The second test problem concerns lift maximization. It is defined as:

$$\boldsymbol{x}^* = \arg\min_{\boldsymbol{x}}[-C_{l.f}(\boldsymbol{x})], \tag{14.6}$$

with the following constraints:

$$C_{d.f}(\boldsymbol{x}) \leq 0.004, \tag{14.7}$$

$$A_f(\boldsymbol{x}) \geq 0.075. \tag{14.8}$$

Additionally, the vector $\boldsymbol{x}$ has to be within the design space determined by the lower/upper bounds, as described in Section 14.1.1.

The operating conditions are the same as test problem 1, i.e. Mach number $M_\infty = 0.75$ and the angle of attack $\alpha = 0$ degrees.

For simplicity, we handle both constraints using the penalty functions, so that the objective function for our test problem is defined as follows:

$$\begin{aligned} H_2(f(\boldsymbol{x})) = -C_{l.f}(\boldsymbol{x}) &+ \beta_1 \cdot [\max\{0, C_{d.f}(\boldsymbol{x}) - 0.004\}/0.004]^2 \\ &+ \beta_2 \cdot [\min\{0, A_f(\boldsymbol{x}) - 0.075\}/0.075]^2. \end{aligned} \tag{14.9}$$

Here, we set $\beta_1 = \beta_2 = 1{,}000$. The original Equation (14.6) is then replaced with the following unconstrained version:

$$\boldsymbol{x}^* = \arg\min_{\boldsymbol{x}} H_2(f(\boldsymbol{x})). \tag{14.10}$$

14.2 CFD Airfoil Models

In order to keep the chapter self-contained, the actual CFD airfoil models are represented here by their response surface approximations (RSA). We consider three models:

- high-fidelity model, f;
- (coarser) low-fidelity model, c_1; and
- (finer) low-fidelity model, c_0.

RSA representations of these models were obtained by means of kriging interpolation of sampled CFD simulation data. In all cases, the simulation was performed using FLUENT (2012). The model setups correspond to what was described in Section 8.2 of the book. The number of grid cells for the high-fidelity model was approximately 400,000 (simulation time ~33 min). Respective figures for the coarser and finer low-fidelity models are ~30,000 grid cells (80 times faster simulation time than the high-fidelity model) and ~100,000 grid cells (40 times faster simulation time than the high-fidelity model). The kriging model is implemented using a DACE toolbox (Lophaven *et al.*, 2002). In each case, 1,000 samples were used to obtain a highly accurate representation of the actual CFD models.

The three RSA models (subsequently referred to as f, c_1, and c_0) can be evaluated using the following syntax:

```
R = airfoil_model_f(x)
R = airfoil_model_c1(x)
R = airfoil_model_c0(x),
```

where R is a column response vector containing lift, drag, and cross-sectional area of the airfoil, whereas x is the input vector of geometry parameters. All models are valid within the design space described in Section 14.1.1.

14.3 Direct Optimization

For the sake of comparison with surrogate-based optimization (SBO) approaches, we include implementation of the direct search using a gradient-based optimizer. Here, we exploit MATLAB's *fmincon* routine. The gradients are obtained internally using finite differentiation.

The script for running direct airfoil optimization is the following:

```
% Direct optimization script
% Initialization
clear all
close all
h1=figure;
h2=figure;
global X; global R; global E; global Ntotal;
X=[]; R=[]; E=[]; Ntotal=[];
% Control variables
problem=2; % Test problem selection: 1 - drag minimization,
% 2 - lift maximization
```

```
f_model=@airfoil_model_f; % model to be optimized
% Lower/upper bounds (do not change - CFD models
% are only valid within these bounds!)
lb=[0 0.4 0.07]';
ub=[0.03 0.8 0.12]';
x0=(lb+ub)/2; % starting point
% options for fmincon routine
options=optimset('Display','iter','Algorithm','active-set',...
    'TolX',1e-8,'TolFun',1e-8,'OutputFcn',@outfun);
% Problem setup
switch(problem)
    case 1 % drag minimization
      plot_data=struct('problem',problem,'Cl',0.5,...
        'A',0.075,'x0',x0,'h1',h1,'h2',h2);
      f_objective=@f_obj_drag_minimization;
    case 2 % lift maximization
      plot_data=struct('problem',problem,'Cd',0.004,...
        'A',0.075,'x0',x0,'h1',h1,'h2',h2);
      f_objective=@f_obj_lift_maximization;
end
% Optimization code (calling fmincon routine)
xopt=fmincon(@f_obj,x0,[],[],[],[],lb,ub,'',options,...
    f_objective,f_model,plot_data);
% Results display code
display_results(xopt,f_model);
```

All auxiliary functions used in the script are provided in the Appendix. The script provides some visualization of the optimization process.

The results of the algorithm run for both test problems are shown in Figs. 14.1–14.4 and in Table 14.1. As can be observed, a gradient search is rather expensive (around 200 and 150 objective function calls for test problems 1 and 2, respectively). Both constraints are active at the optima found.

14.4 SAO

The SAO algorithm implemented here follows the description provided in Section 5.2. There are two approximation models that can be utilized: a linear model (requires $n+1$ training points) or a quadratic model without mixed terms (requires $2n + 1$ training points allocated using star distribution; Cheng *et al.* (2006)), controlled by the *model_order* variable.

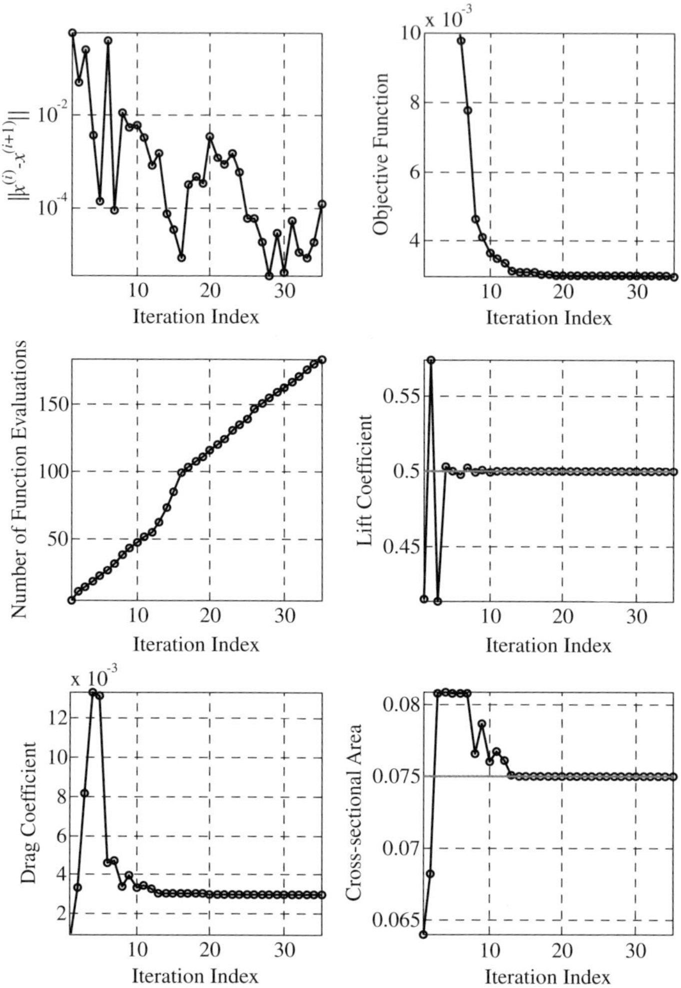

Fig. 14.1. Optimization history for direct airfoil design using MATLAB's *fmincon* (test problem 1).

The script of the SAO algorithm is the following:

```
% Sequential approximate optimization script
% Initialization
clear all
close all
h1=figure('Position',[100 100 700 1000]);
h2=figure('Position',[1000 600 600 400]);
```

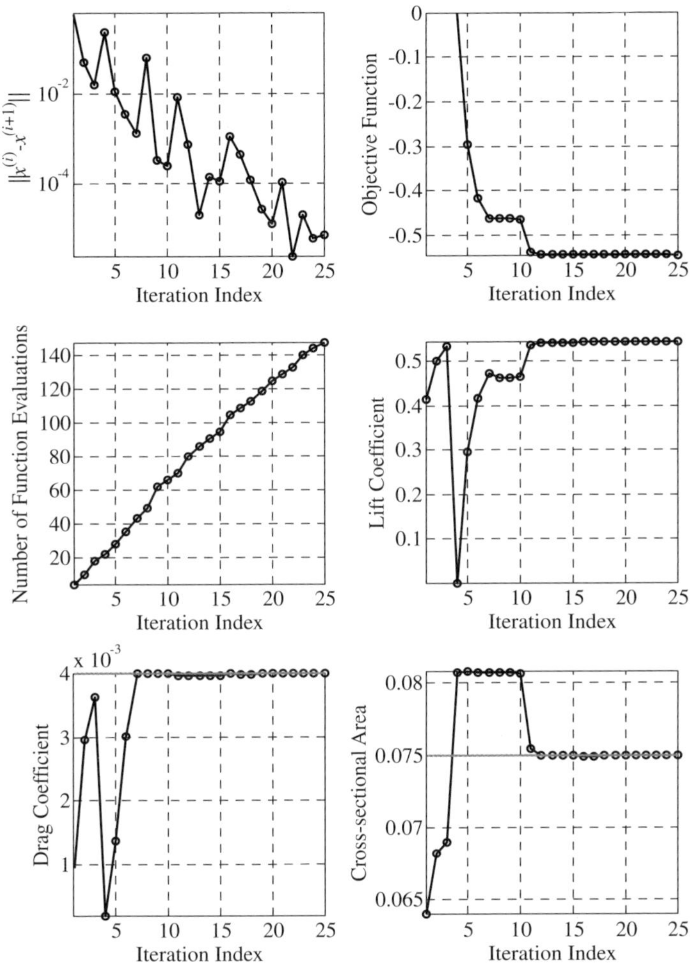

Fig. 14.2. Optimization history for direct airfoil design using MATLAB's *fmincon* (test problem 2).

```
X=[]; R=[]; E=[]; N=[]; Ntotal=0;
% Control variables
problem=2; % Test problem selection: 1 - drag minimization,
% 2 - lift maximization
f_model=@airfoil_model_f; % model to be optimized
model_order=1; % response surface approximation
% model order (1 or 2)
% Lower/upper bounds (do not change - CFD models
% are only valid within these bounds!)
```

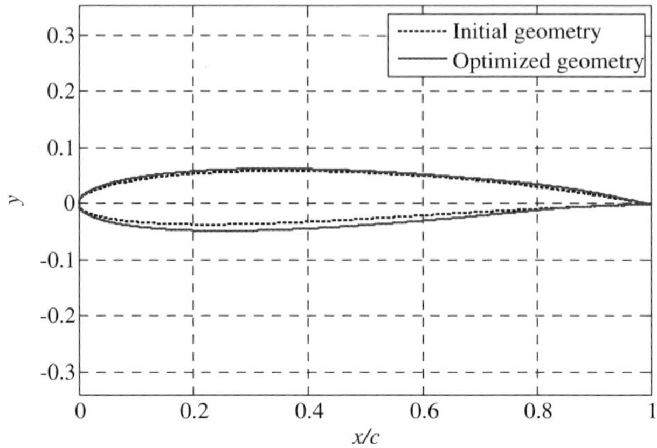

Fig. 14.3. Initial and optimized airfoil geometry for test problem 1 (drag minimization), solved using direct optimization.

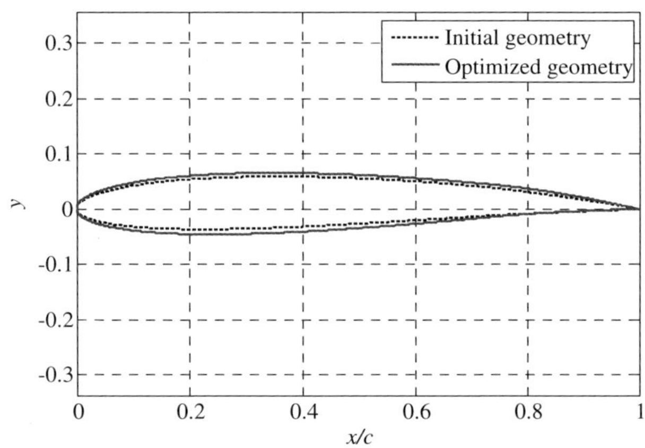

Fig. 14.4. Initial and optimized airfoil geometry for test problem 2 (lift maximization), solved using direct optimization.

```
lb=[0 0.4 0.07]';
ub=[0.03 0.8 0.12]';
d=(ub-lb)/10; % initial region size for setting up RSA model
x0=(lb+ub)/2; % starting point
% options for fmincon routine
options=optimset('Display','iter','Algorithm',...
   'active-set','TolX',1e-8,'TolFun',1e-8);
% Problem setup
switch(problem)
```

Table 14.1 Direct optimization results for test problems 1 and 2.

Parameters & figures of interest	Initial geometry	Test problem 1 optimized geometry	Test problem 2 optimized geometry
m	0.0150	0.01212	0.01512
p	0.6000	0.78964	0.73046
t	0.0950	0.11142	0.11134
$C_{l.f}$	0.4146	0.49999	0.54376
$C_{d.f}$	0.0009	0.00299	0.00400
A_f	0.0640	0.07500	0.07496

```
    case 1 % drag minimization
      plot_data=struct('problem',problem,'Cl',0.5,...
        'A',0.075,'x0',x0,'h1',h1,'h2',h2);
      f_objective=@f_obj_drag_minimization;
    case 2 % lift maximization
      plot_data=struct('problem',problem,'Cd',0.004,...
        'A',0.075,'x0',x0,'h1',h1,'h2',h2);
      f_objective=@f_obj_lift_maximization;
end
% Optimization code
f0=f_model(x0);
E0=f_obj(x0,f_objective,f_model);
Ntotal=Ntotal+1;
% update optimization history
X{1}=x0; E(1)=E0; R{1}=f0; N(1)=Ntotal;
accepted=1;
while 1
    if accepted==1 % RSA model construction
      % (only upon new design acceptance)
      [L,Ntotal]=construct_rsa_model(x0,f0,...
        f_model,d,Ntotal,model_order);
    end
    lb_tmp=max(lb,x0-d); ub_tmp=min(ub,x0+d);
    x1=fmincon(@f_obj,x0,[],[],[],[],lb_tmp,ub_tmp,'',...
      options,f_objective,@rsa_model,L); % RSA model optimization
    if norm(x1-x0)<0.0001 % termination condition
      break;
    end
    f1=f_model(x1);
    E1=f_obj(x1,f_objective,f_model);
    Ntotal=Ntotal+1;
    Es1=f_obj(x1,f_objective,@rsa_model,L);
    rho=(E1-E0)/(Es1-E0); % calculate gain ratio
    if rho>0.75 % RSA model region update
      d=2*d;
    elseif rho<0.25
      d=d/3;
```

```
    end
    if rho>0
      accepted=1;
      x0=x1; f0=f1; E0=E1; % update design
      % update optimization history
      X{length(X)+1}=x0; E(length(E)+1)=E1;
      R{length(R)+1}=f1; N(length(N)+1)=Ntotal;
      plot_history(X,R,E,N,plot_data);
    else
      accepted=0;
    end
    % verify termination condition
    if length(X)>1 && norm(X{end}-X{end-1})<0.0001
      break;
    end
end
display_results(x0,f_model); % Results display code
```

The size of the subregion for setting up and optimizing the response surface model is controlled by the trust region framework.

Table 14.2 and Figs. 14.5–14.8 summarize the results obtained for both test cases using a first order approximation model. It can be observed that while the quality of the final designs are comparable to that found by direct optimization, the optimization cost is lower, particularly for test problem 2 (lift maximization).

14.5 SBO using SM

The SM algorithm considered in this section exploits multi-point output SM, as described in Section 4.5.3. The SM optimization process is embedded in a trust-region framework.

Table 14.2 SAO results.

Parameters & figures of interest	Initial geometry	Test problem 1 optimized geometry	Test problem 2 optimized geometry
m	0.0150	0.01223	0.01299
p	0.6000	0.78620	0.79956
t	0.0950	0.11142	0.11137
$C_{l.f}$	0.4146	0.49999	0.54985
$C_{d.f}$	0.0009	0.00298	0.00399
A_f	0.0640	0.07500	0.07497

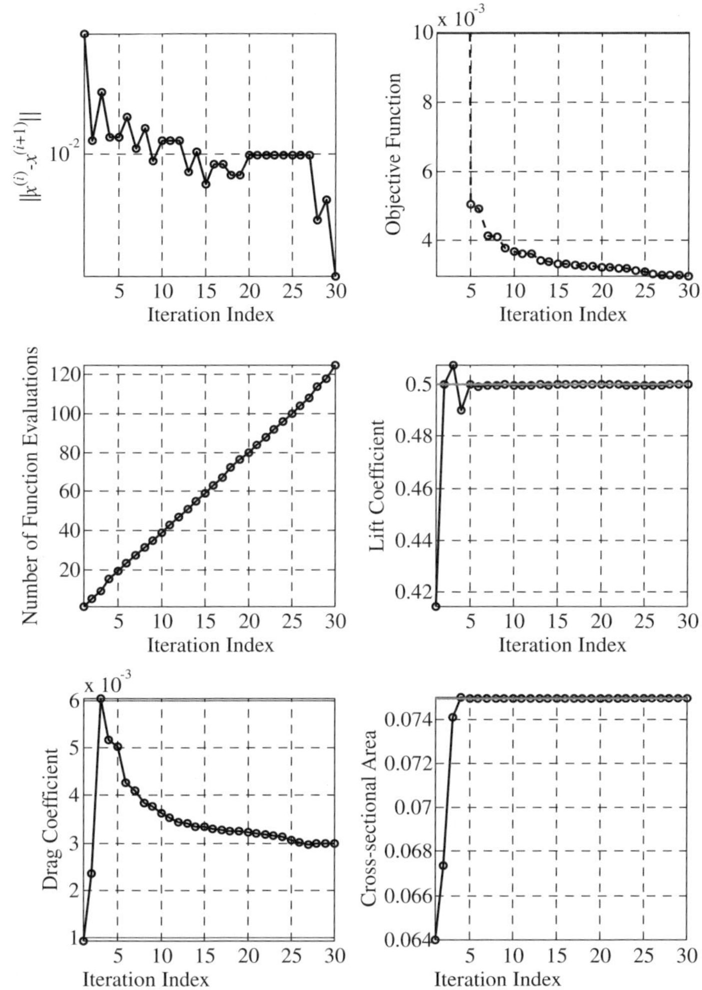

Fig. 14.5. Optimization history for sequential approximate design (test problem 1).

The script of the space mapping algorithm is given below:

```
% Multipoint space mapping script
% Initialization
clear all
close all
h1=figure ('Position',[100 100 700 1000]);
h2=figure ('Position',[1000 600 600 400]);
global X0;
```

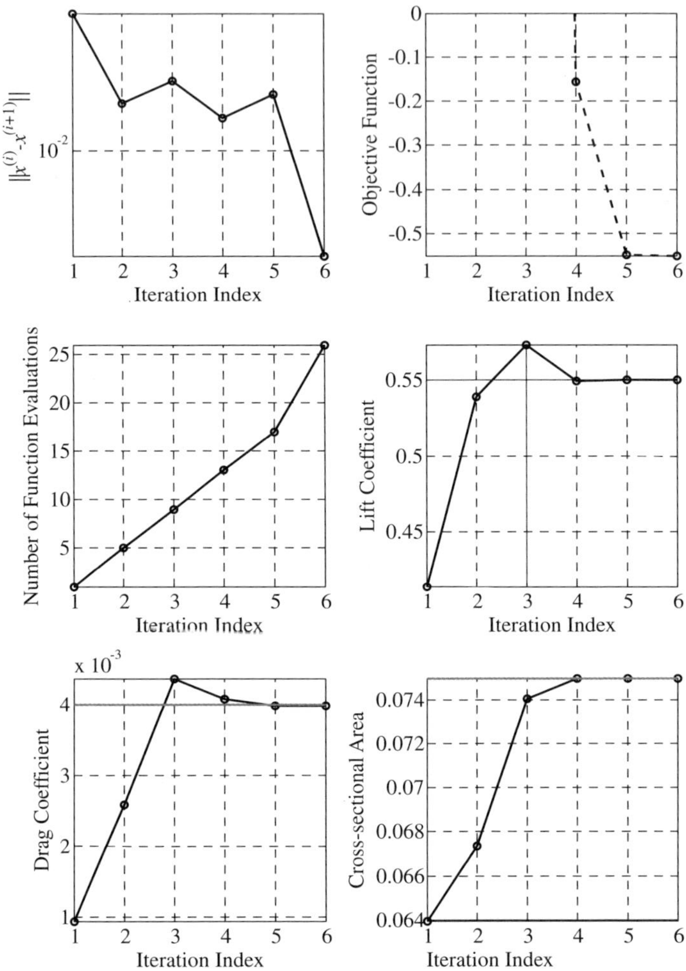

Fig. 14.6. Optimization history for sequential approximate design (test problem 2).

```
global Radius;
X=[]; Rf=[]; Rc=[]; E=[]; N=[]; Ntotal=0;
% Control variables
problem=2; % Test problem selection: 1 - drag minimization,
% 2 - lift maximization
f_model=@airfoil_model_f; % model to be optimized
c_model=@airfoil_model_c1; % low-fidelity model
% Lower/upper bounds (do not change - CFD models
% are only valid within these bounds!)
```

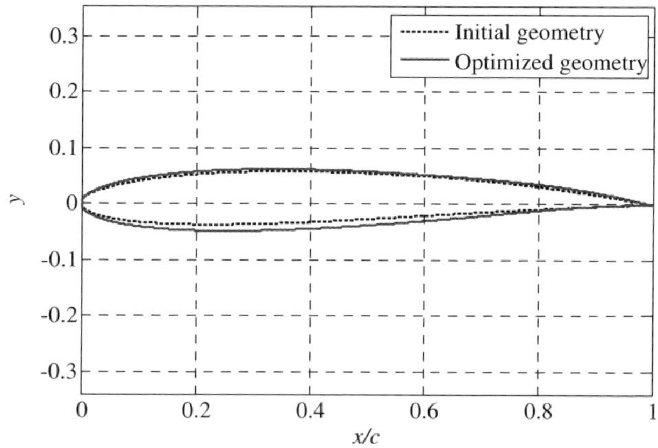

Fig. 14.7. Initial and optimized airfoil geometry for test problem 1 (drag minimization), solved using SAO.

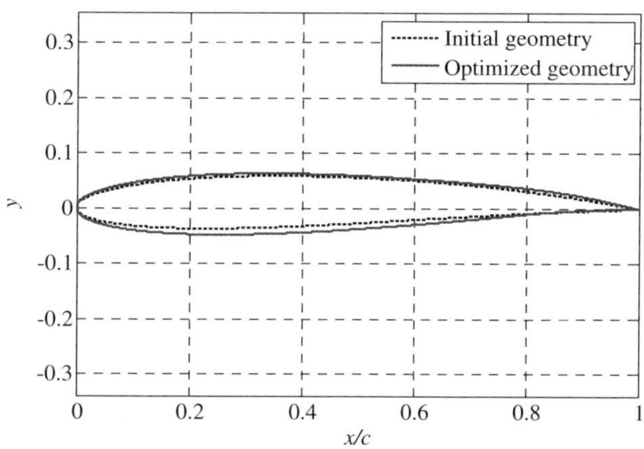

Fig. 14.8. Initial and optimized airfoil geometry for test problem 2 (lift maximization), solved using SAO.

```
lb=[0 0.4 0.07]';
ub=[0.03 0.8 0.12]';
radius=norm((ub-lb)/10); % initial trust-region radius
x0=(lb+ub)/2; % starting point
% options for fmincon routine
options=optimset('Display','iter','Algorithm','active-set',...
    'TolX',1e-8,'TolFun',1e-8);
% Problem setup
switch(problem)
```

```
   case 1 % drag minimization
     plot_data=struct('problem',problem,'Cl',0.5,...
       'A',0.075,'x0',x0,'h1',h1,'h2',h2);
     f_objective=@f_obj_drag_minimization;
   case 2 % lift maximization
     plot_data=struct('problem',problem,'Cd',0.004,...
       'A',0.075,'x0',x0,'h1',h1,'h2',h2);
     f_objective=@f_obj_lift_maximization;
end
% Optimization code
f0=f_model(x0);
c0=c_model(x0);
E0=f_obj(x0,f_objective,f_model);
Ntotal=Ntotal+1;
% update optimization history
[X,E,Rf,Rc,N]=update_storage_sm(X,E,Rf,Rc,N,x0,E0,f0,c0,Ntotal);
while 1
   S=sm_scaling(Rf,Rc);
   S.D=[0 0 0]';
   s0=c_model_scaled(x0,S);
   S.D=f0-s0; % local correction
   X0=x0;
   Radius=radius;
   x1=fmincon(@f_obj,x0,[],[],[],[],lb,ub,@f_constr,options,...
     f_objective,@c_model_scaled,S); % optimize surrogate
   if norm(x1-x0)<0.0001 % termination condition
     break;
   end
   f1=f_model(x1);
   c1=c_model(x1);
   s1=c_model_scaled(x1,S);
   E1=f_obj(x1,f_objective,f_model);
   Ntotal=Ntotal+1;
   % update optimization history
   [X,E,Rf,Rc,N]=update_storage_sm(X,E,Rf,Rc,N,x1,E1,f1,c1,Ntotal);
   plot_history(X,Rf,E,N,plot_data);
   Es1=f_obj(x1,f_objective,@c_model_scaled,S);
   rho=(E1-E0)/(Es1-E0); % calculate gain ratio
   if rho>0.75 % update trust region radius
     radius=2*radius;
   elseif rho<0.25
     radius=radius/3;
   end
   if rho>0 % update design
     x0=x1;
     f0=f1;
     c0=c1;
     E0=E1;
   end
   % verify termination condition
```

Table 14.3 SM optimization results.

Parameters & figures of interest	Initial geometry	Test problem 1 optimized geometry	Test problem 2 optimized geometry
m	0.0150	0.01393	0.01478
p	0.6000	0.73210	0.74227
t	0.0950	0.11141	0.11134
$C_{l.f}$	0.4146	0.50014	0.54462
$C_{d.f}$	0.0009	0.00321	0.00399
A_f	0.0640	0.07500	0.07496

```
    if length(X)>1 && norm(X{end}-X{end-1})<0.0001
      break;
      end
end
display_results(x0,f_model); % Results display code
```

Table 14.3 and Figs. 14.9–14.12 summarize the results obtained for both test cases. It should be emphasized that the computational cost of the design process (while neglecting the low-fidelity model evaluations) is substantially lower than for direct optimization. In practice, as indicated before, the low-fidelity model can be made around 50 times faster than the high-fidelity one. This, in conjunction with the more efficient surrogate model optimization approach, would normally result in the total design cost corresponding to about a dozen high-fidelity model evaluations.

14.6 Airfoil Optimization using Multi-Level Algorithm

The last technique implemented in this chapter is a multi-level optimization algorithm, considered in Section 6.5. Here, we use two low-fidelity models, and the design refinement procedure is embedded in the trust-region framework.

The script of the multi-level algorithm is given below:

```
% Multilevel optimization script
% Initialization
clear all
close all
h1=figure(('Position',[100 100 700 1000]);
h2=figure('Position',[1000 600 600 400]);
```

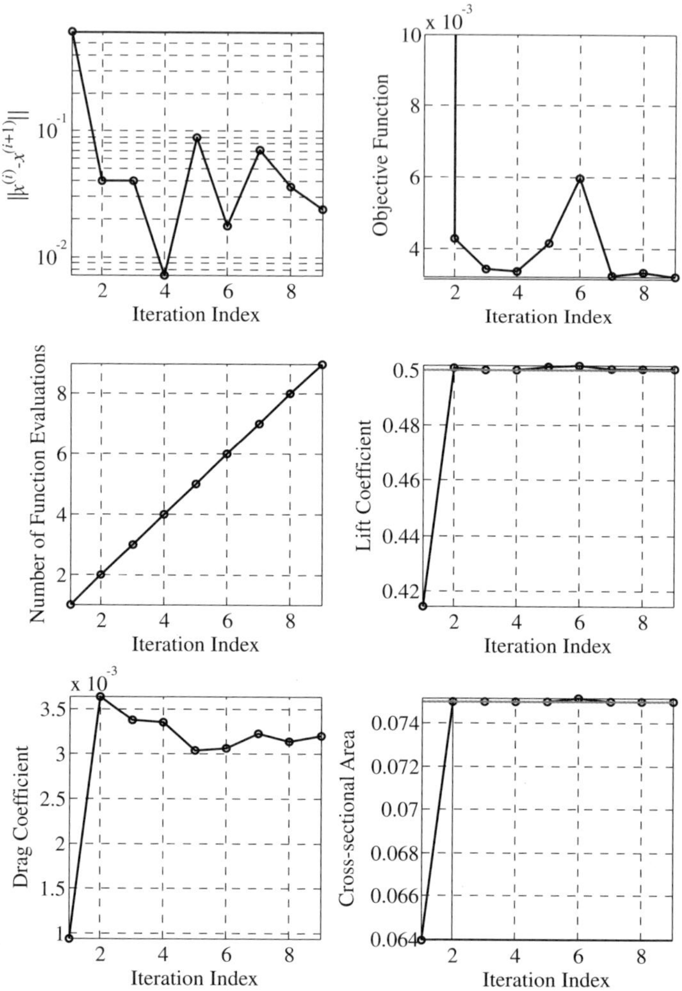

Fig. 14.9. Optimization history for SM design (test problem 1).

```
global X0;
global Radius;
X=[]; R=[]; E=[]; N=[]; Ntotal=0;
% Control variables
problem=2; % 1 - drag minimization, 2 - lift maximization
f_model=@airfoil_model_f; % model to be optimized
c1_model=@airfoil_model_c1; % first (coarser) low-fidelity model
c2_model=@airfoil_model_c0; % second (finer) low-fidelity model
% Lower/upper bounds (do not change - CFD models
```

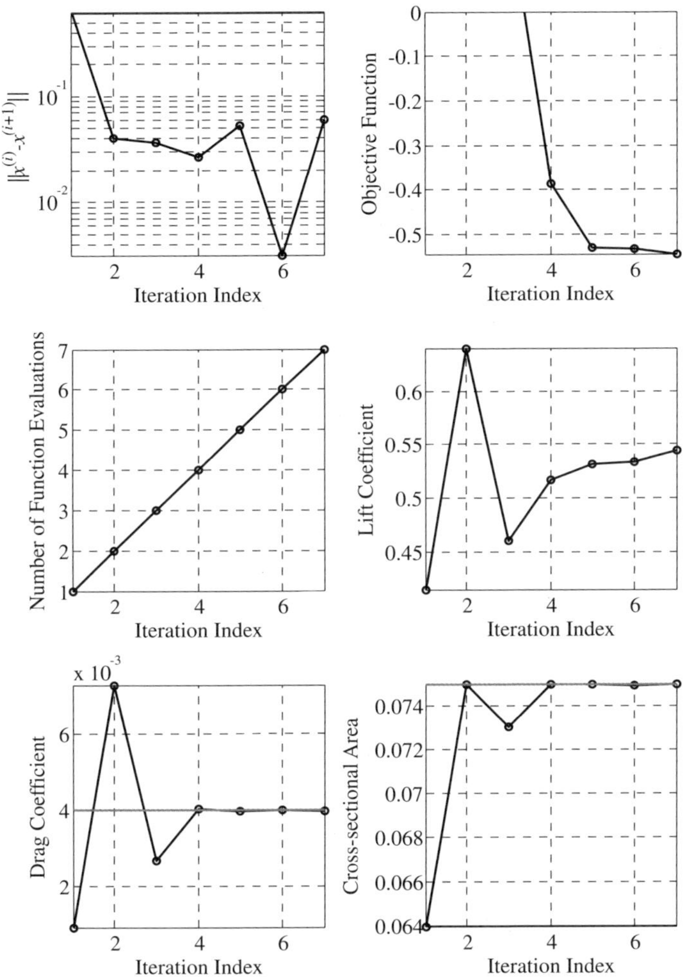

Fig. 14.10. Optimization history for SM design (test problem 2).

```
% are only valid within these bounds!)
lb=[0 0.4 0.07]';
ub=[0.03 0.8 0.12]';
radius=norm((ub-lb)/5); % initial trust-region radius
x0=(lb+ub)/2; % starting point
% options for fmincon routine
options=optimset('Display','iter','Algorithm','active-set',...
    'TolX',1e-8,'TolFun',1e-8);
% Problem setup
switch(problem)
```

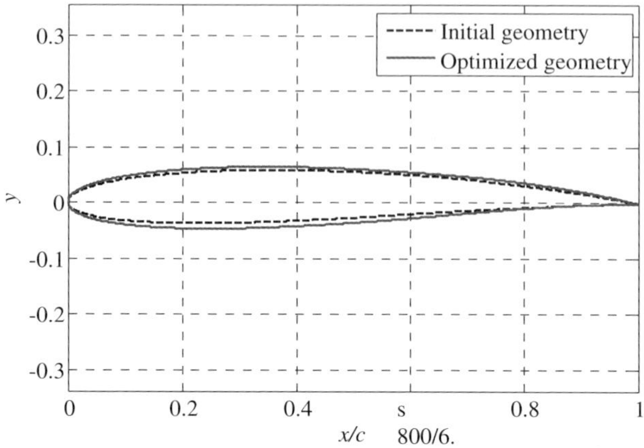

Fig. 14.11. Initial and optimized airfoil geometry for test problem 1 (drag minimization), solved using SM optimization.

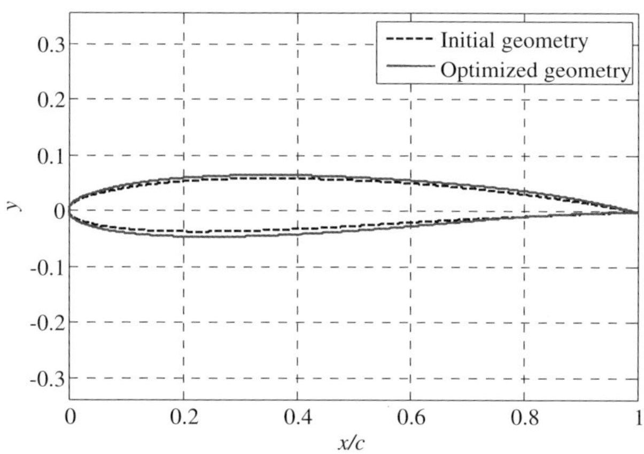

Fig. 14.12. Initial and optimized airfoil geometry for test problem 2 (lift maximization), solved using SM optimization.

```
    case 1 % drag minimization
      plot_data=struct('problem',problem,'Cl',0.5,...
        'A',0.075,'x0',x0,'h1',h1,'h2',h2);
      f_objective=@f_obj_drag_minimization;
    case 2 % lift maximization
      plot_data=struct('problem',problem,'Cd',0.004,...
        'A',0.075,'x0',x0,'h1',h1,'h2',h2);
      f_objective=@f_obj_lift_maximization;
end
% Optimization code
```

```
% Evaluation of the high-fidelity model at x0
f0=f_model(x0);
E0=f_obj(x0,f_objective,f_model);
X{1}=x0; R{1}=f0; E(1)=E0; N=Ntotal; % update optimization history
% Optimization of the first low-fidelity model
x1=fmincon(@f_obj,x0,[],[],[],[],lb,ub,'',options,...
   f_objective,c1_model,plot_data);
f1=f_model(x1);
E1=f_obj(x1,f_objective,f_model);
X{2}=x1; R{2}=f1; E(2)=E1; N(2)=Ntotal; % update optimization history
plot_history(X,R,E,N,plot_data);
% Optimization of the second low-fidelity model
x2=fmincon(@f_obj,x1,[],[],[],[],lb,ub,'',options,...
   f_objective,c2_model,plot_data);
f2=f_model(x2);
E2=f_obj(x2,f_objective,f_model);
% update optimization history
X{3}=x2; R{3}=f2; E(3)=E2; N(3)=Ntotal;
plot_history(X,R,E,N,plot_data);
% Design refinement
model_order=2;
d=(ub-lb)/10;
x0=x2;
f0=f2;
c0=airfoil_model_c0(x0);
% construct RSA model
[L,Ntotal]=construct_rsa_model(x0,c0,c2_model,d,Ntotal,model_order);
while 1
   s0=rsa_model(x0,L); D=f0-s0;
   lb_tmp=max(lb,x0-d);
   ub_tmp=min(ub,x0+d);
   X0=x0; Radius=radius;
   x1=fmincon(@f_obj,x0,[],[],[],[],lb_tmp,ub_tmp,@f_constr,...
     options,f_objective,@rsa_model,L,D); % optimize RSA model
   if norm(x1-x0)<0.0001 % termination condition
     break;
   end
   f1=f_model(x1);
   E1=f_obj(x1,f_objective,f_model);
   Ntotal=Ntotal+1;
   Es1=f_obj(x1,f_objective,@rsa_model,L,D);
   rho=(E1-E0)/(Es1-E0); % calculate gain ratio
   if rho>0.75 % update trust region ratio
     radius=2*radius;
   elseif rho<0.25
     radius=radius/3;
   end
   if rho>0
     x0=x1; f0=f1; E0=E1; % update design
     % update optimization history
```

Table 14.4 Multi-level optimization results.

Parameters & figures of interest	Initial geometry	Test problem 1 optimized geometry	Test problem 2 optimized geometry
m	0.0150	0.01417	0.01362
p	0.6000	0.72292	0.78137
t	0.0950	0.11141	0.11136
$C_{l.f}$	0.4146	0.50006	0.55269
$C_{d.f}$	0.0009	0.00322	0.00401
A_f	0.0640	0.07500	0.07497

```
        X{length(X)+1}=x0; E(length(E)+1)=E1;
        R{length(R)+1}=f1; N(length(N)+1)=Ntotal;
        plot_history(X,R,E,N,plot_data);
    end
    % verify termination condition
    if length(X)>1 && norm(X{end}-X{end-1})<0.0001
        break;
    end
end
display_results(x0,f_model); % Results display code
```

The results are shown in Table 14.4 and Figs. 14.13–14.16. The computational cost of the algorithm is very low compared to direct optimization and comparable to that of SM. It should be noted that the high-fidelity model is evaluated after the first two stages (optimization of both low-fidelity models) for visualization purposes only. Normally, the high-fidelity model is not evaluated until the design refinement stage.

14.7 Summary

In this chapter, we discussed the basic implementations of various SBO techniques. It should be stressed that these implementations are for illustration purposes only and aimed at simplicity rather than efficiency. Still, they can be used by interested readers as starting points for creating more involved implementations. All auxiliary functions used by the scripts are provided in the Appendix. The DACE toolbox (Lophaven *et al.*, 2002) is necessary to run the examples.

It is also worth noting that — because of using response surface approximations to represent actual CFD models — the design problems are somehow simplified: the actual CFD simulation outcome

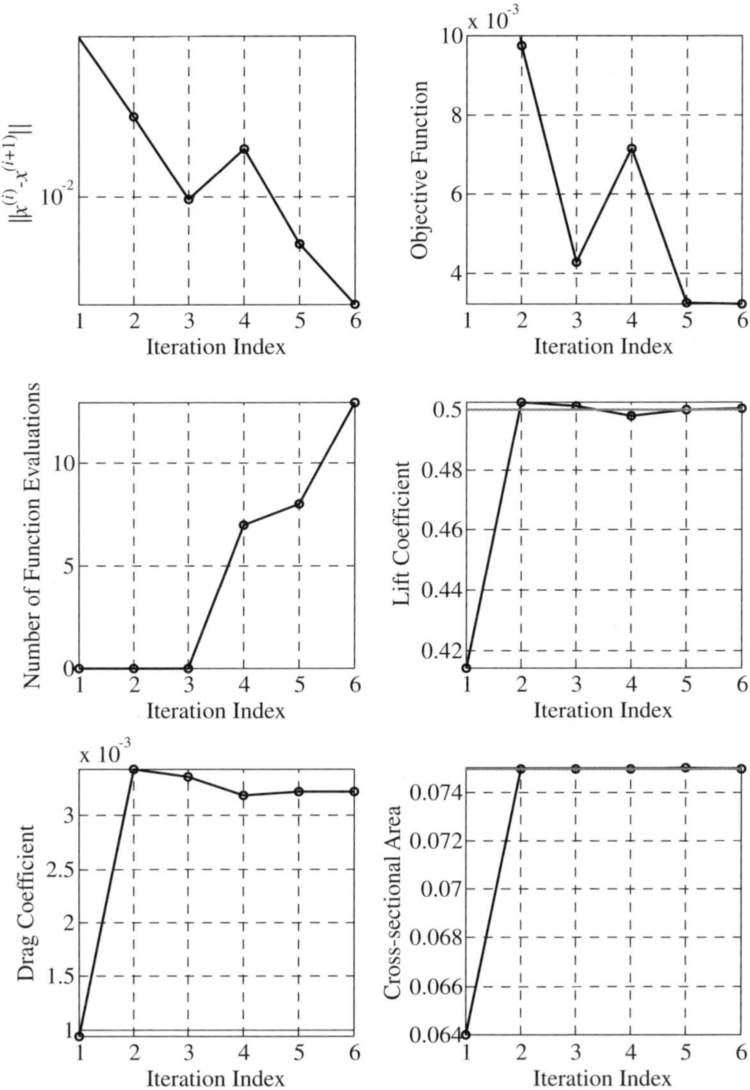

Fig. 14.13. Optimization history for multi-level design (test problem 1).

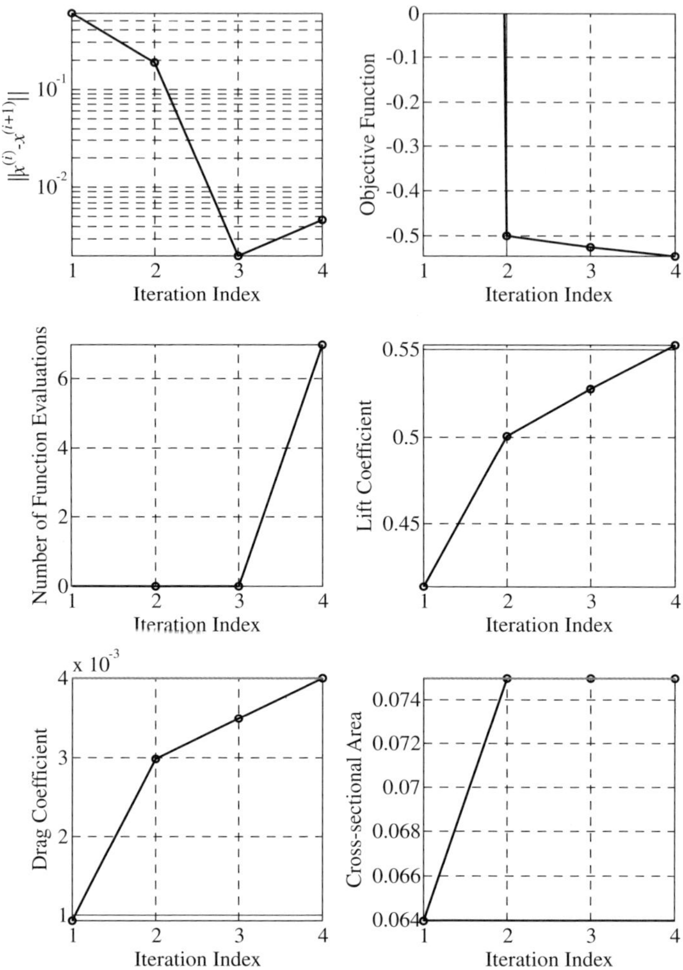

Fig. 14.14. Optimization history for multi-level design (test problem 2).

always contains some level of numerical noise (usually, lower-fidelity models are more noisy), whereas the response surface models are smooth. Thus, applying the discussed algorithms for actual CFD-simulated models would result in a worse performance. For noisy objective functions it is often more advantageous to use derivative-free optimizers (cf. Chapter 3) rather than gradient-based ones, unless adjoint sensitivity is available.

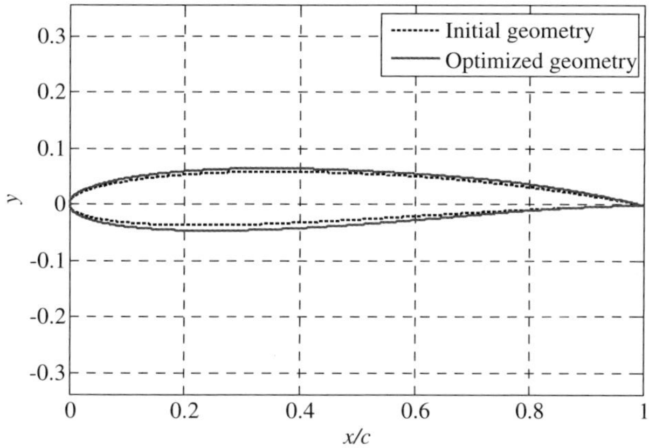

Fig. 14.15. Initial and optimized airfoil geometry for test problem 1 (drag minimization), solved using multi-level optimization.

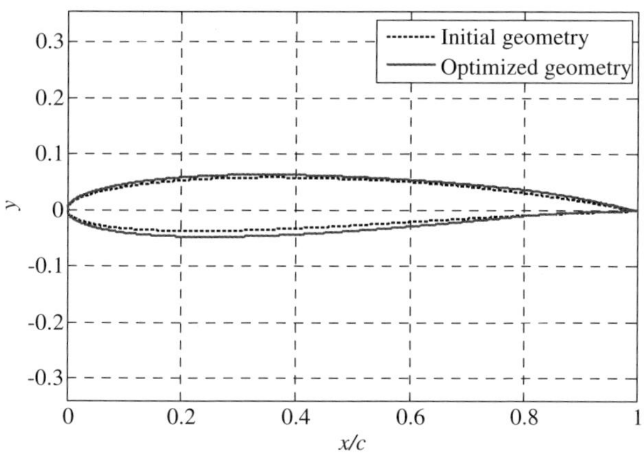

Fig. 14.16. Initial and optimized airfoil geometry for test problem 2 (lift maximization), solved using multi-level optimization.

Part V
Conclusion

Chapter 15

Practical Aspects of Variable-Fidelity Design

As demonstrated in the previous chapters, surrogate-based optimization (SBO) methods exploiting variable-fidelity computational fluid dynamics (CFD) models allow for substantial reduction of the simulation-driven design cost. It is not uncommon that the computational effort of design optimization using SBO corresponds to only a handful of evaluations of the respective high-fidelity model. At the same time, the successful application of such methods requires certain experience from the user regarding, among others, the selection of the particular method and, most importantly, the appropriate setup of the underlying low-fidelity model. Moreover, some variable-fidelity design techniques exhibit potential convergence issues, which is partially because of not using sensitivity data. Other methods may have limitations that arise from the construction of the surrogate model.

In this chapter, we discuss in more detail some practical aspects of variable-fidelity design. In particular, we discuss convergence issues, as well as the importance of the low-fidelity model for the performance of the SBO algorithm, both in terms of the computational cost of the design process and the quality of the final design. We also describe an automated procedure for setting up the low-fidelity models for SBO applications. Finally, some recommendations regarding the selection of a particular SBO and surrogate modeling approach for a given design problem are given.

15.1 Convergence Issues of SBO Algorithms

The generic SBO design scheme (Equation (4.2)) considered in Chapter 4 produces a sequence $\boldsymbol{x}^{(i)}$, $i = 0, 1, \ldots$ of approximate solutions to the original design problem $\boldsymbol{x}^* = \text{argmin}\} \boldsymbol{x}: f(\boldsymbol{x})\}$. Most of the variable-fidelity methods considered in this book are local ones, in the sense that the new approximation, $\boldsymbol{x}^{(i+1)}$ of $\boldsymbol{x}^*$, is obtained by optimizing the current surrogate model, $s^{(i)}$, starting from the previous design, $\boldsymbol{x}^{(i)}$. The surrogate model itself is usually established to meet at least a zero-order consistency condition with the high-fidelity model at $\boldsymbol{x}^{(i)}$, i.e. $s^{(i)}(\boldsymbol{x}^{(i)}) \quad = f(\boldsymbol{x}^{(i)})$.

Formally speaking, this is not sufficient to ensure the convergence of Equation (4.2) to a local optimum of f. However, as explained in Chapter 4, the satisfaction of both zero- and first-order consistency, i.e. $\nabla s^{(i)}(\boldsymbol{x}^{(i)}) = \nabla f(\boldsymbol{x}^{(i)})$ (Alexandrov and Lewis, 2001), is sufficient, provided that the algorithm is embedded in the trust region framework (Conn *et al.* (2000); see also Equation (4.3)) and the functions involved are sufficiently smooth.

Most of the SBO methods considered here do not use sensitivity data. Consequently, first-order consistency is not satisfied. Furthermore, CFD-based objective functions always exhibit a certain level of numerical noise, so that even having $\nabla s^{(i)}(\boldsymbol{x}^{(i)}) = \nabla f(\boldsymbol{x}^{(i)})$ does not guarantee global convergence of the algorithm. On the other hand, practical implementations of the SBO techniques usually use trust-region convergence safeguards to prevent algorithm divergence towards unacceptable designs and to terminate the search process once a design cannot be further improved, e.g. due to the noise. At the same time, because of exploiting knowledge about the system at hand contained in the underlying low-fidelity models (in particular, the trends concerning the change of the objective function under the change of designable parameters), first-order consistency is usually satisfied in an approximate sense, particularly at the initial stages of the design process. In some cases (e.g. output space mapping (OSM), Section 4.5.3), the construction of the surrogate model based on multi-point low-fidelity model correction allows for an improved alignment of the surrogate and the high-fidelity model once the optimization process progresses. As demonstrated in Chapters 10–13, the combination of physics-based modeling and algorithmic measures (trust-region) normally allows for the obtaining of a reasonable practical convergence of the SBO schemes.

It should also be mentioned that virtually all SBO methods considered in this book can be enhanced by using sensitivity data (e.g. obtained through the adjoint approach). Derivative information can be used to improve convergence properties of the SBO search process and to further reduce its computational cost (among others, by speeding up the surrogate model optimization). Examples of such enhancements can be found in the literature (Eldred *et al.*, 2004; Alexandrov and Lewis, 2001; Koziel *et al.*, 2012a).

15.2 Low-Fidelity Models and SBO Performance

A low-fidelity model (or models, in the case of multi-fidelity design algorithms) is the most critical component of a variable-fidelity SBO algorithm. Clearly, an important prerequisite to ensure the efficiency of SBO is that the underlying low-fidelity model is computationally cheap. As the surrogate (and, consequently, the low-fidelity model) is evaluated numerous times during the optimization run, a related computational overhead should be reduced as much as possible. At the same time, the low-fidelity model should be reasonably accurate so that it adequately captures relevant features of the physical phenomena being described (e.g. a pressure shock in the case of transonic airfoils) and can be aligned with the high-fidelity model using a surrogate modeling technique chosen by the designer. While in some cases it is possible to develop low-fidelity models with negligible evaluation costs (e.g. models based on simple analytical formulas), in general relaxed accuracy CFD simulation may be necessary. Unfortunately, such models may be relatively expensive. Consequently, their evaluation cost cannot be neglected and may contribute considerably to the overall design expenses (Koziel and Ogurtsov, 2011, 2013c).

Thus, the proper setup of the low-fidelity model (controlled, among others, by the mesh density) is of great significance. On one hand, using a coarser model allows us to reduce its evaluation time. On the other hand, the coarser models are less accurate. As a result, a larger number of iterations of the SBO algorithm may be necessary to yield a satisfactory design in order that the total cost may be about the same or even higher than the total cost of an optimization algorithm employing the finer model. Also, the SBO process may simply fail if the underlying low-fidelity model is not sufficiently accurate. For finer models, the individual evaluation time may be higher, but this is not directly translated into a higher total design cost, because a smaller number of iterations may be sufficient to find a good design. In general, finding a good trade-off between the low-fidelity model speed and accuracy is not obvious.

The computational expenses of the low-fidelity models that are built from coarse-mesh simulations can be alleviated to some extent on the algorithmic level. For example, in some versions of space mapping (SM), the surrogate model parameters are repeatedly extracted with nonlinear regression at every iteration of the optimization algorithm (Bandler *et al.*, 2004a), which results in a large number of low-fidelity model evaluations and, consequently, high total costs. Unlike SM, response correction techniques, e.g. manifold mapping (Echeverría and Hemker, 2005) or shape-preserving response prediction (Koziel, 2010a), do not have these issues because no extractable parameters are utilized there.

For the sake of illustration, let us consider SBO of transonic airfoils. The CFD analysis is performed using ICEM CFD (2013; mesh generation) and FLUENT (2012; fluid flow simulations) using the computational domain shown in Fig. 15.1 (Leifsson *et al.*, 2012). A grid convergence study was performed using the NACA (National Advisory Committee for Aeronautics) 0012 airfoil at Mach number $M_\infty = 0.75$ and angle of attack $\alpha = 1$ degree. The results are shown in Fig. 15.2. First the fine mesh was constructed (this is mesh number 2 in Fig. 15.3); then the mesh density was doubled in each direction to produce mesh number 1. The lift and drag coefficients changed by less than one lift count ($\Delta C_l = 10^{-2}$) and one drag count ($\Delta C_d = 10^{-4}$), respectively. The remaining meshes were generated by halving the mesh density in each direction. We conclude that the fine mesh (number 2) is adequate, and that 407,676 cells are needed for mesh convergence. We use that particular mesh for

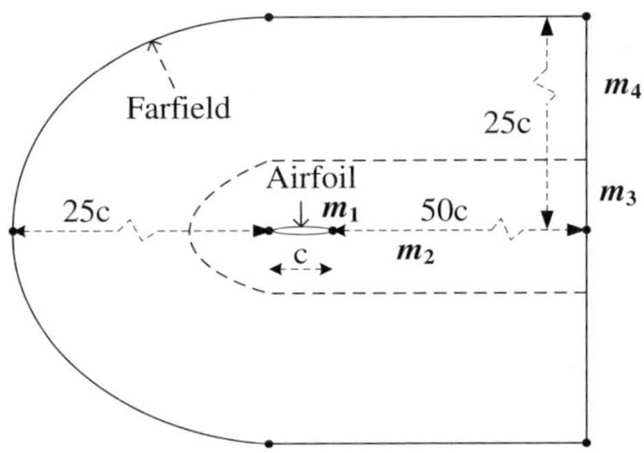

Fig. 15.1. A sketch of the computational domain.

the high-fidelity model, i.e. mesh number 2 in Fig. 15.2. The overall simulation time for the case considered with the fine mesh is around 33 min (Fig. 15.2(b)). The solver reached a fully converged solution (a reduction in residuals by six orders) in this case after 275 iterations. The solver was able to reach a fully converged solution in all cases, except in the very last one, mesh number 7, which the solver ran for 1,000 iterations.

In order to investigate the effect of low-fidelity model setup on the performance of the SBO process, we consider transonic airfoil design using a multi-point OSM technique (Koziel and Leifsson (2012a); see also Section 4.5.3). The SM algorithm is run several times using different underlying low-fidelity models that utilize meshes 3–7 in Fig. 15.2. On top of using a coarser mesh, the maximum number of CFD solver iterations is set to 100 for all low-fidelity models, as this number is sufficient for convergence of both lift and drag coefficients (Leifsson *et al.*, 2012). A comparison of the pressure distributions, shown in Fig. 15.3, indicates that the low-fidelity model, in spite of being based on much coarser mesh and reduced flow solver iterations, captures the main features of the high-fidelity model pressure distribution quite well. The biggest discrepancy in the distributions is around the shock on the upper surface, leading to an overestimation of both lift and drag (Fig. 15.2).

As mentioned before, in order to optimize the airfoils, we use the SM technique (Section 4.5.3). The surrogate model optimization

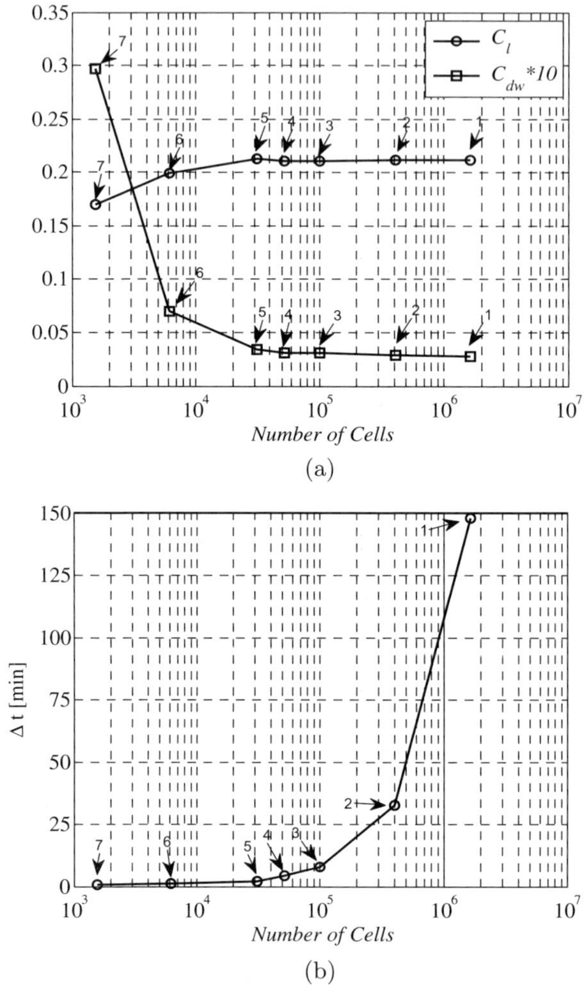

Fig. 15.2. Grid convergence study using the NACA 0012 airfoil at Mach number $M_\infty = 0.75$ and angle of attack $\alpha = 1$ degree. (a) Lift and drag coefficients versus the number mesh cells; (b) the simulation time versus the number of mesh cells.

is performed using a pattern-search algorithm (Koziel, 2010a). The termination condition for the algorithm is $\|x^{(i)} - x^{(i-1)}\| < 10^{-3}$. The airfoil shapes are parameterized using the NACA four-digit method with three design variables, $x = [m\ p\ t/c]^T$, where the chord is set to $c = 1$. We consider two design cases, one case of lift maximization and

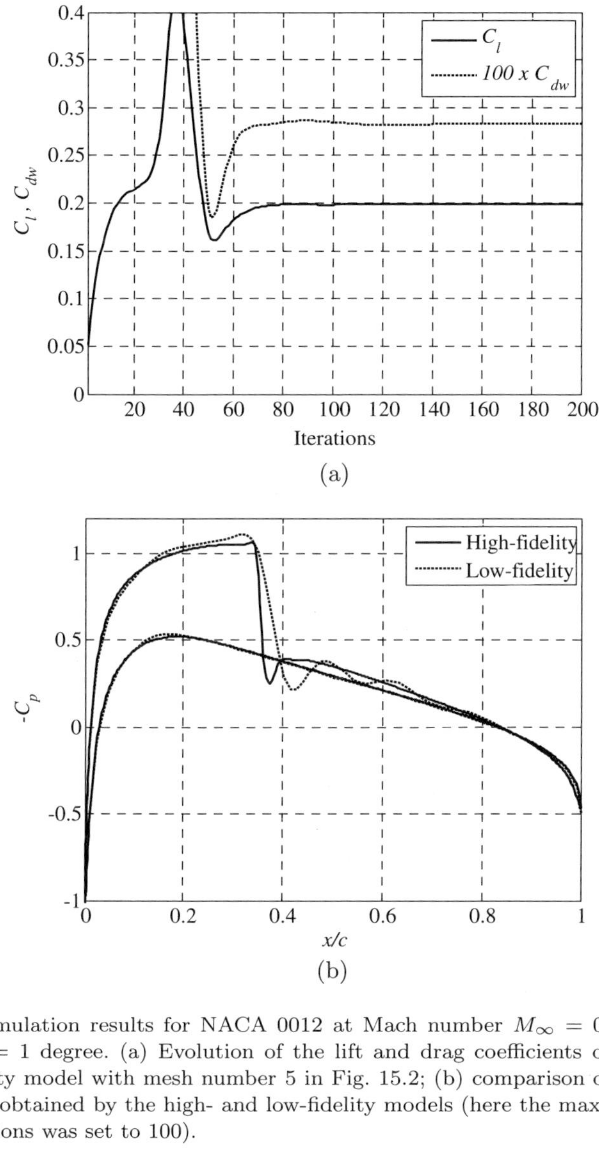

(b)

Fig. 15.3. Simulation results for NACA 0012 at Mach number $M_\infty = 0.75$ and angle of attack $\alpha = 1$ degree. (a) Evolution of the lift and drag coefficients obtained using the low-fidelity model with mesh number 5 in Fig. 15.2; (b) comparison of the pressure distributions obtained by the high- and low-fidelity models (here the maximum number of flow iterations was set to 100).

one of drag minimization. Details of the test cases and optimization results are given in Table 15.1. The bounds on the design variables are set as $0 \le m \le 0.1, 0.2 \le p \le 0.8$, and $0.05 \le t/c \le 0.2$. Here, the low-fidelity models are denoted by numbers 1–5. Low-fidelity model 1

Table 15.1 Numerical results of lift maximization (case 1) and drag minimization (case 2) for five different low-fidelity models of varying mesh density.*

Variables	Case 1 (lift maximization) $M_\infty = 0.75$, $\alpha = 1°$, $C_{d,max} = 41\ dc$, $A_{min} = 0.065$					
	Initial	LF 1	LF 2	LF 3	LF 4	LF 5
m	0.02	0.0090	0.0090	0.0090	0.0090	0.0113
p	0.40	0.6119	0.6333	0.7185	0.7165	0.6111
t/c	0.10	0.1019	0.1014	0.0979	0.0980	0.0947
$C_l(l.c.)$	66.01	46.38	47.12	50.82	50.66	52.40
$C_{dw}(d.c.)$	166.55	40.99	41.00	40.95	40.91	39.98
A	0.0674	0.0686	0.0683	0.0659	0.0660	0.0638
N_c	—	300	500	400	550	750
N_f	—	6	10	8	11	15
Time ratio	—	22	44	80	186	248
Total cost	—	<20	<22	13	<14	<19

Variables	Case 2 (drag minimization) $M_\infty = 0.75$, $\alpha = 1°$, $C_{l.min} = 50\ lc$, $A_{min} = 0.065$					
	Initial	LF 1	LF 2	LF 3	LF 4	LF 5$
m	0.01	0.0149	0.0141	0.0150	0.0138	—
p	0.50	0.5400	0.5815	0.5296	0.5941	—
t/c	0.12	0.0963	0.0963	0.0963	0.0958	—
$C_l(l.c.)$	33.29	49.98	49.98	49.97	49.72	—
$C_{dw}(d.c.)$	182.34	154.75	139.68	158.65	131.46	—
A	0.0808	0.0648	0.0648	0.0648	0.0645	—
N_c	—	300	550	300	700	—
N_f	—	6	11	6	14	—
Time ratio	—	22	44	80	186	—
Total cost	—	<20	<24	<10	<18	—

*The lift coefficient values are given in lift counts ($\Delta C_l = 10^{-2}$), and the drag coefficient values in drag counts ($\Delta C_{dw} = 10^{-4}$). N_c and N_f are the number of low- and high-fidelity model evaluations, respectively. The total optimization cost is expressed in the equivalent number of high-fidelity model evaluations. All the numerical values are from the high-fidelity model.
$The flow solver failed for the low-fidelity model.

uses mesh number 3 in Fig. 15.2, low-fidelity model 2 uses mesh number 4, and so forth up to low-fidelity model 5, which uses mesh number 7. The time evaluation ratios between high- and a given low-fidelity model are 22 (for the low-fidelity model 1), 43 (model 2), 80 (model 3), 186 (model 4), and 248 (model 5). While in practice the

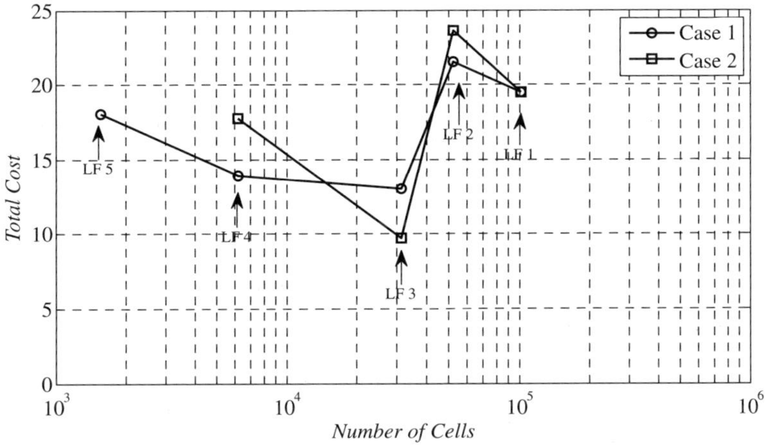

Fig. 15.4. Total design cost versus mesh density of the low-fidelity model for design cases 1 and 2. Shown are the results from the five different low-fidelity models. The algorithm working with the low-fidelity model 3 exhibits the lowest computational cost for both test cases.

ratio may vary from one design to another, we use fixed values that are typical for a specific mesh setup.

The optimization results for both cases are shown in Table 15.1, as well as in Fig. 15.4. In this study, we are mostly concerned about the computational cost of the design process. As expected, the algorithm working with the finer low-fidelity models generally requires a smaller number of iterations (as the surrogate is more accurate). On the other hand, each iteration is more expensive because of the higher evaluation cost of the low-fidelity model itself. The cost of each iteration is lower for coarser low-fidelity models; however, a larger number of iterations is necessary for the algorithm to converge. Moreover, the algorithm may have problems in finding a satisfactory design either because of the low-fidelity model inaccuracy or because the CFD simulation fails for certain designs considered during the optimization run (which is what happened for the low-fidelity model 5 in case 2). Thus, the numerical results generally confirm common-sense expectations. It seems that using the middle-fidelity models (such as model 3) could be the most beneficial. While the overall trend is quite consistent for lower-fidelity models (an increase of the design cost and lack of reliability for the coarsest models), one can also observe some fluctuations for higher-fidelity models. In order to formulate more definite recommendations, more comprehensive study seems to be necessary, with the algorithm performance averaged over the large number of test cases. This will be the subject of future work.

15.3 Low-Fidelity Model Setup through Optimization

As indicated in the previous section, the low-fidelity model setup is an important factor determining the performance of the SBO process. Typically, the selection of the appropriate model parameters, such as the number of grid points in specific regions of the computational domain as well as the maximum number of flow solver iterations, is only guided by engineering experience. The objective is to reduce the model simulation cost while maintaining reasonable accuracy. In this section, we demonstrate that automation of this process by using numerical optimization of model parameters may lead to substantially better results and greater computational savings.

We consider again the transonic airfoil with the high-fidelity model, set up in the same manner as in Section 15.2. The grid density is controlled by the following parameters (shown in Fig. 15.1): m_1 = number of grid points on the upper and lower airfoil surfaces; m_2 = number of grid points on the horizontal line behind the airfoil from the trailing-edge to the farfield boundary; m_3 = number of grid points on the vertical line from the airfoil surface to one-quarter of the distance to the farfield; m_4 = number of grid points on the three-quarters of the vertical line from the farfield down to the airfoil surface; and m_5 = distance from the airfoil surface to the first grid point.

Local clustering on the airfoil surface is also controlled, but not parameterized. The number of flow solver iterations is denoted by N. The vector of all combined parameters will be referred to as $\boldsymbol{z} = [m_1\ m_2\ m_3\ m_4\ m_5\ N]$.

The study below concerns airfoil optimization using a multi-level optimization algorithm, described in Section 6.5. We consider three different methods of setting up these parameters in order to set up the low-fidelity models for the multi-level optimization algorithm, as well as their influence on the algorithm performance.

15.3.1 *Low-fidelity model setup based on grid independence study*

The most common strategy for setting up the low-fidelity models is by using the results of a grid independence study. The process is typically done in "reverse", meaning that a high-fidelity grid is developed by an experienced engineer and then the number of grid points in each direction is reduced by half. The distance to the grid point closest to the surface is doubled as well. Figure 15.2 is an example of such a study.

The flow solution history, shown in Fig. 15.3(a) for a specific grid setup, for a low-fidelity model indicates that the lift and drag coefficients are nearly converged after 80–100 iterations. The maximum number of iterations is set to $N = 100$ for the low-fidelity model. This reduces the overall simulation time further. A comparison of the pressure distributions of the high- and low-fidelity models, shown in Fig. 15.3(b), indicates that the low-fidelity model, in spite of being based on much coarser mesh and reduced flow-solver iterations, captures the main features of the high-fidelity model pressure distribution quite well. The biggest discrepancy in the distributions is around the shock on the upper surface, leading to an overestimation of both lift and drag (cf. Fig. 15.2).

15.3.2 *Low-fidelity model setup based on insight*

An alternative method of setting up the low-fidelity models is by modifying the grid parameters based on the insight of the engineer. The objective would be to reduce the simulation time, but at the same time retain the accuracy of the high-fidelity model. For example, regions with large gradients need to be resolved better than other regions. With that in mind, one can reduce the number of grid points in the outer regions more than in the region close to the surface. In our case, the grid parameters m_2 and m_4 can be reduced more rapidly than the other grid parameters. The number of flow solvers is set in the same way as described in Section 15.3.1.

15.3.3 *Low-fidelity model setup based on numerical optimization*

The last low-fidelity model setup methodology considered here exploits numerical optimization. More specifically, the grid parameters as well as the number of iterations N, i.e. the vector $\boldsymbol{z}$, are optimized in order to reduce the discrepancy between the drag coefficients predicted by the low-fidelity model and the high-fidelity one, assuming given simulation time ratios between the models. Let us denote the drag coefficient predicted by a CFD model simulated using the grid/iteration parameters $\boldsymbol{z}$ as $C_d(\boldsymbol{x}, M, \alpha, \boldsymbol{z})$, where $\boldsymbol{x}$ represents the airfoil geometry and M and α are operating conditions (Mach number and angle of attack) for a reference airfoil for which the low-fidelity model is being set up. The optimization problem is defined as follows:

$$\boldsymbol{z}^* = \arg\min_{\boldsymbol{z}} H_z(C_d(\boldsymbol{x}, M, \alpha, \boldsymbol{z})), \tag{15.1}$$

with the objective function defined as:

$$\begin{aligned} &H_z(C_d(\boldsymbol{x}, M, \alpha, \boldsymbol{z})) \\ &\quad = \left(\frac{C_d(\boldsymbol{x}, M, \alpha, \boldsymbol{z}) - C_d(\boldsymbol{x}, M, \alpha, \boldsymbol{z}_f)}{C_d(\boldsymbol{x}, M, \alpha, \boldsymbol{z}_f)}\right)^2 \\ &\qquad + \gamma\left(\frac{t(\boldsymbol{z}_f)}{t(\boldsymbol{z})} - R_{\text{target}}\right)^2. \end{aligned} \tag{15.2}$$

Here, $\boldsymbol{z}_f$ are the grid parameters of the high-fidelity model, $C_d(\boldsymbol{x}, M, \alpha, \boldsymbol{z}_f)$ is the drag coefficient predicted by the high-fidelity model for the reference airfoil and operating conditions, and $t(\boldsymbol{z})$ represents the CFD model simulation time for given grid parameters $\boldsymbol{z}$. The objective function contains a penalty factor with the proportionality coefficient γ (here, we use $\gamma = 1{,}000$) that forces the optimization process to obtain a given simulation time ratio, R_{target}.

The low-fidelity model setup through the optimization Equations (15.1) and (15.2) allows us to obtain the best possible grid setup for a required simulation time ratio that can be controlled much more precisely than for the typical methods of Sections 15.3.1 and 15.3.2. As demonstrated below, this is directly translated into a better efficiency of the multi-level optimization algorithm.

15.3.4 *Numerical results*

Here we present an application of the multi-level optimization algorithm of Section 6.5 to the design of transonic airfoils. Using three test cases involving both drag minimization and lift maximization, we compare the performance of the algorithm with respect to the quality of the final design, as well as the computational complexity of the design process for various setups of the low-fidelity models, as described in Sections 15.3.1–15.3.3.

The low-fidelity models were configured using the NACA 0012 airfoil shape as a baseline and assuming the following operating conditions: $M_\infty = 0.75$ and $\alpha = 1$ degree. Three sets of low-fidelity models were considered: LFM 1 = models created using linear variation of all grid parameters (cf. Section 15.3.1); LFM 2 = models created using a variation of a set of grid parameters (cf. Section 15.3.2); and LFM 3 = models obtained by numerical optimization of the grid parameters (cf. Section 15.3.3).

The characteristics of all three low-fidelity model sets are presented in Figs. 15.5–15.7. It can be observed that the lift coefficient values (Fig. 15.5) diverge from the high-fidelity value upon reducing the grid cells for all the low-fidelity model families. The behavior of the drag coefficient values (Fig. 15.6) is different. For the sets LFM 1 (constructed by a typical grid independence study) and LFM 2 (constructed by insight), a deviation from the high-fidelity model drag due to the reduction of the cell number is much more pronounced when compared to LFM 3 (constructed by numerical optimization). LFM 3 can be characterized by the drag coefficient being nearly constant with respect to the number of cells. What is even more important, the time ratio (Fig. 15.7) increases more rapidly with a reduced number of cells for the LFM 3 family than the other two. It can be concluded, therefore, that constructing the low-fidelity model

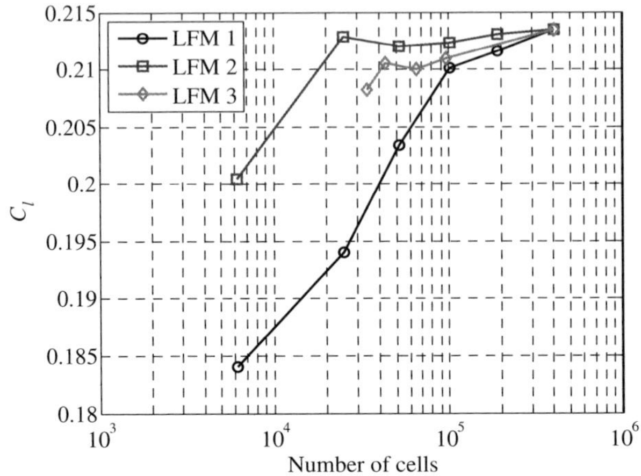

Fig. 15.5. Comparison of low-fidelity model set characteristics LFM 1–LFM 3 for NACA 0012 at $M_\infty = 0.75$, $\alpha = 1$ degree: variations of the lift coefficient.

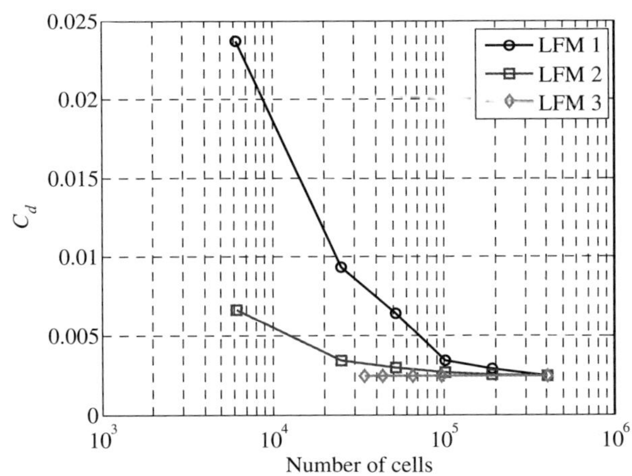

Fig. 15.6. Comparison of low-fidelity model set characteristics LFM 1–LFM 3 for NACA 0012 at $M_\infty = 0.75$, $\alpha = 1$ degree: variations of the drag coefficient.

families using numerical optimization yields both more accurate and faster models, at least for our baseline airfoil shape.

For each design case, three optimization studies are performed using the three different low-fidelity model sets, constructed as described in the previous section. Two low-fidelity models are used by

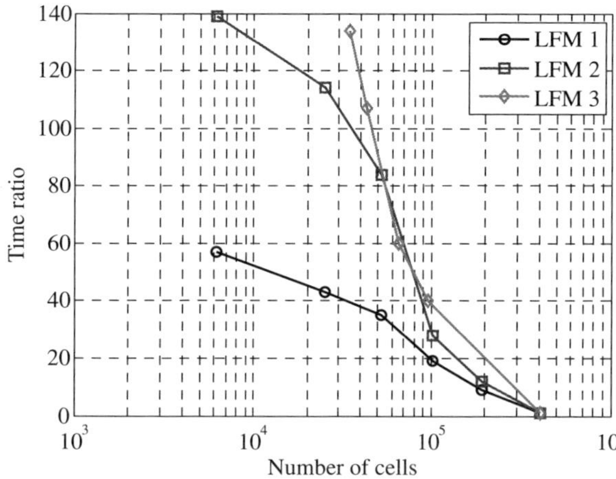

Fig. 15.7. Comparison of low-fidelity model set characteristics LFM 1–LFM 3 for NACA 0012 at $M_\infty = 0.75$, $\alpha = 1$ degree: ratio of the high-to-low-fidelity simulation time with the number of grid cells.

Table 15.2 Simulation time ratio for the high- to low-fidelity models for each low-fidelity model used from different families.

Model	LFM 1	LFM 2	LFM 3
c_1	25	84	100
c_2	3	12	40

the multi-level optimization algorithm in each case. Table 15.2 shows the time ratio of the two low-fidelity models used from each family. Using faster models for LFM 1 and LFM 2 either resulted in failed simulations during the optimization run, i.e. the grids were simply too coarse for the flow solver to handle, or the optimizer would not yield improved designs.

The design cases are described in Table 15.3. In each case, the airfoil shape is parameterized using the NACA four-digit airfoil with m (the maximum ordinate of the mean camberline as a fraction of chord), p (the chordwise position of the maximum ordinate), and t/c (the thickness-to-chord ratio) as the design variables (see Abbott and von Doenhoff (1959) for details). The bounds on the design variables

Table 15.3 Design case formulations.

Case	M_∞	α	Objective	Constraint 1	Constraint 2
1	0.70	$1°$	min C_{dw}	$C_l \geq 0.6000$	$A \geq 0.075$
2	0.75	$0°$	max C_l	$C_{dw} \leq 0.0050$	$A \geq 0.075$
3	0.80	$0°$	min C_{dw}	$C_l \geq 0.5000$	$A \geq 0.065$

Table 15.4 Numerical results for case 1.

Variable	Initial	LFM 1	LFM 2	LFM 3
m	0.0200	0.0175	0.0166	0.0180
p	0.4000	0.5500	0.5800	0.5233
t	0.1200	0.1114	0.1164	0.1114
C_l	0.5963	0.6001	0.6000	0.6000
C_{dw}	0.0047	0.0016	0.0018	0.0017
A	0.0808	0.0750	0.0784	0.0751
N_{c1}	—	51	54	52
N_{c2}	—	38	38	38
N_f	—	3	3	2
Cost	—	<18	<7	<4

Table 15.5 Numerical results for case 2.

Variable	Initial	LFM 1	LFM 2	LFM 3
m	0.0200	0.0152	0.0146	0.0151
p	0.4000	0.7433	0.7656	0.7661
t	0.1200	0.1140	0.1140	0.1120
C_l	0.4745	0.5676	0.5702	0.5900
C_{dw}	0.0115	0.0050	0.0050	0.0050
A	0.0808	0.0767	0.0768	0.0754
N_{c1}	—	52	51	51
N_{c2}	—	38	38	38
N_f	—	3	4	3
Cost	—	<18	<8	<5

are $0 \leq m \leq 0.1$, $0.2 \leq p \leq 0.8$, and $0.05 \leq t/c \leq 0.2$. The Mach number is $M_\infty = 0.75$ and the angle of attack $\alpha = 1$ degree. Details of the optimization results are given in Table 15.3.

The results presented in Tables 15.4–15.6 indicate that the performance of the multi-level algorithm is consistent throughout all the test cases. The results show that the optimized designs obtained

Table 15.6 Numerical results for case 3.

Variable	Initial	LFM 1	LFM 2	LFM 3
m	0.0000	0.0150	0.0153	0.0151
p	0.0000	0.5300	0.5150	0.5200
t	0.1000	0.0966	0.0965	0.0966
C_l	0.0006	0.5002	0.5004	0.4999
C_{dw}	0.0016	0.0156	0.0162	0.0159
A	0.0673	0.0651	0.0650	0.0650
N_{c1}	—	51	51	51
N_{c2}	—	38	38	38
N_f	—	3	3	2
Cost	—	<18	<7	<4

by using the three different low-fidelity model families are similar. The thickness-to-chord ratio is nearly the same for all designs, but the maximum camber and the location of maximum camber differ slightly. All designs satisfy both constraints, but they differ slightly in the objective function.

The algorithm using the low-fidelity model set LFM 3 yields designs similar to those produced with the low-fidelity model sets LFM 1 and LFM 2, but at a considerably lower CPU cost (4–5 equivalent evaluations of the high-fidelity model on average). At the same time, the quality of the final designs produced with all sets is similar. It should be emphasized that because of the algorithm setup, the number of evaluations of particular models are similar ($N_{c1} \sim 50$, $N_{c2} \sim 40$, $N_f \sim 3$), so that the computational benefit mostly comes from the fact that the models of LFM 3 are faster than those of LFM 1 and LFM 2. However, the fact that the LFM 3 models are also more accurate than LFM 1 and LFM 2 has its contribution to the computational savings: the average number of the refinement iterations of the multi-level algorithm (and, consequently, the number of high-fidelity model evaluations) equals 3 for LFM 1, 3.3 for LFM 2, but only 2.3 for LFM 3.

15.4 Discussion and Recommendations

In this book, a number of SBO techniques suitable for solving computationally expensive aerodynamic shape optimization problems have been discussed and demonstrated. The majority of these techniques are based on variable-fidelity CFD simulations. The relaxed-accuracy CFD models serve, upon suitable correction, as fast prediction tools that guide the search process towards an improved design. It has been demonstrated — using both aerodynamic and hydrodynamic examples, as well as problems from other engineering disciplines such as electrical engineering — that the typical computational cost of a surrogate-based design process is substantially lower than conventional methods.

At the same time, physics-based surrogate modeling and optimization requires some user experience and, at the current stage of research, it may not always be possible to implement SBO techniques as fully automated design procedures. Perhaps the two most important issues are — as already discussed in this chapter — potential convergence problems and the necessity of finding a proper trade-off between the speed and the accuracy of the low-fidelity model. On the other hand, these difficulties can be alleviated by using trust-region frameworks as convergence safeguards, as well as automation of the low-fidelity model setup, e.g. through numerical optimization of grid parameters (cf. Section 13.3). It seems that a combination of physics-based SBO techniques with adjoint sensitivity approaches has a lot of potential, not only in improving the robustness of the optimization process but also a further reduction of its computational cost (Koziel *et al.*, 2012a).

Another important aspect is a design method selection. On the one hand, a variety of optimization techniques are available that utilize both physics-based and approximation-based surrogates, as well as the mixture of the two. On the other hand, the performance of these methods is often comparable. In general, finding the most suitable design optimization approach for a given design problem is not straightforward. It should take into account the following factors:

- user experience regarding numerical modeling and optimization techniques (perhaps less efficient but simpler techniques are recommended for less experienced users);

- problem-specific knowledge and visual inspection of available model responses (visual assessment of the differences between the low- and high-fidelity model responses may help in determining the most suitable method of constructing the surrogate); and
- available low-fidelity models (in particular, selection of the optimization method may be limited if the low-fidelity model is relatively expensive).

It should be indicated that simple methods are often very efficient (e.g. basic multiplicative response correction, cf. Sections 12.1, 12.3, and 13.6). For inexperienced users, methods such as multi-point output SM (Section 4.5.3) or multi-level design optimization (Section 6.5) may be the most suitable because of the fact that they are simple in implementation. Methods exploiting the knowledge embedded in the low-fidelity model to the fullest extent, such as shape-preserving response prediction (Section 6.3) are more difficult to implement but, at the same time, very efficient for specific tasks (e.g. inverse design).

It is also worth mentioning that combining physics-based surrogate modeling with approximation methods is often very efficient. In particular, in case the low-fidelity model is relatively expensive, it might be beneficial to replace it with its response surface approximation model at later stages of the optimization process to reduce the overall design cost.

15.5 Variable-Fidelity Design Optimization: Prospective Look

Surrogate-based aerodynamic optimization using variable-fidelity models has been the main theme of this book. It is a relatively young yet promising field of research. As demonstrated using a number of application examples, variable-fidelity SBO methods can be very efficient, leading to satisfactory design at low computational cost corresponding to a handful of evaluations of the high-fidelity CFD model of the structure/component of interest. It seems that there is a lot of potential in this type of design method, particularly from the point of view of efficient handling of computationally expensive design problems. However, their wide acceptance by the engineering community depends on resolving the difficulties mentioned in this and other chapters, in particular improving the robustness and making them less dependent on user experience. Addressing these and other issues of variable-fidelity SBO are the most urgent objectives of future research in this area.

Appendix

MATLAB Code

In this chapter, the auxiliary MATLAB function necessary to run the SBO algorithms discussed in Chapter 14 are provided. The codes and relevant material to download are available from: http://www.worldscientific.com/r/p987-supp; you will be required to register and sign in to download the material.

The rest of this chapter provides the MATLAB code for the auxiliary functions. The MATLAB scripts for running the four SBO algorithms can be found in Chapter 14.

```
%-----------------------------------------------------------------%
%-----------------------------------------------------------------%

% CFD coarse model c0
%
% Input Arguments:
% ----------------
% x - geometry parameters
%
% Output Arguments:
% -----------------
% R - model response
function R=airfoil_model_c0(x,varargin)
% kriging model evaluation
load airfoil_model_c0.mat
for q=1:length(dmodel)
   Y{q}=predictor(x',dmodel{q});
   R(q,1)=Y{q};
end
```

```
%-----------------------------------------------------------------%
%-----------------------------------------------------------------%

% CFD coarse model c1
%
% Input Arguments:
% ----------------
% x - geometry parameters
%
% Output Arguments:
% -----------------
% R - model response
function R=airfoil_model_c1(x,varargin)
% kriging model evaluation
load airfoil_model_c1.mat
for q=1:length(dmodel)
   Y{q}=predictor(x',dmodel{q});
   R(q,1)=Y{q};
end

%-----------------------------------------------------------------%
%-----------------------------------------------------------------%

% CFD fine model f
%
% Input Arguments:
% ----------------
% x - geometry parameters
%
% Output Arguments:
% -----------------
% R - model response
function R=airfoil_model_f(x,varargin)
% kriging model evaluation
load airfoil_model_f.mat
for q=1:length(dmodel)
   Y{q}=predictor(x',dmodel{q});
   R(q,1)=Y{q};
end

%-----------------------------------------------------------------%
%-----------------------------------------------------------------%

% Scaled CFD coarse model c1
%
% Input Arguments:
% ----------------
% x - geometry parameters
% varargin{1}- scaling parameters
```

```
%
% Output Arguments:
% -----------------
% R - model response
function R=c_model_scaled(x,varargin)
load airfoil_model_c1.mat
for q=1:length(dmodel)
   Y{q}=predictor(x',dmodel{q});
   R(q,1)=Y{q};
end
% model scaling
S=varargin{1};
R=[S.Al*R(1)+S.Dl S.Ad*R(2)+S.Dd R(3)]'+S.D;

%-----------------------------------------------------------------%
%-----------------------------------------------------------------%

% Construction of the response surface approximation model
%
% Input Arguments:
% ----------------
% x0 - center of the model validity region
% f0 - corresponding CFD model response
% f_model - CFD model handle
% d - model validity region size
% N - number of CFD model evaluations so far
% model_order - order of the RSA model (1 - first-order, 2 - second-order)
%
% Output Arguments:
% -----------------
% L - model coefficients
% N - (updated) number of CFD model evaluations
function [L,N]=construct_rsa_model(x0,f0,f_model,d,N,model_order)
n=length(x0);
switch(model_order)
    case 1 % matrices for first-order model construction
      A(1,:)=[1 x0'];
      F(1,:)=f0';
      for j=1:n
         xtmp=x0;
         xtmp(j)=xtmp(j)+d(j);
         A(j+1,:)=[1 xtmp'];
         ftmp=feval(f_model,xtmp);
         F(j+1,:)=ftmp';
      end
      N=N+n;
    case 2 % matrices for second-order model construction
      A(1,:)=[1 x0' x0'.^2];
      F(1,:)=f0';
      for j=1:n
```

```
        xtmp=x0;
        xtmp(j)=xtmp(j)+d(j);
        A(2*j,:)=[1 xtmp' xtmp'.^2];
        ftmp=feval(f_model,xtmp);
        F(2*j,:)=ftmp';
        xtmp(j)=xtmp(j)-2*d(j);
        A(2*j+1,:)=[1 xtmp' xtmp'.^2];
        ftmp=feval(f_model,xtmp);
        F(2*j+1,:)=ftmp';
      end
      N=N+2*n;
    end
L=inv(A)*F; % calculating coefficients (through linear regression problem)

%-------------------------------------------------------------%
%-------------------------------------------------------------%

% Results display function
function display_results(xopt,f_model)
Ropt=f_model(xopt);
fprintf('----------------------\n');
fprintf('Optimization completed\n');
fprintf('----------------------\n');
fprintf('Final geometry: xopt=[%1.5f %1.5f
%1.5f]\n\n',xopt(1),xopt(2),xopt(3));
fprintf('Final figures of
interest:\nCl=%1.5f\nCd=%1.5f\nA=%1.5f\n',Ropt(1),Ropt(2),Ropt(3));
fprintf('----------------------\n');

%-------------------------------------------------------------%
%-------------------------------------------------------------%

% Constraint function for trust-region implementation
%
% Input Arguments:
% ----------------
% x - geometry parameters
%
% Output Arguments:
% -----------------
% cineq - inequality constraints
% ceq - equality constraints
function [cineq,ceq]=f_constr(x,varargin)
global X0;
global Radius;
ceq=[];
cineq(1)=norm(X0-x)-Radius;

%-------------------------------------------------------------%
```

```
%----------------------------------------------------------------%

% Objective function
%
% Input Arguments:
% ----------------
% x - geometry parameters
% f_obj - objective function handle
% f_model - CFD model handle
%
% Output Arguments:
% ----------------
% E - objective function value
function E=f_obj(x,f_obj,f_model,varargin)
R=feval(f_model,x,varargin{:});
E=feval(f_obj,R);

%----------------------------------------------------------------%
%----------------------------------------------------------------%

% Objective function for drag minimization test case
%
% Input Arguments:
% ----------------
% R - model response
%
% Output Arguments:
% ----------------
% E - objective function value
function E=f_obj_drag_minimization(R,varargin)
E=R(2); % main objective: drag
Cltarget=0.5; % target lift
Amin=0.075; % minimum allowed crosseactiona area
Cl=R(1);
A=R(3);
E=E+1000*(Cl-Cltarget)^2; % penalty factor for lift constraint violation
if A<Amin % penalty factor for area constraint violation
   E=E+1000*((A-Amin)/Amin)^2;
end

%----------------------------------------------------------------%
%----------------------------------------------------------------%

% Objective function for lift maximization test case
%
% Input Arguments:
% ----------------
% R - model response
%
```

```
% Output Arguments:
% ----------------
% E - objective function value
function E=f_obj_lift_maximization(R,varargin)
E=-R(1); % main objective: negative lift
Cdmax=0.004; % maximum allowed drag
Amin=0.075; % minimum allowed crosseactiona area
Cd=R(2);
A=R(3);
if Cd>Cdmax % penalty factor for drag constraint violation
   E=E+1000*((Cd-Cdmax)/Cdmax)^2;
end
if A<Amin % penalty factor for area constraint violation
   E=E+1000*((A-Amin)/Amin)^2;
end

%---------------------------------------------------------------%
%---------------------------------------------------------------%

% Description:
%    Airfoil shape generation using the NACA 4-digit shape. The TE is
%    closed and the airfoil is scaled to chord 1.
%    Equations from "Theory of Wing Sections" by Abbott and Von Doenhof.
%
% Input:
%    x: Airfoil parameters
%       x(1): = m; Max camber
%       x(2): = p; Position of max camber
%       x(3): = t; Max thickness-to-chord ratio
%    N: Number of points
%
% Output:
%    XU:       x-coordinates on upper surface
%    YU:       y-coordinates on upper surface
%    XL:       x-coordinates on lower surface
% YL: y-coordinates on lower surface
function [XU,YU,XL,YL] = naca4digit(x,N)

% Parameters
    m = x(1);
p = x(2);
t = x(3);

% Airfoil generation
xi=1:N;
x = 1.0 - cos((xi-1.0)*pi/2.0/(N - 1.0)); % concentrated at LE
I = find(x<p);
J = find(x>=p);
yc(I) = m/p^2*(2*p*x(I)-x(I).^2);
```

```
dyc(I) = 2*m/p^2*(p-x(I));
yc(J) = m/(1-p)^2*((1-2*p)+2*p*x(J)-x(J).^2);
dyc(J) = 2*m/(1-p)^2*(p-x(J));
yt = t/0.2*(0.2969*sqrt(x)-0.126*x-0.3516*x.^2+0.2843*x.^3-0.1015*x.^4);
theta = atan(dyc);
xu=x-yt.*sin(theta);
xl=x+yt.*sin(theta);
yu=yc+yt.*cos(theta);
yl=yc-yt.*cos(theta);

% Close TE
au=(yu(end-1)-yu(end))/(xu(end-1)-xu(end));
bu=yu(end-1)-au*xu(end-1);
al=(yl(end-1)-yl(end))/(xl(end-1)-xl(end));
bl=yl(end-1)-al*xl(end-1);
xi=-(bu-bl)/(au-al);
yi=au*xi+bu;
c=xi;

% Prepare x- and y-coordinates
XU=[xu xi]/c;
XL=[xl xi]/c;
YU=[yu yi]/c;
YL=[yl yi]/c;

%-----------------------------------------------------------------%
%-----------------------------------------------------------------%

% Output function for direct optimization visualization
% Input/output arguments consistent with Matlab requirements
function stop=outfun(x,optimValues,state,varargin)
stop=false;
global X;
global R;
global E;
global Ntotal;
if strcmp(state,'iter')
   X{length(X)+1}=x;
   R{length(R)+1}=feval(varargin{2},x);
   E(length(E)+1)=optimValues.fval;
   Ntotal(length(Ntotal)+1)=optimValues.funccount;
   plot_history(X,R,E,Ntotal,varargin{3});
end

%-----------------------------------------------------------------%
%-----------------------------------------------------------------%

% Plotting optimization history
function plot_history(Xh,Rh,Eh,Ntotal,plot_data)
```

```
N=length(Xh);
if N<2
   return;
end
D(1)=norm(Xh{1});
for j=1:N-1
   D(j+1)=norm(Xh{j}-Xh{j+1});
end
fontsize=14;

figure(plot_data.h1);
% plot convergence
subplot(3,2,1);
semilogy(1:N,D,'ko-','LineWidth',2,'MarkerSize',5);
xlabel('Iteration Index','FontSize',fontsize,'FontName','Times New
Roman','FontAngle','normal');
ylabel('||\itx^{\rm(\iti\rm)}-x^{(\iti\rm+1)}||','FontSize',fontsize,
'FontName','Times New Roman','FontAngle','normal');
set(gca,'FontSize',fontsize,'FontName','Times New Roman');
axis([1 N 0 Inf]);
grid on; zoom on
% plot objective function
subplot(3,2,2);
plot(1:N,Eh,'ko-','LineWidth',2,'MarkerSize',5);
xlabel('Iteration Index','FontSize',fontsize,'FontName','Times New
Roman','FontAngle','normal');
ylabel('Objective Function','FontSize',fontsize,'FontName','Times New
Roman','FontAngle','normal');
set(gca,'FontSize',fontsize,'FontName','Times New Roman');
axis([1 N -Inf Inf]);
switch(plot_data.problem)
    case 1
      if Eh(end)<0.01
         axis([1 N -Inf 0.01]);
      end
    case 2
      if Eh(end)<0
         axis([1 N -Inf 0]);
      end
    end
grid on; zoom on

% plot objective function
subplot(3,2,2);
plot(1:N,Eh,'ko-','LineWidth',2,'MarkerSize',5);
xlabel('Iteration Index','FontSize',fontsize,'FontName','Times New
Roman','FontAngle','normal');
ylabel('Objective Function','FontSize',fontsize,'FontName','Times
New Roman','FontAngle','normal');
```

```
set(gca,'FontSize',fontsize,'FontName','Times New Roman');
axis([1 N -Inf Inf]);
switch(plot_data.problem)
    case 1
      if Eh(end)<0.01
        axis([1 N -Inf 0.01]);
      end
    case 2
      if Eh(end)<0
        axis([1 N -Inf 0]);
      end
    end
grid on; zoom on

% plot total number of function evaluations
subplot(3,2,3);
plot(1:N,Ntotal,'ko-','LineWidth',2,'MarkerSize',5);
xlabel('Iteration Index','FontSize',fontsize,'FontName','Times New
Roman','FontAngle','normal');
ylabel('Number of Function
Evaluations','FontSize',fontsize,'FontName','Times New
Roman','FontAngle','normal');
set(gca,'FontSize',fontsize,'FontName','Times nEW Roman');
axis([1 N -Inf Inf]);
grid on; zoom on

% plot lift coefficient
subplot(3,2,4);
for j=1:N
    Cl(j)=Rh{j}(1);
end
plot(1:N,Cl,'ko-','LineWidth',2,'MarkerSize',5);
hold on
if plot_data.problem==1
    plot([1 N],[plot_data.Cl
plot_data.Cl],'r','LineWidth',2,'MarkerSize',5);
end
hold off
xlabel('Iteration Index','FontSize',fontsize,'FontName','Times New
Roman','FontAngle','normal');
ylabel('Lift Coefficient','FontSize',fontsize,'FontName','Times
New Roman','FontAngle','normal');
set(gca,'FontSize',fontsize,'FontName','Times New Roman');
axis([1 N -Inf Inf]);
grid on; zoom on;

% plot drag coefficient
subplot(3,2,5);
for j=1:N
    Cd(j)=Rh{j}(2);
```

```
end
plot(1:N,Cd,'ko-','LineWidth',2,'MarkerSize',5);
hold on
if plot_data.problem==2
   plot([1 N],[plot_data.Cd plot_data.Cd],'r','LineWidth',2,'MarkerSize',5);
end
hold off
xlabel('Iteration Index','FontSize',fontsize,'FontName','Times New
Roman','FontAngle','normal');
ylabel('Drag Coefficient','FontSize',fontsize,'FontName','Times
New Roman','FontAngle','normal');
set(gca,'FontSize',fontsize,'FontName','Times New Roman');
axis([1 N -Inf Inf]);
grid on; zoom on;

% plot crossectional area
subplot(3,2,6);
for j=1:N
   A(j)=Rh{j}(3);
end
plot(1:N,A,'ko-','LineWidth',2,'MarkerSize',5);
hold on
if plot_data.problem==1
    plot([1 N],[plot_data.A plot_data.A],'r','LineWidth',2,'MarkerSize',5);
elseif plot_data.problem==2
plot([1 N],[plot_data.A plot_data.A],'r','LineWidth',2,'MarkerSize',5);
end
hold off
xlabel('Iteration Index','FontSize',fontsize,'FontName','Times New
Roman','FontAngle','normal');
ylabel('Crossectiona Area','FontSize',fontsize,'FontName','Times
New Roman','FontAngle','normal');
set(gca,'FontSize',fontsize,'FontName','Times
New Roman');
axis([1 N -Inf Inf]);
grid on; zoom on;

drawnow;

figure(plot_data.h2);
[XU0,YU0,XL0,YL0] = naca4digit(plot_data.x0,500);
[XU,YU,XL,YL] = naca4digit(Xh{end},500);
plot(XU0,YU0,'k--',XU,YU,'b',XL0,YL0,'k--',XL,YL,'b','LineWidth',2);
xlabel('\itx/c','FontSize',fontsize,'FontName','Times New
Roman','FontAngle','normal');
ylabel('\ity','FontSize',fontsize,'FontName','Times New
Roman','FontAngle','normal');
legend('Initial geometry','Optimized geometry');
set(gca,'FontSize',fontsize,'FontName','Times New Roman');
grid on
```

```
zoom on
axis equal

%---------------------------------------------------------------%
%---------------------------------------------------------------%
% Response surface model. The model is either linear or quadratic
% (without mixed terms). Model order recognized by the length of the
% coefficient table L
%
% Input Arguments:
% ----------------
% x - geometry parameters
% L - model coefficients
% varargin{1}- optional parameter (additive response correction term)
%
% Output Arguments:
% -----------------
% R - model response: [lift drag area]'
function R=rsa_model(x,L,varargin)
switch(size(L,1))
   case 4
     y=[1 x'];
   case 7
     y=[1 x' x'.^2];
end
R=L'*y'; % calculate model response
if nargin>2 % optional additive response correction
   R=R+varargin{1};
end
% Function for multipoint additive/multiplicative low-fidelity model scaling
% Linear regression problem to be solved (Cl.f, Cl.c - high-,low-fidelity
lift:
%    min{Al,Dl : sum_k(Cl.f{k}- Al*Cl.c{k}- Dl)^2} % similarly for drag.
%
% Input Arguments:
% ----------------
% f - cell array containing available high-fidelity model response vectors
% c - cell array containing available low-fidelity model response vectors
%
% Output Arguments:
% -----------------
% S - structure with scaling parameters
function S=sm_scaling(f,c)
n=length(f); % number of response vectors
% scaling
if n==1 % only one point available =>no additive correction
   S.Al=f{1}(1)/c{1}(1);
   S.Dl=0;
   S.Ad=f{1}(2)/c{1}(2);
   S.Dd=0;
```

```
else
   for j=1:n
      Cl(j,1)=c{j}(1);
      Cl(j,2)=1;
      Fl(j,1)=f{j}(1);
      Cd(j,1)=c{j}(2);
      Cd(j,2)=1;
      Fd(j,1)=f{j}(2);
   end
   Xl=inv(Cl'*Cl)*Cl'*Fl; % least square solution to the regression problem
at hand (lift)
   S.Al=Xl(1);
   S.Dl=Xl(2);
   Xd=inv(Cd'*Cd)*Cd'*Fd; % least square solution to the regression problem
at hand (drag)
   S.Ad=Xd(1);
   S.Dd=Xd(2);
end

%---------------------------------------------------------------%
%---------------------------------------------------------------%

% Function for storing accumulated optimization results
%
% Input Arguments:
% ----------------
% X - previous iteration points
% E - previous objective function values
% Rf - previous high-fidelity model responses
% Rc - previous low-fidelity model responses
% N - previous optimization costs
% x,e,f,c,n - new iteration point, objective function value, model
% responses and cost
%

% Output Arguments:
% -----------------
% X - updated iteration points
% E - updated objective function values
% Rf - updated high-fidelity model responses
% Rc - updated low-fidelity model responses
% N - updated optimization costs
function [X,E,Rf,Rc,N]=update_storage_sm(X,E,Rf,Rc,N,x,e,f,c,n)
found=0;
for j=1:length(X) % check for existence of the new item in the tables
   if norm(X{j}-x)<1e-6
      found=1;
      break;
   end
end
```

```
if found==0 % update tables if the new item is not there yet
   if isempty(X)
      X{1}=x;
      Rf{1}=f;
      Rc{1}=c;
   else
      X=[X x];
      Rf=[Rf f];
      Rc=[Rc c];
   end
   E=[E e];
   N=[N n];
end
```

References

Abbott, I.H. and Von Doenhoff, A.E. (1959) *Theory of Wing Sections.* Dover Publications, New York, NY.

Agilent ADS. (2011) Agilent Technologies, Santa Rosa, CA.

Aidala, P.V., Davis, Jr., W.H., and Mason, W.H. (1983) Smart aerodynamic optimization. AIAA-83-1863. *AIAA Applied Aerodynamics Conference*, Danvers, MA, July 13–15.

Aitken, A.C. (1935) On least squares and linear combinations of observations. *Proceedings of the Royal Society of Edinburgh*, 55, 42–48.

Alba, E. and Marti, R. (Eds) (2006) *Metaheuristic Procedures for Training Neural Networks.* Springer, New York, NY.

Alexandrov, N.M. and Lewis, R.M. (2001) An overview of first-order model management for engineering optimization. *Optimization and Engineering*, 2, 413–430.

Alexandrov, N.M., Dennis, J.E., Lewis, R.M., and Torczon, V. (1998) A trust region framework for managing use of approximation models in optimization. *Structural Multidisciplinary Optimization*, 15, 16–23.

Alexandrov, N.M., Nielsen, E.J., Lewis, R.M., and Anderson, W.K. (2000) First-order model management with variable-fidelity physics applied to multi-element airfoil optimization. *AIAA/USAF/NASA/ISSMO Symposium on Multidisciplinary Design and Optimization*, Long Beach, CA, September 6–8.

Alexandrov, N.M., Lewis, R.M., Gumbert, C.R., Green, L.L., and Newman, P.A. (2001) Approximation and model management in aerodynamic optimization with variable-fidelity models. *AIAA Journal of Aircraft*, 38, 1093–1101.

Allaire, D. and Willcox, K. (2014) A mathematical and computational framework for multifidelity design and analysis with computer models. *International Journal on Uncertainty Quantification*, 4, 1–20.

Allen, B., Vorus, W.S., and Prestero, T. (2000) Propulsion system performance enhancements on REMUS AUVs. *Proceedings MTS/IEEE Oceans 2000*, Providence, RI, September 11–14.

Alvarez, A., Bertram, V., and Gualdesi, L. (2009) Hull hydrodynamic optimization of autonomous underwater vehicles operating at snorkelling depth. *Ocean Engineering*, 36, 105–112.

Anderson, J.D. (2007) *Modern Compressible Flow*, 5th ed. McGraw Hill Inc., New York, NY.

Anderson, J.D. (2010) *Fundamentals of Aerodynamics*, 5th ed. McGraw-Hill Inc., New York, NY.

Anderson. W.K. and Venkatakrishnan, V. (1999) Aerodynamic design optimization on unstructured grids with a continuous adjoint formulation. *Computers & Fluids*, 28, 443–480.

Anderson, W.K. and Venkatakrishnan, V. (1999) Aerodynamic design optimization on unstructured grids with a continuous adjoint formulation. *Computers & Fluids*, 28, 443–480.

Andrés, E., Salcedo-Sanz, S., Monge, F., and Pérez-Bellido, A.M. (2012) Efficient aerodynamic design through evolutionary programming and support vector regression algorithms. *International Journal of Expert Systems with Applications*, 39, 10700–10708.

Angiulli, G., Cacciola, M., and Versaci, M. (2007) Microwave devices and antennas modeling by support vector regression machines. *IEEE Transactions on Magnetics*, 43, 1589–1592.

Back, T., Fogel, D.B., and Michalewicz Z. (Eds) (2000) *Evolutionary Computation 1: Basic Algorithms and Operators*. Taylor & Francis Group, Abingdon.

Bandler, J.W., Biernacki, R.M., Chen, S.H., Grobelny, P.A., and Hemmers, R.H. (1994) Space mapping technique for electromagnetic optimization. *IEEE Transactions on Microwave Theory and Techniques*, 42, 2536–2544.

Bandler, J.W., Biernacki, R.M., Chen, S.H., Hemmers, R.H., and Madsen, K. (1995) Electromagentic optimization exploiting aggressive space mapping. *IEEE Transactions on Microwave Theory and Techniques*, 43, 2874–2882.

Bandler, J.W., Cheng, Q.S., Gebre-Mariam, D.H., Madsen, K., Pedersen, F., and Søndergaard, J. (2003) EM-based surrogate modeling and design exploiting implicit, frequency and output space mappings. *IEEE International Microwave Symposiun Digest,* Philadelphia, PA, June 6–13.

Bandler, J.W., Cheng, Q.S., Dakroury, S.A., Mohamed, A.S., Bakr, M.H., Madsen, K., and Søndergaard, J. (2004a) Space mapping: the state of the art. *IEEE Transactions on Microwave Theory and Techniques*, 52, 337–361.

Bandler, J.W., Cheng, Q.S., Nikolova, N.K., and Ismail, M.A. (2004b) Implicit space mapping optimization exploiting preassigned parameters. *IEEE Transactions on Microwave Theory and Techniques*, 52, 378–385.

Banks, H.T. and Kunisch, K. (1989). *Estimation Techniques for Distributed Parameter Systems.* Birkhauser, Basel.

Barrett, T.R., Bressloff, N.W., and Keane, A.J. (2006) Airfoil shape design and optimization using multifidelity analysis and embedded inverse design. *AIAA Journal*, 44, 2051–2060.

Barlow, J.B., Rae, W.H., and Pope, A. (1999) *Low-Speed Wind Tunnel Testing*, 3rd ed. John Wiley & Sons, New York, NY.

Baysal, O. and Eleshaky, M.E. (1991) Aerodynamic sensitivity analysis methods for the compressible Euler equations. *Journal of Fluids Engineering*, 113, 681–688.

Baysal, O. and Eleshaky, M.E. (1992) Aerodynamic design optimization using sensitivity analysis and computational fluid dynamics. *AIAA Journal*, 30, 718–725.

Beachkofski, B. and Grandhi, R. (2002) Improved distributed hypercube sampling. AIAA Paper 2002-1274, *43rd AIAA/ASME/ASCE/AHS/ASC Structures, Structural Dynamics, and Materials Conference*, Denver, CO, April 22–25.

Bischof, C., Bücker, H.M., Hovland, P.D., Naumann, U., and Utke, J. (Eds) (2008) *Advances in Automatic Differentiation. Lecture Notes in Computational Science and Engineering Vol 64*. Springer, Berlin.

Bischof, C.H., Mauer, A., Jones, W.T., and Samareh, J. (1998) Experiences with automatic differentiation applied to a volume grid generation code. *Journal of Aircraft*, 35, 569–573.

Booker, A.J., Dennis, J.E., Frank, P.D., Serafini, D.B., Torczon, V., and Trosset, M.W. (1999) A rigorous framework for optimization of expensive functions by surrogates. *Structural Optimization*, 17, 1–13.

Brooks, C.J., Forrester, A.I., Keane, A.I.J, and Shahpar, S. (2011) Multi-Fidelity Design Optimization of a Transonic Compressor Rotor. *9th European Conference on Turbomachinery Fluid Dynamics and Thermodynamics*, Istanbul, Turkey, 21–25 March.

Burgreen, G.W. and Baysal, O. (1996) Three-dimensional aerodynamic shape optimization using discrete sensitivity analysis. *AIAA Journal*, 34, 1761–1770.

Ceperic, V. and Baric, A. (2004) Modeling of analog circuits by using support vector regression machines. *Proceedings of the 11th International Conference on Electronics, Circuits, and Systems*, Tel Aviv, Israel, December 13–15, 391–394.

Chakraborty, U. (2008) *Advances in Differential Evolution. Studies in Computational Intelligence*. Springer-Verlag, Berlin.

Chen, Z.N. (2008) Wideband microstrip antennas with sandwich substrate. *IET Microwaves Antennas & Propagation*, 2, 538–546.

Cheng, Q.S., Koziel, S., and Bandler, J.W. (2006) Simplified space mapping approach to enhancement of microwave device models. *International Journal of RF and Microwave Computer-Aided Engineering*, 16, 518–535.

Cheng, Q.S., Rautio, J.C., Bandler, J.W., and Koziel, S. (2010a) Progress in simulator-based tuning — the art of tuning space mapping. *IEEE Microwave Magazine*, 11, 96–110.

Cheng, Q.S., Bandler, J.W., and Koziel, S. (2010b) Space mapping design framework exploiting tuning elements. *IEEE Transactions on Microwave Theory and Techniques*, 58, 136–144.

Clerc, M. and Kennedy, J. (2002) The particle swarm — explosion, stability, and convergence in a multidimensional complex space. *IEEE Transactions on Evolutionary Computation*, 6, 58–73.

Cliff, S.E., Reuther, J.J., Saunders, D.A., and Hicks, R.M. (2001) Single-point and multipoint aerodynamic shape optimization of high-speed civil transport. *Journal of Aircraft*, 38, 997–1005.

Coello Coello, C.A. (2013) *Evolutionary Algorithms for Solving Multi-Objective Problems*, 2nd ed. Springer, New York, NY.

Conn, A.R., Gould, N.I.M., and Toint, P.L. (2000) Trust region methods. *MPS-SIAM Series on Optimization, Society for Industrial and Applied Mathematics (SIAM).* Philadelphia, PA.

Conn, A.R., Scheinberg, K., and Vicente, L.N. (2009) Introduction to derivative-free optimization. *MPS-SIAM Series on Optimization, Society for Industrial and Applied Mathematics (SIAM)*, Philadelphia, PA.

Cook, P.H., McDonald, M.A., and Firmin, M.C.P. (1979) *Aerofoil RAE 2822 — Pressure Distributions, and Boundary Layer and Wake Measurements.* Experimental Data Base for Computer Program Assessment by Fluid Dynamics Panel, Working Group 04, AGARD Report AR 138. NATO Science and Technology Organization, Brussels.

Couckuyt, I. (2013) Forward and inverse surrogate modeling of computationally expensive problems. PhD Thesis, Ghent University.

CST Microwave Studio (2012) CST AG, Darmstadt, Germany.

Dalton, C. and Zedan, M.F. (1980) Design of low-drag axisymmetric shapes by the inverse method. *Journal of Hydronautics*, 15, 48–54.

Deb, K. (2009) *Multi-Objective Optimization Using Evolutionary Algorithms.* Wiley, Chichester.

de Barros, E.A., Dantas, J.L.D., Pascoal, A.M., and de Sá, E. (2008) Investigation of normal force and moment coefficients for an AUV at nonlinear angle of attack and sideslip angle. *IEEE Journal of Oceanic Engineering*, 33, 538–549.

Derksen, R.W. and Rogalsky, T. (2010) Bezier-PARSEC: an optimized aerofoil parameterization for design. *Advances in Engineering Software*, 41, 923–930.

Desai, M. and Ito, K. (1994) Optimal controls of Navier-Stokes equations. *SIAM Journal of Control and Optimization*, 32, 1428–1446.

Doolan, C.J and Morgans, R.C. (2007) Numerical evaluation and optimization of low speed wind tunnel contractions. *18th AIAA Computational Fluid Dynamics Conference*, Miami, FL, June 25–27.

Dorigo, M. and Stutzle, T. (2004) *Ant Colony Optimization.* MIT Press, Cambridge.

Drela, M. and Giles, M.B. (1987) Viscous-inviscid analysis of transonic and low Reynolds number airfoils. *AIAA Journal*, 25, 1347–1355.

Duvigneau, R. and Visonneau, M. (2001) Shape optimization of incompressible and turbulent flows using the simplex method. AIAA Paper 2001-2533, *15th AIAA Computational Fluid Dynamics Conference*, Anaheim, CA, June 11–14.

Echeverría, D. and Hemker, P.W. (2005) Space mapping and defect correction. *CMAM International Mathematical Journal Computational Methods in Applied Mathematics*, 5, 107–136.

Echeverría, D. and Hemker, P.W. (2008) Manifold mapping: a two-level optimization technique. *Computing and Visualization in Science*, 11, 193–206.

Eldred, M.S., Giunta, A.A., and Collis S.S. (2004) Second-order corrections for surrogate-based optimization with model hierarchies. *10th AIAA/ISSMO multidisciplinary analysis and optimization conference.* Albany, NY, August 30–September 1.

Elliott, J. and Peraire, J. (1997) Practical three-dimensional aerodynamic design and optimization using unstructured meshes. *AIAA Journal*, 35, 1479–1485.

em^{TM} (2012) Sonnet Software, Inc., North Syracuse, NY.

Epstein, B. and Peigin, S. (2006) Optimization of 3D wings based on Navier-Stokes solutions and genetic algorithms. *International Journal of Computational Fluid Dynamics*, 20, 75–92.

Eyi, S., K.D. Lee, K.D., Rogers, S.E. and Kwak, D. (1996) High-lift design optimization using Navier-Stokes equations. *Journal of Aircraft*, 33(3), 499–504.

Fennel, W. and Neumann, T. (2004). *Introduction to the Modelling of Marine Ecosystems*. Elsevier, Amsterdam.

FLUENT (2012) ANSYS Inc., Canonsburg, PA.

Fonseca, C.M. (1995) Multiobjective genetic algorithms with applications to control engineering problems. PhD Thesis, University of Sheffield.

Forrester, A.I.J. (2010) Black-box calibration for complex systems simulation. *Philosophical Transactions of the Royal Society of London, Series A: Mathematical and Physical Sciences*, 368, 1924.

Forrester, A.I.J. and Keane, A.J. (2009) Recent advances in surrogate-based optimization. *Progress in Aerospace Sciences*, 45, 50–79.

Forrester, A.I.J., Bressloff, N.W., and Keane, A.J. (2006) Optimization using surrogate models and partially converged computationally fluid dynamics simulations. *Proceedings of the Royal Society A: Mathematical, Physical and Engineering Sciences*, 462(2071), 2177–2204.

Forrester, A.I.J., Sóbester, A., and Keane, A.J. (2007) Multi-fidelity optimization via surrogate modeling. *Proceedings of the Royal Society A*, 463, 3251–3269.

Frank, P.D. and Shubin, G.R. (1992) A comparison of optimization-based approaches for a model computational aerodynamic design problem. *Journal of Computational Physics*, 98, 74–89.

Garner, J. (1967) *Botnvarpan og bunadur hennar* (Fishingnets and Equipment). Fiskifelag Islands, Reykjavik.

Gatsis, J. and Zingg, D.W. (2003) A fully-coupled Newton-Krylov algorithm for aerodynamic optimization. AIAA Paper 2003-3956, *16th Computational Fluid Dynamics Conference*, June 23–26.

Geem, Z.W., Kim, J.H., and Loganathan, G.V. (2001) A new heuristic optimization algorithm: harmony search. *Simulation*, 76, 60–68.

Geisser, S. (1993) *Predictive Inference*. Springer-Science+Business Media, B.V., Dordrecht.

Giannakoglou, K.C. (2002) Design of optimal aerodynamic shapes using stochastic optimization methods and computational intelligence. *Progress in Aerospace Sciences*, 38(2), 43–76.

Gill, P.E., Murray, W., Saunders, M.A., and Wright, W.A. (1986) User's guide for NPSOL: A FORTRAN package for nonlinear programming. Stanford University, Stanford, CA.

Gill, P.E., Murray, W., and Saunders, M.A. (2005) SNOPT: an SQP algorithm for large-scale constrained optimization. *SIAM Review*, 47, 99–131.

Giunta, A.A. (1997) Aircraft multidisciplinary design optimization using design of experiments theory and response surface modeling methods. PhD Thesis, Virginia Polytechnic Institute and State University.

Giunta, A.A. and Eldred, M.S. (2000) Implementation of a trust region model management strategy in the DAKOTA optimization toolkit. *Proceedings of the AIAA/USAF/NASA/ISSMO Symposium on Multidisciplinary Analysis and Optimization*, Long Beach, CA, September 6–8.

Giunta, A.A., Wojtkiewicz, S.F., and Eldred, M.S. (2003) Overview of modern design of experiments methods for computational simulations. AIAA Paper 2003-0649, *41st AIAA Aerospace Sciences Meeting and Exhibit*, Reno, NV, January 6–9.

Goldberg, D.E. (1989) *Genetic Algorithms in Search, Optimization & Machine Learning*. Pearson Education, Upper Saddle River, NJ.

Goldschmied, F.R. (1966) Integrated hull design, boundary-layer control, and propulsion of submerged bodies. *Journal of Hydronautics*, 1, 2–11.

Golub, G.H. and Van Loan, Ch.F. (1996) *Matrix Computations*, 3rd ed. The Johns Hopkins University Press, Baltimore, MD.

Gorissen, D., Crombecq, K., Couckuyt, I., Demeester, P., and Dhaene, T. (2010) A surrogate modeling and adaptive sampling toolbox for computer based design. *Journal of Machine Learning Research*, 11, 2051–2055.

Grasmeyer, J.M., (1997) A Discrete Vortex Method for Calculating the Minimum Induced Drag and Optimum Load Distribution for Aircraft Configurations with Noncoplanar Surfaces. VPI-AOE-242, Department of Aerospace and Ocean Engineering, Virginia Polytechnic Institute and State University. Blacksburg, VA, 24061.

Grasmeyer, J.M. (1999) Multidisciplinary Design Optimization of a Transonic Strut-Braced Wing Aircraft. AIAA-99-001. *37th AIAA Aerospace Sciences Meeting and Exhibit*, Reno, NV, January 12–13.

Greenman, R.M. (1998) Two-dimensional high-lift aerodynamic optimization using neural networks. NASA/TM-1998-112233, NASA, Washington, DC, USA.

Griewank, A. (2000) *Evaluating Derivatives: Principles and Techniques of Algorithmic Differentiation*. Society for Industrial and Applied Mathematics (SIAM), Philadelphia, PA.

Guan, X., Ma, Z., Cai, P., Anada, T., and Hagiwara, G. (2008) A microstrip dual-band bandpass filter with reduced size and improved stopband characteristics, *Microwave and Optical Technology Letters*, 50, 618–620.

Gunn, S.R. (1998) Support vector machines for classification and regression. Technical Report. School of Electronics and Computer Science, University of Southampton.

Gunzburger, M.D. (1997) Inverse design and optimization methods. *Lecture Series 1997-05*, The Von Karman Institute for Fluid Dynamics, Saint-Gilles, Belgium.

Han, S.P. (1977) A globally convergent method for nonlinear programming. *Journal of Optimization Theory and Applications*, 22, 297–309.

Haraldsson, H.O. (1996) Fluid dynamics simulation of fishing gear. MSc Thesis, University of Iceland.

Haykin, S. (1998) *Neural Networks: A Comprehensive Foundation*, 2nd ed. Prentice Hall, Upper Saddle River, NJ.

Haykin, S. (2008) *Neural Networks and Learning Machines.* Persons Education, Upper Saddle River, NJ.

Hazra, S.B. and Jameson, A. (2007) One-shot pseudo-time method for aerodynamic shape optimization using the Navier-Stokes equations. *International Journal for Numerical Methods in Fluids*, 68(5), 564–581.

Hicken, J. E. and Zingg, D. W., (2007) A parallel Newton-Krylov flow solver for the Euler equations on multi-block grids. AIAA-2007-4333, *18th AIAA Computational Fluid Dynamics Conference*, Miami, FL, June 25–28.

Hicks, R.M. and Henne, P.A (1978) Wing design by numerical optimization. *Journal of Aircraft*, 15, 407–412.

Hicks, R.M., Murman, E.M., and Vanderplaats, G.N. (1974) An assessment of airfoil design by numerical optimization. NASA Technical Memorandum X-3092, NASA, Washington, DC.

Hirsch, C. (2007) *Numerical Computation of Internal & External Flows. Fundamentals of Computational Fluid Dynamics*, 2nd ed. Elsevier, Burlington, MA.

Holst, T.L. and Pulliam, T.H. (2003) Transonic wing shape optimization using a genetic algorithm. *Fluid Mechanics and its Applications*, 73, 245–252.

Hosder, S., Watson, L.T., Grossman, B., Mason, W.H., and Kim, H. (2001) Polynomial response surface approximations for the multidisciplinary design optimization of a high speed civil transport. *Optimization and Engineering*, 2, 431–452.

Huan, J.C. and Modi, V. (1994) Optimum design for drag minimizing bodies in incompressible flow. *Inverse Problems in Engineering*, 1, 1–25.

Huang, L. and Gao, Z. (2012) Wing-body optimization based on multi-fidelity surrogate model. *28th Int. Congress of the Aeronautical Sciences*, Brisbane, Australia, September 23–28.

Huang, T.T., Liu, H.L., Groves, N.C., Forlini, T.J., Blanton, J.N., and Gowing, S. (1992) Measurements of flows over an axisymmetric body with various appendages (DARPA SUBOFF Experiments). *Proceedings of the 19th Symposium on Naval Hydrodynamics*, Seoul, Korea, August 23–28.

ICEM CFD (2013) ANSYS Inc., Canonsburg, PA.

Iollo, A., Kuruvila, G., and Ta'asan, S. (2005) Pseudo-time method for optimal shape design using the Euler equations. NASA CR 198205, NASA, Washington, DC.

Jacobs, J.H., Etman, L.F.P., van Keulen, F., and Rooda, J.E. (2004) Framework for sequential approximate optimization. *Structural and Multidisciplinary Optimization*, 27, 384–400.

Jagadeesh, P., Murali, K., and Idichandy, V.G. (2009) Experimental investigation of hydrodynamic force coefficients over AUV hull form. *Ocean Engineering*, 36, 113–118.

Jameson, A. (1988). Aerodynamic design via control theory. *Journal of Scientific Computing*, 3, 233–260.

Jameson, A., Martinelli, L., and Pierce, N. (1998) Optimum aerodynamic design using the Navier-Stokes equations. *Theoretical Fluid Dynamics*, 10, 213–237.

Jameson, A., Martinelli, L., and Haimes, B. (2004) Aerodynamic shape optimization of complete aircraft configurations using a unstructured grids. AIAA Paper 2004-0533, *42nd AIAA Aerospace Science Meeting*, Reno, NV, January 5–8.

Jones, D.R. (2001) A taxonomy of global optimization methods based on response surfaces. *Global Optimization*, 21, 345–383.

Jones, D.R., Perttunen, C.D., and Stuckman, B.E. (1998) Efficient global optimization of expensive black-box functions. *Journal of Global Optimization*, 13, 455–492.

Jouhoud, J.-C., Sagaut, P., Montagnac, M., and Laurenceau, J. (2007) A Surrogate-Model Based Multidisciplinary Shape Optimization Method with Applications to a 2D Subsonic Airfoil. *Computers & Fluids*, 36, 520–529.

Journel, A.G. and Huijbregts, Ch.J. (1981) *Mining Geostatistics.* Academic Press/Harcourt Brace & Company, San Diego, CA.

Kabir, H., Wang, Y., Yu, M., and Zhang, Q.J. (2008). Neural network inverse modeling and applications to microwave filter design. *IEEE Transactions on Microwave Theory and Techniques*, 56, 867–879.

Kanazaki, M., Tanaka, K., Jeong, S., and Yamamoto, K. (2006) Multi-objective aerodynamic optimization of elements' setting for high-lift airfoil using kriging model. *44th AIAA Aerospace Sciences Meeting and Exhibit*, Reno, NV, January 9–12.

Katz, J. and Plotkin, J. (2001) *Low-speed Aerodynamics (Vol. 13).* Cambridge University Press, Cambridge.

Kennedy, J., Eberhart, R.C., and Shi, Y. (2001) *Swarm Intelligence.* Academic Press/Harcourt Brace & Company, San Diego, CA.

Kennedy, M.C. and O'Hagan, A. (2000) Predicting the output from complex computer code when fast approximations are available. *Biometrika*, 87, 1–13.

Kim, H., Obayashi, S., and Nakahashi, K. (2001) Aerodynamic optimization of supersonic transport wing using unstructured adjoint method. *AIAA Journal*, 39, 1011–1020.

Kim, S., Alonso, J.J., and Jameson, A. (2000) Two-dimensional high-lift aerodynamic optimization using the continuous adjoint method. *8th AIAA/USAF/NASA/ISSMO Symposium on Multidisciplinary Analysis and Optimization.* Long Beach, CA, September 6–8.

Kim, S., Alonso, J.J., and Jameson, A. (2002) Design optimization of high-lift configurations using a viscous continuous adjoint method. *40th AIAA Aerospace Sciences Meeting and Exhibit*, Reno, NV, January 14–17.

Kim, S., Hosseini, K., Leoviriyakit, K., and Jameson, A. (2005) Enhancement of adjoint design methods via optimization of adjoint parameters. AIAA Paper 2005-448, *43rd AIAA Aerospace Sciences Meeting and Exhibit*, Reno, NV, January 10–13.

Kinsey, D.W. and Barth, T.J. (1984) Description of a hyperbolic grid generation procedure for arbitrary two-dimensional bodies. *AFWAL Technical Memorandum No. 84-191-FIMM*, NASA, Washington, DC.

Kirkpatrick, S., Gelatt, C.D., and Vecchi, M.P. (1983) Optimization by simulated annealing. *Science*, 220, 671–680.

Kleijnen, J. (2008) *Design and Analysis of Simulation Experiments.* Springer Science + Business Media, LLC, New York, NY.

Kleijnen, J.P.C. (2009) Kriging metamodeling in simulation: a review. *European Journal of Operational Research*, 192, 707–716.

Koehler, J.R., and Owen, A.B. (1996). 'Computer experiments', in Ghosh, S. and Rao, C.R. (eds), *Handbook of Statistics*, Elsevier Science B.V., Amsterdam, pp. 261–308.

Kolda, T.G., Lewis, R.M., and Torczon, V. (2003) Optimization by direct search: new perspectives on some classical and modern methods. *SIAM Review*, 45, 385–482.

Korivi, V.M., Newman, P.A., and Taylor, A.C. (1998) Aerodynamic optimization using sensitivity derivatives from a three-dimensional supersonic Euler code. *Journal of Aircraft*, 35, 405–411.

Koziel, S. (2010a) Shape-preserving response prediction for microwave design optimization. *IEEE Transactions on Microwave Theory and Techniques*, 58, 2829–2837.

Koziel, S. (2010b) Adaptively adjusted design specifications for efficient optimization of microwave structures. *Progress in Electromagnetic Research B (PIER B)*, 21, 219–234.

Koziel, S. (2010c) Multi-fidelity multi-grid design optimization of planar microwave structures with Sonnet. *International Review of Progress in Applied Computational Electromagnetics*, Tampere, Finland, April 26–29, pp. 719–724.

Koziel, S. and Bandler, J.W. (2007a) Space-mapping optimization with adaptive surrogate model. *IEEE Transactions on Microwave Theory and Techniques*, 55, 541–547.

Koziel, S. and Bandler, J.W. (2007b) Coarse and surrogate model assessment for engineering design optimization with space mapping. *IEEE MTT-S International Microwave Symposium Digest*, Honolulu, HI, June 3–8, pp. 107–110.

Koziel, S. and Bandler, J.W. (2010) Recent advances in space-mapping-based modeling of microwave devices. *International Journal of Numerical Modelling*, 23, 425–446.

Koziel, S. and Leifsson, L. (2011) Transonic airfoil shape optimization using variable-resolution models and pressure distribution alignment. *29th AIAA Applied Aerodynamics Conference*, Honolulu, HI, June 27–30.

Koziel, S. and Leifsson, L. (2012a) Knowledge-Based Airfoil Shape Optimization Using Space Mapping. AIAA paper 2012-3016, *30th AIAA Applied Aerodynamics Conference*, New Orleans, LA, June 25–28.

Koziel, S. and Leifsson, L. (2012b) Adaptive response correction for surrogate-based airfoil shape optimization. AIAA Paper 2012-3121, *30th AIAA Applied Aerodynamics Conference*, New Orleans, LA.

Koziel, S. and Leifsson, L., (2012c) Multi-fidelity airfoil shape optimization with adaptive response prediction. AIAA Paper 2012-5454, *14th AIAA/ISSMO Multidisciplinary Analysis and Optimization Conference*, Indianapolis, IN, September 17–19.

Koziel, S. and Leifsson, L., (2013a) Surrogate-based aerodynamic shape optimization by variable-resolution models. *AIAA Journal*, 51, 94–106.

Koziel, S. and Leifsson, L., (2013b) Multi-level surrogate-based airfoil shape optimization. *AIAA Aerospace Sciences Meeting*, Grapevine, TX, January 7–10.

Koziel, S. and Ogurtsov, S. (2010) Robust multi-fidelity simulation-driven design optimization of microwave structures. *IEEE MTT-S International Microwave Symposium Digest*, Anaheim, CA, May 23–28, pp. 201–204.

Koziel, S. and Ogurtsov, S. (2011). Rapid design optimization of antennas using space mapping and response surface approximation models. *International Journal of RF and Microwave Computer-Aided Engineering*, 21, 611–621.

Koziel, S. and Ogurtsov, S. (2012) CPU-budget-driven automated microwave design optimization using variable-fidelity electromagnetic simulations. *IEEE MTT-S International Microwave Symposium Digest*, Montreal, QC, June 17–22.

Koziel, S. and Ogurtsov, S. (2013a) Computational-budget-driven automated microwave design optimization using variable-fidelity electromagnetic simulations. *International Journal of RF and Microwave Computer-Aided Engineering*, 23, 349–356.

Koziel, S. and Ogurtsov, S. (2013b) Multi-level microwave design optimization with automated model fidelity adjustment. *International Journal of RF and Microwave Computer-Aided Engineering*, 24, 281–288.

Koziel, S. and Ogurtsov, S. (2013c) 'Simulation-driven design of broadband antennas using surrogate-based optimization', in Koziel, S., Yang, X.S. and Zhang, Q.J. (eds), *Simulation-Driven Design Optimization and Modeling for Microwave Engineering*, Imperial College Press, London, pp. 159–190.

Koziel, S. and Ogurtsov, S. (2013d) Multi-objective design of UWB antennas using surrogate-based optimization. *IEEE International Symposium on Antennas and Propagation*, 61, 210–211.

Koziel, S. and Szczepanski, S. (2011) Accurate modeling of microwave structures using shape-preserving response prediction. *IET Microwaves, Antennas & Propagation*, 5, 1116–1122.

Koziel, S., Bandler, J.W., Mohamed, A.S., and Madsen, K. (2005) Enhanced surrogate models for statistical design exploiting space mapping technology. *IEEE MTT-S International Microwave Symposium Digest*, Long Beach, CA, June 12–17, 1609–1612.

Koziel, S., Bandler, J.W., and Madsen, K. (2006) A space mapping framework for engineering optimization: theory and implementation. *IEEE Transactions on Microwave Theory and Techniques*, 54, 3721–3730.

Koziel, S., Bandler, J.W., and Madsen, K. (2008a) Quality assessment of coarse models and surrogates for space mapping optimization. *Optimization and Engineering*, 9, 375–391.

Koziel, S., Cheng, Q.S., and Bandler, J.W. (2008b) Space mapping. *IEEE Microwave Magazine*, 9, 105–122.

Koziel, S., Bandler, J.W., and Madsen, K. (2009a) Space mapping with adaptive response correction for microwave design optimization. *IEEE Transactions on Microwave Theory and Techniques*, 57, 478–486.

Koziel, S., Meng, J., Bandler, J.W., Bakr, M.H., and Cheng, Q.S. (2009b) Accelerated microwave design optimization with tuning space mapping. *IEEE Transactions on Microwave Theory and Techniques*, 57, 383–394.

Koziel, S., Bandler, J.W., and Cheng, Q.S. (2010a) Adaptively constrained parameter extraction for robust space mapping optimization of microwave circuits. *IEEE MTT-S International Microwave Symposium Digest*, Anaheim, CA, May 23–28, 205–208.

Koziel, S., Bandler, J.W., and Cheng, Q.S. (2010b) Robust trust-region space-mapping algorithms for microwave design optimization. *IEEE Transactions on Microwave Theory and Techniques*, 58, 2166–2174.

Koziel, S., Cheng, Q.S., and Bandler, J.W. (2010c) Implicit space mapping with adaptive selection of preassigned parameters. *IET Microwaves, Antennas & Propagation*, 4, 361–373.

Koziel, S., Cheng, Q.S., and Bandler, J.W. (2011a) Rapid design optimization of microwave structures through automated tuning space mapping. *IET Microwaves, Antennas & Propagation*, 4, 1892–1902.

Koziel, S., Echeverría-Ciaurri, D., and Leifsson, L. (2011b) 'Surrogate-based methods', in Koziel, S. and Yang, X.S. (eds), *Computational Optimization, Methods and Algorithms, Series: Studies in Computational Intelligence*, Springer-Verlag, Berlin, pp. 33–60.

Koziel, S., Ogurtsov, S., Bandler, J.W., and Cheng, Q.S. (2012a) Robust space mapping optimization exploiting EM-based models with adjoint sensitivity. *IEEE MTT-S International Microwave Symposium Digest*, Montreal, Canada, June 17–22, pp. 1301–1308.

Koziel, S., Ogurtsov, S., Couckuyt, I., and Dhaene, T. (2012b) Variable-fidelity electromagnetic simulations and co-kriging for accurate modeling of antennas. *IEEE Transactions on Antennas and Propagation*, 1–3.

Koziel, S., Bandler, J.W., and Cheng, Q.S. (2013) Enhanced fidelity modeling of microwave structures combining shape-preserving response prediction with space mapping. *IEEE International Microwave Symposium*, Seattle, WA, June 2–7, pp. 481–493.

Kuhn, G.H. and Tucker, A.W. (1951) 'Nonlinear programming', in Neyman, J. (ed), *Proceedings of the 2nd Berkeley Symposium on Mathematical Statistics and Probability*, University of California Press, Berkeley, pp. 481–493.

Laurenceau, J. and Sagaut, P. (2008) Building efficient response surfaces of aerodynamic functions with kriging and cokriging. *AIAA Journal*, 46, 498–507.

Leary, S.J., Bhaskar, A., and Keane, A.J. (2001) A constraint mapping approach to the structural optimization of an expensive model using surrogates. *Optimization and Engineering*, 2, 385–398.

Leary, S., Bhaskar, A., Keane, A. (2003) Optimal orthogonal-array-based Latin hypercubes. *Journal of Applied Statistics*, 30, 585–598.

Leifsson, L., Koziel, S., and Ogurtsov, S. (2012) Low-fidelity model mesh density and the performance of variable-resolution shape optimization algorithms. *International Conference on Computational Science*, Omaha, NE, June 4–6.

Le Moigne, A. and Qin, N. (2004) Variable-fidelity aerodynamic optimization for turbulent flows using a discrete adjoint formulation. *AIAA Journal*, 42, 1281–1292.

Lepine, J., Guibault, F., Trepanier, J.-Y., and Pepin, F. (2001) Optimized nonuniform rational b-spline geometrical representation for aerodynamic design of wings. *AIAA Journal*, 39, 2033–2041.

Leung, T. (2010) A Newton-Krylov approach to aerodynamic shape optimization in three dimensions. PhD Thesis, University of Toronto.

Levin, D. (1998) The approximation power of moving least-squares. *Mathematics of Computation*, 67, 1517–1531.

Leoviriakit, K. and Jameson, A. (2005) Multipoint wing planform optimization via control theory. AIAA Paper 2005-0450, *43rd AIAA Aerospace Sciences Meeting and Exhibit*, Reno, NV, January 10–13.

Liao, C.-K., Chi, P.-L., and Chang, C.-Y. (2007) Microstrip realization of generalized Chebyshev filters with box-like coupling schemes. *IEEE Transactions on Microwave Theory and Techniques*, 55, 147–153.

Liu, J., Han, Z., and Song, W. (2012) Comparison of infill sampling criteria in kriging-based aerodynamic optimization. *28th Int. Congress of the Aeronautical Sciences*. Brisbane, Australia, September 23–28.

Lophaven, S.N., Nielsen, H.B., and Søndergaard, J. (2002) *DACE: A MATLAB kriging toolbox*. Technical University of Denmark, Copenhagen.

Lutz, Th. and Wagner, S. (1998) Numerical shape optimization of natural laminar flow bodies. *Proceedings of the 21st ICAS Congress*. Melbourne, Australia, September 13–18.

Majda, A. (2003). *Introduction to PDE's and Waves for the Atmosphere and Ocean*. American Mathematical Society and Courant Institute of Mathematical Sciences at New York University, New York, NY.

March, A. and Willcox, K. (2012) A provably convergent multifidelity optimization algorithm not requiring high-fidelity derivatives. *AIAA Journal*, 50, 1079–1089.

March, A., Wang, Q. and Willcox, K. (2011) Gradient-based multifidelity optimization for aircraft design using bayesian model calibration, *The Aeronautical Journal*, Vol. 115, No. 1173 December.

Marchuk, G.I. (1982). *Methods of Numerical Mathematics*, 2nd ed. Springer-Verlag, New York, NY.

Marheineke, N., Pinnau, R., and Reséndiz, E. (2012) Space mapping-focused control techniques for particle dispersions in fluids. *Optimization and Engineering*, 13, 101–120.

Marsden, A.L., Wang, M., Dennis, J.E., and Moin, P. (2004) Optimal aeroacoustic shape design using the surrogate management framework. *Optimization and Engineering*, 5, 235–262.

Martins, J.R.R.A, Alonso, J.J., and Reuther, J.J. (2005) A coupled-adjoint sensitivity analysis method for high-fidelity aero-structural design. *Optimization and Engineering*, 6, 33–62.

Martins, J.R.R.A., Sturdza, P., and Alonso, J.J. (2003) The complex-step derivativeapproximation. *ACM Transactions on Mathematical Software*, 29, 245–262.

Mathew, J., Bahr, C., Carroll, B., Sheplak, M., and Cattafesta, L. (2005) Design, fabrication, and characterization of an anechoic wind tunnel facility. *11th AIAA/CEAS Aeroacoustics Conference*, Monterey, CA, May 23–25.

Mason, W.H., (1990) Analytic Models for Technology Integration in Aircraft Design. AIAA-90-3262. *AIAA/AHS/ASEE Aircraft Design, Systems and Operations Conference*, Dayton, OH, September 17–19.

MATLAB (2012) Mathworks Inc., Natick, MA.

Maute, K., Nikbay, M., and Farhat, C. (2003) Sensitivity analysis and design optimization of three-dimensional non-linear aeroelastic systems by the adjoint method. *International Journal of Numerical Methods in Engineering*, 56, 911–933.

Mavriplis, D.J. (2007) Discrete adjoint-based approach for optimization problems on three-dimensional unstructured meshes. *AIAA Journal*, 45, 741–750.

McGuffie, K. and Henderson-Sellers, A., (2005). *A Climate Modelling Primer*, 3rd ed. John Wiley & Sons Ltd., West Sussex.

McKay M, Conover W, and Beckman R. (1979) A comparison of three methods for selecting values of input variables in the analysis of output from a computer code. *Technometrics*, 21, 239–245.

Meng, J. and Xia, L. (2007) Support-vector regression model for millimeter wave transition. *Int. J. Infrared and Milimeter Waves*, 28, 413–421.

Menter, F. (1994) Two-equation eddy-viscosity turbulence models for engineering applications. *AIAA Journal*, 32, 1598–1605.

Meunier, M. (2007) Simulation and optimization of flow control strategies for novel high-lift configurations. *25th AIAA Applied Aerodynamic Conference*, Miami, FL, June 25–28.

Michalewicz, Z. (1996) *Genetic Algorithms + Data Structures = Evolution Programs*, 3rd ed. Springer-Verlag, Berlin.

Minsky, M.I. and Papert, S.A. (1969) *Perceptrons: An Introduction to Computational Geometry.* The MIT Press, Cambridge, MA.

Morris, M.D., Mitchell, T.J., and Ylvisaker, D. (1993) Design and analysis of computer experiments: use of derivatives in surface prediction. *Technometrics*, 35, 243–255.

Myring, D.F. (1976) A theoretical study of body drag in subcritical axisymmetric flow. *Aeronautical Quarterly*, 28, 186–194.

Nadarajah, S. and Jameson, A. (2000) A comparison of the continuous and discrete adjoint approach to automatic aerodynamic optimization. AIAA Paper 2000-0667, *38th Aerospace Sciences Meeting & Exhibit*, Reno, NV, January 10–13.

Nelder, J.A. and Mead, R. (1965) A simplex method for function optimization. *Computer Journal*, 7, 308–313.

Nemec, M. and Zingg, D.W. (2002) Newton-Krylov algorithm for aerodynamic design using the Navier-Stokes equations. *AIAA Journal*, 40, 1146–1154.

Nemec, M. and Zingg, D.W. (2003) Optimization of high-lift configurations using a Newton-Krylov algorithm. *16th AIAA Computational Fluid Dynamics Conference*, Orlando, FL, June 23–26.

Nemec, M., Zingg, D.W., and Pulliam, T. (2004) Multipoint and multi-objective aerodynamic shape optimization. *AIAA Journal*, 42, 1057–1065.

Nielsen, H.B. (1999) Damping parameter in Marquardt's method. IMM DTU. Report IMM-REP-1999-05, Danish Technical Institute, Copenhagen.

Nocedal, J. and Wright, S.J. (2006) *Numerical Optimization*, 2nd ed. Springer-Verlag, Berlin.

Obayashi, S. (1997) Aerodynamic optimization with evolutionary algorithms. Technical Report. Lecture Notes for the von Kármán Institute for Fluid Dynamics Lecture Series: Inverse Design and Optimizations, Brussels, Belgium.

O'Hagan, A. (1978) Curve fitting and optimal design for predictions. *Journal of the Royal Statistical Society*, 40, 1–42.

Oschlies, A. and Garcon, V. (1999). An eddy-permitting coupled physical-biological model of the North Atlantic. 1. Sensitivity to advection numerics and mixed layer physics. *Global Biogeochemical Cycles*, 13, 135–160

Palmer, K. and Tsui, K.-L. (2001) A minimum bias Latin hypercube design. *IIE Transactions*, 33, 793–808.

Papadimitriou, D.I. and Giannakoglou, K.C. (2008) Aerodynamic shape optimization using first and second order adjoint and direct approaches. *Archives of Computational Methods in Engineering*, 15, 447–488.

Petosa, A. (2007). *Dielectric Resonator Antenna Handbook.* Artech House, Boston, MA.

Pérez, V.M., Renaud, J.E., and Watson, L.T. (2002) Interior point sequential approximate optimization methodology. *Proceedings of the 9th AIAA/ISSMO Symposium of Multidisciplinary Analysis and Optimization*, Atlanta, GA, September 4–6.

Pironneau, O. (1974) On optimum design in fluid mechanics. *Journal of Fluid Mechanics*, 64, 97–110.

Pironneau, O. (1983) *Optimal Shape Design for Elliptic Systems.* Springer-Verlag, Berlin.

Pozar, D.M. (2012) *Microwave Engineering*, 4th ed. John Wiley & Sons, Inc., Hoboken, New Jersey.

Price K., Storn R., and Lampinen J. (2005). *Differential Evolution: A Practical Approach to Global Optimization.* Springer-Verlag, Berlin.

Priess, M., Koziel, S., and Slawig, T. (2011a) Surrogate-based optimization of climate model parameters using response correction. *Journal of Computational Science*, 2, 335–344.

Priess, M., Slawig, T., and Koziel, S. (2011b) Improved surrogate-based optimization of climate model parameters using response correction. *International*

Conference on Simulation and Modeling Methodologies, Technologies and Applications (SIMULTECH) 2011, Noordwijkerhout, The Netherlands, July 29–31, pp. 449–457.

Priess, M., Piwonski, J., Koziel, S., Oschlies, A., and Slawig, T. (2013) Accelerated parameter identification in a 3D marine biogeochemical model using surrogate-based optimization. *Ocean Modeling*, 68, 22–36.

Queipo, N.V., Haftka, R.T., Shyy, W., Goel, T., Vaidynathan, R., and Tucker, P.K. (2005) Surrogate-based analysis and optimization. *Progress in Aerospace Sciences*, 41, 1–28.

Rao, S.S. (1996) *Engineering Optimization: Theory and Practice*, 3rd ed. John Wiley & Sons, Inc., Hoboken, NJ.

Rasmita, A., Yong-Xin, G., and Alphones, A. (2009) Constrained explicit knowledge embedded space mapping using circuit tuning based on physical augmentation as parameter extraction. *Asia Pacific Microwave Conference*, Singapore, December 7–9, pp. 1431–1434.

Rasmussen, C.E. and Williams, C.K.I. (2006) *Gaussian Processes for Machine Learning*. MIT Press, Cambridge.

Rautio, J.C. (2006) RF design closure — companion modeling and tuning methods. *IEEE MTT IMS Workshop: Microwave Component Design using Space Mapping Technology*, San Francisco, CA, June 11–16.

Rayas-Sanchez, J.E. (2004) EM-based optimization of microwave circuits using artificial neural networks: the state-of-the-art. *IEEE Transactions on Microwave Theory and Techniques*, 52, 420–435.

Redhe, M. and Nilsson, L. (2004) Optimization of the new Saab 9-3 exposed to impact load using a space mapping technique. *Structural and Multidisciplinary Optimization*, 27, 411–420.

Reuther, J.J., Jameson, A., Alonso, J.J., Rimlinger, M.J., and Saunders, D. (1999a) Constrained multipoint aerodynamic shape optimization using an adjoint formulation and parallel computers, part 1. *Journal of Aircraft*, 26, 51–60.

Reuther, J.J., Jameson, A., Alonso, J.J., Rimlinger, M.J., and Saunders, D. (1999b) Constrained multipoint aerodynamic shape optimization using an adjoint formulation and parallel computers, part 2. *Journal of Aircraft*, 26, 61–74.

Robinson, T.D., Eldred, M.S., Willcox, K.E., and Haimes, R. (2008) Surrogate-based optimization using multifidelity models with variable parameterization and corrected space mapping. *AIAA Journal*, 46, 2316–2326.

Rojo-Alvarez, J.L., Camps-Valls, G., Martinez-Ramon, M., Soria-Olivas, E., Navia-Vazquez, A., and Figueiras-Vidal, A.R. (2005) Support vector machines framework for linear signal processing. *Signal Processing*, 85, 2316–2326.

Roux, W.J., Stander, N., Haftka, R.T. (1998) Response surface approximations for structural optimization. *International Journal of Numerical Methods in Engineering*, 42, 517–534.

Rumpfkeil, M. and Zingg, D.W. (2008) The optimal control of unsteady flows with a discrete adjoint method. *Optimization and Engineering*, 11, 5–22.

Rumsey, C.L. and Ying, S.X. (2002) Prediction of high lift: review of present CFD capability. *Progress in Aerospace Sciences*, 38(2), 145–180.

Samareh, J.A. (2001) Survey of shape parametrization techniques for high-fidelity multidisciplinary shape optimization. *AIAA Journal*, 39, 877–884.

Santner, T.J., Williams, B., and Notz, W. (2003) *The Design and Analysis of Computer Experiments*. Springer-Verlag, Berlin.

Sarmiento, J. and Gruber, N. (2006). *Ocean Biogeochemical Dynamics*. Princeton University Press, Princeton, NJ.

Schmitt, V. and Charpin, F. (1979) Pressure distributions on the ONERA-M6-wing at transonic Mach numbers. Experimental Data Base for Computer Program Assessment. Report of the Fluid Dynamics Panel, Working Group 04, AGARD AR 138, NATO Science and Technology Organization, Brussels, Belgium.

Selig, M. (2014) *The University of Illinois at Urbana-Champaing Airfoil Coordinates Database*. [Database] Available at: http://aerospace.illinois.edu/m-selig/ads/coord_database.html. Accessed 30 July 2014.

Sherman, L.L., Taylor A.C., Green, L.L., Newman, P.A., Hou, G.W., and Korivi, V.M. (1996) First- and second-order aerodynamic sensitivity derivatives via automatic differentiation with incremental iterative methods. *Journal of Computational Physics*, 129, 307–331.

Simpson, T.W., Peplinski, J., Koch, P.N., and Allen, J.K. (2001) Metamodels for computer based engineering design: survey and recommendations. *Engineering with Computers*, 17, 129−150.

Smola, A.J. and Schölkopf, B. (2004) A tutorial on support vector regression. *Statistics and Computing*, 14, 199−222.

Sobieczky, H. (1998) 'Parametric airfoils and wings', in Fuji, K. and Dulikravich, G.S. (eds), *Notes on Numerical Fluid Mechanics, Vol. 68*, Wiesbaden, Vieweg..

Sobieszczanski-Sobieski, J. and Haftka, R.T. (1997). Multidisciplinary aerospace design optimization: survey of recent developments. *Structural Optimization*, 14, 1–23.

Solov'ev, S.A. (2009) Determining the shape of an axisymmetric body in a viscous incompressible flow on the basis of the pressure distribution on the body surface. *Journal of Applied Mechanics and Technical Physics*, 50, 927–935.

Søndergaard, J. (2003) Optimization using surrogate models — by the space mapping technique. PhD Thesis, Technical University of Denmark.

Spalart, P. R. and Allmaras, S. R. (1992) A one equation turbulence model for aerodynamic flows. AIAA-92-0439. *38th AIAA Aerospace Sciences Meeting and Exhibit*, Reno, NV, January 6–9.

Su, Y. (1991) Flow analysis and design of three-dimensional wind tunnel contractions. *AIAA Journal*, 29, 1912–1920.

Staum, J. (2009) Better simulation metamodeling: the why, what, and how of stochastic kriging. *Proceedings of the 2009 Winter Simulation Conference*, Austin, TX, December 13–16, pp. 119–133.

Storn R. and Price, K. (1997) Differential evolution — a simple and efficient heuristic for global optimization over continuous spaces. *Journal of Global Optimization*, 11, 341–359.

Swanson, D.G. and Wenzel, R.J. (2001) Fast analysis and optimization of combline filters using FEM. *IEEE MTT-S International Microwave Symposium Digest*, Boston, MA, May 20–24, pp. 1159–1162.

Swanson, D. and Macchiarella, G. (2007) Microwave filter design by synthesis and optimization. *IEEE Microwave Magazine*, 8, 55–69.

Takemiya, T. (2008) Aerodynamic design applying automatic differentiation and using robust variable fidelity optimization. PhD Thesis, Georgia Institute of Technology.

Takemiya, T. and Mavris, D.N., (2009) Aerodynamic optimization through automatic differentiation and CFD. *Japan Society of Aeronautical Space Sciences*, 57, 89–97.

Tarantola, A. (2005). *Inverse Problem Theory and Methods for Model Parameter Estimation.* Society for Industrial and Applied Mathematics (SIAM), Philadelphia, PA.

Tannehill, J.A., Anderson, D.A., and Pletcher, R.H. (1997) *Computational Fluid Mechanics and Heat Transfer*, 2nd ed. Taylor & Francis, New York, NY.

Toal, D.J.J. and Keane, A.J. (2011) Efficient multipoint aerodynamic design optimization via cokriging. *Journal of Aircraft*, 48, 1685–1695.

Toal, D.J.J., Bressloff, N.W., and Keane, A.J. (2008) Geometric filtration using POD for aerodynamic design optimization. *26th AIAA Applied Aerodynamics Conference*, Honolulu, HI, August 18–21.

Toropov, V.V., Filatov, A.A., and Polynkin, A.A. (1993). Multiparameter structural optimization using FEM and multipoint explicit approximations. *Structural Optimization*, 6, 7–14.

Tu, S., Cheng, Q.S., Zhang, Y., Bandler, J.W., and Nikolova, N.K. (2013) Space mapping optimization of handset antennas exploiting thin-wire models. *IET Microwaves Antennas & Propagation*, 61, 3797–3807.

van Dam, C.P. (2002) The aerodynamic design of multi-element high-lift systems for transport airplanes. *Progress in Aerospace Sciences*, 38(2), 101–144.

Vapnik, V.N. (1995) *The Nature of Statistical Learning Theory.* Springer–Verlag, New York, NY.

Watmuff, J. (1986) Wind tunnel contraction design. *9th Australasian Fluid Mechanics Conference*, Auckland, New Zealand, December 8–12.

Wild, S.M., Regis, R.G, and Shoemaker, C.A. (2008) ORBIT: Optimization by radial basis function interpolation in trust-regions. *SIAM Journal on Scientific Computing*, 30, 3197–3219.

Wrenn, G.A. (1989) An indirect method for numerical optimization using the kreissel-meier-steinhauser function. NASA CP-4220, NASA, Washington, DC.

Wu, K.-L., Zhao, Y.-J. Wang, J., and Cheng, M.K.K. (2004) An effective dynamic coarse model for optimization design of LTCC RF circuits with aggressive space mapping. *IEEE Transactions on Microwave Theory and Techniques*, 52, 393–402.

Xia, L., Meng, J., Xu, R., Yan, B., and Guo, Y. (2006) Modeling of 3-D vertical interconnect using support vector machine regression. *IEEE Microwave and Wireless Components Letters*, 16, 639–641.

Yamamoto, I. (2006) "Research and development of past, present, and future AUV technologies," Masterclass in AUV Technology for Polar Science, Southampton, UK, March 28–30.

Yang, C. and Löhener, R. (2003) Prediction of flows over an axisymmetric body with appendages. 8th International Conference on Numerical Ship Hydrodynamics, Busan, South Korea, September 22–25.

Yang, X.S. (2005) Engineering optimization via nature-inspired virtual bee algorithms. IWINAC 2005. *Lecture Notes in Computer Science*, 3562, 317–323.

Yang, X.S. (2008) *Nature-Inspired Metaheuristic Algorithms*. Luniver Press, Frome.

Yang, X.S. (2010) *Engineering Optimization: An Introduction with Metaheuristic Applications*. John Wiley & Sons Ltd., West Sussex.

Yang, Y., Hu, S.M., and Chen, R.S. (2005) A combination of FDTD and least-squares support vector machines for analysis of microwave integrated circuits. *Microwave and Optical Technology Letters*, 44, 296–299.

Ye, K.Q. (1998) Orthogonal column Latin hypercubes and their application in computer experiments. *Journal of the American Statistical Association*, 93, 1430–1439.

Zhang, Q.J. and Gupta, K.C. (2000) *Neural Networks for RF and Microwave Design*. Artech House, Norwood.

Zhang, K. and Han, Z. (2013) Support vector regression-based multidisciplinary design optimization in aircraft conceptual design. AIAA paper 2013-1160, *51st AIAA Aerospace Sciences Meeting*, Grapevine, TX, January 7–10.

Zingg, D.W., Nemec, M., and Pulliam, T. (2008) A comparative evaluation of genetic and gradient-based algorithms applied to aerodynamic optimization. *Revue Européenne de Mcanique Numérique*, 17, 103–126.

Index